Biomolecular Structure and Dynamics

NATO ASI Series

Advanced Science Institutes Series

A Series presenting the results of activities sponsored by the NATO Science Committee, which aims at the dissemination of advanced scientific and technological knowledge, with a view to strengthening links between scientific communities.

The Series is published by an international board of publishers in conjunction with the NATO Scientific Affairs Division

A	**Life Sciences**	Plenum Publishing Corporation
B	**Physics**	London and New York
C	**Mathematical and Physical Sciences**	Kluwer Academic Publishers
D	**Behavioural and Social Sciences**	Dordrecht, Boston and London
E	**Applied Sciences**	
F	**Computer and Systems Sciences**	Springer-Verlag
G	**Ecological Sciences**	Berlin, Heidelberg, New York, London,
H	**Cell Biology**	Paris and Tokyo
I	**Global Environmental Change**	

PARTNERSHIP SUB-SERIES

1.	**Disarmament Technologies**	Kluwer Academic Publishers
2.	**Environment**	Springer-Verlag / Kluwer Academic Publishers
3.	**High Technology**	Kluwer Academic Publishers
4.	**Science and Technology Policy**	Kluwer Academic Publishers
5.	**Computer Networking**	Kluwer Academic Publishers

The Partnership Sub-Series incorporates activities undertaken in collaboration with NATO's Cooperation Partners, the countries of the CIS and Central and Eastern Europe, in Priority Areas of concern to those countries.

NATO-PCO-DATA BASE

The electronic index to the NATO ASI Series provides full bibliographical references (with keywords and/or abstracts) to more than 50000 contributions from international scientists published in all sections of the NATO ASI Series.
Access to the NATO-PCO-DATA BASE is possible in two ways:

– via online FILE 128 (NATO-PCO-DATA BASE) hosted by ESRIN,
Via Galileo Galilei, I-00044 Frascati, Italy.

– via CD-ROM "NATO-PCO-DATA BASE" with user-friendly retrieval software in English, French and German (© WTV GmbH and DATAWARE Technologies Inc. 1989).

The CD-ROM can be ordered through any member of the Board of Publishers or through NATO-PCO, Overijse, Belgium.

Series E: Applied Sciences - Vol. 342

Biomolecular Structure and Dynamics

edited by

Gérard Vergoten

CRESIMM,
Department of Chemistry,
University of Science and Technology,
Lille, Villeneuve d'Ascq, France

and

Theophile Theophanides

National Technical University of Athens,
Chemical Engineering Department,
Zografou,
Athens, Greece

Springer Science+Business Media, B.V.

Proceedings of the NATO Advanced Study Institute on
Biomolecular Structure and Dynamics: Experimental and Theoretical Advances
Loutraki, Greece
May 27–June 6, 1996

A C.I.P. Catalogue record for this book is available from the Library of Congress

ISBN 978-94-010-6307-4   ISBN 978-94-011-5484-0 (eBook)
DOI 10.1007/978-94-011-5484-0

Printed on acid-free paper

TABLE OF CONTENTS

PREFACE vii

Part I.   **MODELING AND COMPUTER SIMULATIONS**   1

The Physical Chemistry of Specific Recognition 3
J. Janin

Approaches to Protein-Ligand Binding from Computer Simulations 21
William L. Jorgensen, Erin M. Duffy, Jonathan W. Essex, Daniel L. Severance, James F. Blake,
Deborah K. Jones-Hertzog, Michelle L. Lamb and Julian Tirado-Rives

Dynamics of Biomolecules: Simulation versus X-Ray and Far-Infrared Experiments 35
S. Hery, M. Souaille and J.C. Smith

Semiempirical and *ab initio* Modeling of Chemical Processes: From Aqueous solution
to Enzymes 47
Richard P. Muller, Jan Florián and Arieh Warshel

Professional Gambling 79
Rolando Rodriguez and Gerrit Vriend

Molecular Modeling of Globular Proteins: Strategy ID $\Rightarrow$ 3D: Secondary Structures and
Epitopes 121
Alain J.P. Alix

Physiochemical Properties in Vacuo and in Solution of Some Molecules with Biological
Significance from Density Functional Computations 151
Tiziana Marino, T. Mineva, Nino Russo and Marirosa Toscano

GMMX Conformation Searching and Prediction of MNR Proton-Proton Coupling Constants 179
Fred L. Tobiason and Gérard Vergoten

PART II.   **PROTEINS AND LIPIDS**   187

Biomolecular Structure and Dynamics: Recent Experimental and Theoretical Advances   189
R. Kaptein, A.M.J.J. Bonvin and R. Boelens

What Drives Associations of α-Helical Peptides in Membrane Domains of Proteins? Role of
Hydrophobic Interactions 211
Roman G. Efremov and Gérard Vergoten

Infrared Spectroscopic Studies of Membrane Lipids 229
José Luis R. Arrondo and Félix M. Goñi

Time-Resolved Infrared Spectroscopy of Biomolecules 243
H. Georg, K. Hauser, C. Rödig, O. Weidlich and F. Siebert

UV Resonance Raman Determination of α-Helix Melting During the Acid Denaturation of
Myoglobin 263
Sanford A. Asher and Zhenhuan Chi

vi

PART III.   **NUCLEIC ACIDS**   271

The use of Fourier Transform Infrared (FT-IR) Spectroscopy in the Structural Analysis of
Nucleic Acids 273
Theo Theophanides and Jane Anastassopoulou

Geometries and Stabilities of G.GC, T.AT, A.AT and C.GC Nucleic Acid Base Triplets   285
Tiziana Marino, Nino Russo, Anna Sarubbo and Marirosa Toscano

Vibrational Circular Dichroism of Nucleic Acids: *Survey of Techniques, Theoretical Background
and Example Applications* 299
Timothy A. Keiderling

Subject/Author Index 319

PREFACE

This book contains the formal lectures and contributed papers presented at the NATO Advanced Study Institute on Biomolecular Structure and Dynamics : Recent Experimental and Theoretical Advances. The meeting convened at the city of Loutraki, Greece on 27 May 1996 and continued to 6 June 1996.

The material presented describes the fundamental and recent advances in experimental and theoretical aspects of molecular dynamics and stochastic dynamics simulations, X-ray crystallography and NMR of biomolecules, structure prediction of proteins, time resolved Fourier transform infrared spectroscopy of biomolecules, computation of free energy, applications of vibrational circular dichroism of nucleic acids and solid state NMR spectroscopy.

In addition, recent advances in UV resonance Raman spectroscopy of biomolecules semiempirical molecular orbital methods, empirical force fields, quantitative studies of the structure of proteins in water by Fourier transform infrared spectroscopy, density function theory (DPT) were presented.

Metal-ligand interactions, DFT treatment of organometallic and biological systems, simulation versus X-ray and far-infrared experiments are also discussed in some detail. In addition, a large proportion of program was devoted to current experimental and theoretical studies of the structure of biomolecules and intramolecular dynamic processes.

The purpose of the proceedings is to provide the reader with a rather broad perspective on the current theoretical aspects and recent experimental findings in the field of biomolecular dynamics. Moreover, the material presented in the proceedings should make apparent the future trends for research in this field, as well as could provide grants for collaborative research between theoreticians and experimentalists in areas of importance to the understanding of biomolecular structure and dynamics.

The proceedings should be of interest to graduate and postgraduate students who are involved or starting research in these areas, and to scientists who are actively pursuing research in biomolecular structure and dynamics.

Appreciable part of the information contained in the proceeding has not yet been published in books on biomolecular structure and dynamics.

G. Vergoten

T. Theophanides

ACKNOWLEDGEMENTS

We express our appreciation to the local organizing committee . Special thanks are due to all lecturers and authors for their cooperation in preparing the manuscripts, excellent delivery of lectures and dedication to the meeting. Mrs Françoise Bailly provided invaluable secretarial assistance in preparation for the meeting and during the meeting.

Last, but not least, we should like to express our gratitude to the NATO Scientific Affairs Division for granting financial support for the meeting.

MODELING   AND COMPUTER SIMULATIONS

THE PHYSICAL CHEMISTRY OF SPECIFIC RECOGNITION

J. JANIN

Laboratoire d'Enzymologie and Biochimie Structurales, CNRS UPR 9063
Bât. 34, 91198 Gif-sur-Yvette, France

1. Introduction

It is our opinion that the processes of synthesis and folding of highly complexes molecules
in living cells involve, in addition to covalent bonds, only the intermolecular interactions of
van der Waals attraction and repulsion, electrostatic interactions, hydrogen-bond formation,
etc., which are now well understood. These interactions are such as to give stability to a
system of two molecules with *complementary* structures in juxtaposition...
In order to achieve maximum stability, the two molecules must have complementary
surfaces, like die and coin, and also a complementary distribution of active groups.
L. Pauling & M. Delbrück (1940) [1]

The reader of these remarkable sentences should remember that they were
composed four years before Avery, McLeod & McCarty showed DNA to be the
molecule genes are made of, fifteen years before Fred Sanger sequenced insulin, and
twenty years before Max Perutz & John Kendrew obtained the X-ray structure of
myoglobin. In 1940, Linus Pauling and Max Delbrück had no experimental evidence
whatsoever to support their statements. They were addressing colleagues in physics and
chemistry rather than biologists who, in these times, seldomly spoke in terms of atomic
interactions. Still, in the US at least, biologists were ready to consider physical
chemistry as a partner science in the study of the mechanisms that rule the cell and the
organism. Figures like Pauling and Delbrück were in the lead, and they were so fully
right in this particular case that we find not a word must be changed in their definition
of complementarity, which makes it possible for two (macro)molecules to assemble into
a specific stable complex. The only question we may ask at the end of this century, is

3

G. Vergoten and T. Theophanides (eds.), Biomolecular Structure and Dynamics, 3–19.
© 1997 *Kluwer Academic Publishers.*

whether we can make Pauling and Delbrück's definition quantitative and find numbers that express specificity and stability.

2. Affinity and the law of mass action: equilibrium and rate constants

For stability, the answer seems an easy yes. A non-covalent complex being ruled by the Gulberg-Waage law of mass action, the reaction formula:

$$A + B \xleftarrow[k_d]{k_a} AB \tag{1}$$

implies a relationship between the equilibrium concentrations of components A and B and complex AB:

$$K_d = \frac{1}{K_a} = \frac{k_d}{k_a} = \frac{[A][B]}{[AB]} \tag{2}$$

K_a and K_d are the two equilibrium constants, k_a and k_d the two rate constants for association and dissociation. K_d (or its reciprocal K_a) measure the stability of complex AB and the affinity of A and B for each other. K_d values in the micro- or nanomolar range can be derived by measuring concentrations at equilibrium. This is no longer possible when the affinity is much higher. Then, it is more practical to measure the two rate constants, the ratio of which yields K_d values in the picomolar range or below.

The second order rate constant k_a has an upper value that comes from the stochastic diffusion of molecules in solution: $k_a \approx 10^9$ $M^{-1}.s^{-1}$ in water at 25°C. Table 1 quotes rate constants for typical specific protein-protein complexes: two enzyme-inhibitor and one antigen-antibody complexes. The enzyme is bovine trypsinin one case, a bacterial ribonuclease, barnase, in the other. The antigen is hen lysozyme, the antibody, a covalent pair of *E. coli* expressed variable domains (single chain Fv). Affinities cover six orders of magnitude with $K_d = 10^{-8} - 10^{-14}$ M, mostly due to k_d, k_a being $10^6 - 10^8$ $M^{-1}.s^{-1}$. Barnase, barstar and the Fv fragment have been subjected to site-directed mutagenesis. In mutant R59A of barnase, part of a long series analyzed by Schreiber & Fersht [2-3], the point substitution makes the affinity drop by a factor of 10^4. Variant M3 of the Fv fragment has been selected by phage display to raise the affinity for lysozyme by a factor of 5 [4]. Similar changes are observed upon point substitution in other systems.

Table 1: Experimental rate and equilibrium constants in some protein-protein complexes

Complex	k_a $(M^{-1}.s^{-1})$	k_d (s^{-1})	K_d (M)	ΔG_d	$\Delta\Delta G_d$
				(kcal.mol^{-1})	
Trypsin-PTI [a]	$1.1.10^6$	$6.6.10^{-8}$	6.10^{-14}	18.1	-
Barnase-barstar [b]	$3.7.10^8$	$3.7.10^{-6}$	1.10^{-14}	19.0	-
R59A variant	$3.4.10^7$	$2.4.10^{-3}$	7.10^{-11}	13.8	5.4
Lysozyme-Fv D1.3 [c]	$1.8.10^6$	6.10^{-3}	3.10^{-9}	11.7	-
M3 variant	$1.6.10^6$	1.10^{-3}	6.10^{-10}	12.6	-0.9

Values near 25° taken from:
(a) Vincent & Lazdunski [21]; (b) Schreiber & Fersht [2]; (c) Hawkins and al. [4]

The higher affinity of variant M3 is entirely due to the lower rate of dissociation. In contrast, barnase mutation R59A both increases k_d by a factor of 10^3 and lowers k_a by a factor of 10. It should be stressed that barnase-barstar association is extremely fast, with k_a near the diffusion limit for molecules having $M_r \approx 10$ kDa. Nearly every collision between barnase and barstar must yield a specific stable complex. This may seem absurd if we consider that the contact region (covering the enzyme active site) is no more than 10-15% of each component surface. A mutation such as R59A that modifies the net electric charge of barnase as well as the k_a value, shows that the association between barnase and barstar is electrostatically assisted [2-3]. At very high ionic strength, long-range electrostatic interactions are shielded and k_a drops by over four orders of magnitude to $\approx 10^5$ M^{-1}.s^{-1}, a value compatible with the precise geometry observed in the complexe.

3. Enthalpies, free enthalpies and entropies

Affinity may also be defined in terms of the usual thermodynamic parameters, the enthalpy H (internal energy at constant pressure), the entropy S and the free enthalpy G (Gibbs energy). Changes in these parameters are quoted in reference to a 'standard' state, per mole of product of reaction (1) and in either direction. We choose to quote values for dissociation, and signs must be changed for association:

$$\Delta G_d = - RT \ln \frac{K_d}{c_\emptyset} \qquad (3)$$

Here, R is the gas constant ($\approx$2 cal.mol^{-1}.K^{-1}), T the temperature and $c_\emptyset$ the concentration taken to be the standard state. For solution studies, the usual convention is $c_\emptyset$=1 M, yet this is an arbitrary choice and $c_\emptyset$=55,5 M, the molar concentration of pure water, is sometimes used. Moreover, tabulated values almost never use this convention: they relate to the pure liquid or solid chemical species, not to aqueous solution. The $c_\emptyset$ convention is unimportant when comparing the affinity of two different ligands for the same site or, as in Table 1, the affinity of a mutant and and the wild type of the same protein. The dissociation changes from K_d to K'_d, the free enthalpy change from ΔG_d to $\Delta G_d+\Delta\Delta G_d$:

$$\Delta\Delta G_d = RT \ln \frac{K'_d}{K_d} \qquad (4)$$

The free enthalpy of dissociation ΔH_d does not depend on $c_\emptyset$. It can be derived from K_d measurements made at several temperatures by applying Van t'Hoff law:

$$\Delta H_d = \frac{d(\Delta G_d/T)}{d(1/T)} = - R \frac{d(\ln K_d)}{d(1/T)} \qquad (5)$$

Then, the entropy of dissociation ΔS_d (which does depend on $c_\emptyset$) is derived from:

$$\Delta G_d = \Delta H_d - T\Delta S_d \qquad (6)$$

In recent years, a direct determination of ΔH_d can be made by isothermal mixing calorimetry as the heat evolved when two solutions are mixed [5]. By performing measurements at several temperatures, the heat capacity of dissociation ΔC_d comes out as:

$$\Delta C_d = \frac{d(\Delta H_d)}{dT} = T \frac{d(\Delta S_d)}{dT} \qquad (7)$$

Assuming ΔC_d to be a constant in the temperature range under study, one may integrate Eq. 7 and predict ΔH_d, ΔS_d and ΔG_d at all temperatures knowing K_d and ΔH_d at 25°C (T_0=298K) only. Fig. 1 shows the result for a lysozyme-antibody HyHEL5 complex [6]. In this particular case, ΔC_p, ΔH_d and ΔS_d all have positive values: association releases heat, a favourable enthalpy stabilizes the complex and a unfavourable entropy fights it. In other systems, negative values of ΔH_d or ΔS_d can be observed at 25°C. Moreover, ΔC_d is high. Therefore, both the enthalpy and the entropy vary quickly and change sign with temperature. In the lysozyme-HyHEL5 complex, ΔS_d is negative below 0°C. Then, entropy favours complex formation - but that statement is valid only at concentrations above $c_{\emptyset}$=1M !

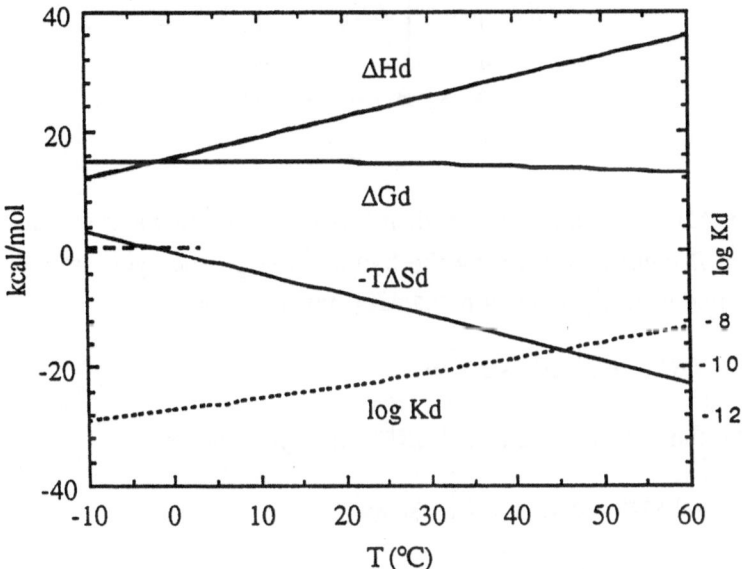

Figure 1: Temperature dependence of thermodynamic parameters for the lysozyme-antibody HyHEL5 complex. ΔH_d was measured by isothermal mixing calorimetry at several temperatures between 10° and 37°C yielding ΔC_d=0.34 kcal.mol^{-1}.K^{-1} [6] and the dissociation constant at T_0=278K (25°C). Assuming ΔC_d to be temperature-independent, we have at all temperatures:

$$\Delta H_d\,(T) = \Delta H_d\,(T_0) + (T-T_0)\,\Delta C_d \quad \text{and} \quad \Delta S_d\,(T) = \Delta S_d\,(T_0) + \Delta C_d \ln \frac{T}{T_0}$$

The temperature dependence of enthalpy and entropy almost exactly compensate each other; thus, ΔG_d varies by <1 kcal.mol^{-1} between 0 and 37°C, whereas K_d (dashes) changes by a factor of 100.

4. A physical model of affinity

An excellent test of our understanding of protein-protein interaction is our capacity to calculate the thermodynamic parameters ΔH_d and ΔS_d in systems where structural data are available from X-ray studies. Here is my particular way of doing it [7]. Let us draw a thermodynamic cycle:

$$\Delta X^{gaz}$$

Gas state $\qquad\qquad$ A + B $\longleftrightarrow$ AB

Hydration $\qquad \Delta X_A \Big\updownarrow \quad \Big\updownarrow \Delta X_B \qquad \Big\updownarrow \Delta X_{AB}$

Water solution $\qquad$ A + B $\longleftrightarrow$ AB

$$\Delta X_d$$

Complex AB forms in gas phase, then it is transfered to water solution together with its A and B components during the hydration step. The cycle implies that any extensive thermodynamic parameter noted ΔX_d may be derived as:

$$\Delta X_d = \Delta X^{gaz} + \Delta X^{hyd} \tag{8}$$

where ΔX^{gaz} is the gas phase value and ΔX^{hyd} the hydration term:

$$\Delta X^{hyd} = \Delta X_A + \Delta X_B - \Delta X_{AB} \tag{9}$$

An advantage of this presentation is that protein-protein interactions are evaluated separately from protein-solvent interactions. The first are evaluated in gas phase, the second, by applying empirical coefficients to changes in the solvent accessible surface area. Such coefficients are obtained from calorimetric and solubility studies performed on small molecules [8-9]. They multiply the solvent accessible surface area A of the various atom types to yield physical parameters for hydration; X^{hyd} can be the heat capacity, the enthalpy or the free enthalpy. A being evaluated on atomic coordinates of the complex AB and of the two components A and B, the area B buried at the interface (the interface area) is:

$$B = A_A + A_B - A_{AB} \tag{10}$$

Eq. 9 and 10 show that the coefficients yielding X^{hyd} from A also yield ΔX^{hyd} from B. Alternative thermodynamic cycles have the complex form in a non-polar organic solvent, followed with a solvent-water transfer. The transfer coefficients of Eisenberg & McLachlan [10] then multiply the same B values as ours to yield thermodynamic quantities.

Our calculation goes through the following five steps:

- in gas phase:

     (a) make one molecule out of two;

     (b) establish van der Waals and electrostatic interactions;

     (c) adjust the conformations of A and B;

     (d) immobilize side chains and part of the main chain;

- then:

     (e) hydration.

During steps (a) and (b), the molecules behave as rigid bodies; in (c) and (d), deviations from the rigid-body approximation are considered. The enthalpy ΔH^{gaz} change upon (b) and (c) is equal to the potential energy for the interaction of A and B *in vacuo*. It includes terms for electrostatic and van der Waals interactions, and the energy cost for changing the conformation of A and B. Any conformation change is assumed to be the same in gas phase and in solution. Molecular mechanics yields a value for ΔH^{gaz}, but not for the corresponding entropy change ΔS^{gaz}. Entropy is essential at step (a): six degrees of rotational/translational freedom are lost when the number of molecules changes. The ·vibrational modes which replace them have a lower entropy, and the difference is ΔS^{rt}, the single term in the whole calculation that explicitly depends on concentration and leads to the law of mass action. Other gas phase entropy terms concern internal degrees of freedom such as side-chain rotation and main-chain mobility.

5. A structural interpretation of affinity

Table 2 quotes results of our calculation applied to three systems. Gas phase electrostatic and van der Waals interactions are intense and yield a large ΔH^{gaz}. Upon transfer to water, protein-solvent interactions are lost where the protein surface is buried at the protein-protein interface. This explains the large negative ΔH^{hyd}. Compensation of protein-protein by protein-solvent interactions is incomplete in the case of non-polar interactions (van der Waals forces). Its extent changes with the system considered for polar interactions. The subtilisin-eglin protease-inhibitor interface has many neutral polar groups forming relatively weak hydrogen bonds. Burying these polar groups costs more enthalpy than the hydrogen bonds are worth. In contrast, the barnase-barstar interface is rich in charged groups. Burying them costs 82 kcal.mol^{-1}, whereas the hydrogen bonds and salt bridges they form are worth 100 kcal.mol^{-1}. Overall, polar interactions favour association.

A point of major importance should be made here: the calculation has a large error bar. E^{elec} is extremely sensitive to both the position of electric charges in space and to the modelling of dielectric effects. Here, I used a continuous dielectric model with a uniform dielectric constant $\varepsilon=3$. This is a plausible gas phase value, yet it is uncertain and E^{elec} changes by 100 kcal.mol^{-1} between $\varepsilon=2$ and $\varepsilon=4$. Therefore, the good agreement observed in Table 2 between the calorimetric value of the enthalpy of lysozyme-antibody dissociation and the calculated value is largely spurious. Nevertheless, we can accept it as qualitative and extrapolate the calculated values to other temperatures than 25°C, since the heat capacities and entropies are also fitted.

In any case, Table 2 fully vindicates the ideas of Pauling & Delbrück [1]. Van der Waals interactions do stabilize protein-protein complexes, the reason being that the atomic packing at a protein-protein interface is more compact on the average than at a water-protein interface. As predicted by Pauling & Delbrück, the interface forms between surfaces having complementary shapes 'like die and coin'. Electrostatic interactions play a more ambiguous part. We find them to favour either association or dissociation depending on the system, but also on the way the calculation is made. Experimental data from site-directed mutagenesis come to the rescue of theory in this case, but here again, the answer is ambiguous. In the barnase-barstar complex, residues Arg 59 of barnase and Glu 76 of barstar make a salt bridge. The R59A substitution raises K_d by four orders of magnitude, equivalent to $\Delta\Delta G_d = 5.2$ kcal.mol^{-1} (Table 1). Is that the salt bridge energy? No, because mutation E76A of barstar has only a minor effect on affinity. Arg 59 makes other interactions and the effective value of the salt bridge must be much less than 5.2 kcal.mol^{-1}.

Table 2: Calculated thermodynamic parameters for three complexes at 25°C.

Complex (kcal.mol⁻¹) $(kcal.mol^{-1})$	Lysozyme-HyHEL5	Subtilisin-eglin	Barnase-barstar
Gas phase			
E^{vdw}	59	64	82
E^{elec} (ε=3)	66	31	100
ΔH^{gas} (a)	125	95	182
$-T\Delta S^{rt}$ (b)	⁻15	-15	-15
$-T\Delta S^{sc}$ (c)	⁻24	-15	-18
$-T\Delta S^{vib}$ (d)	⁻25	(⁻25)	(⁻25)
$-T\Delta S^{gas}$	-64	(-55)	(-58)
ΔG^{gas}	61	(40)	(124)
Hydration (e)			
ΔC (kcal.mol⁻¹.K⁻¹)	0.26	0.29	0.21
ΔH^{npol}	-15	-13	-15
ΔH^{pol}	-91	-66	-82
ΔH^{hyd}	--106	-79	-97
ΔG^{npol}	16	18	12
ΔG^{pol}	-69	-50	-63
ΔG^{hyd}	-53	-32	-51
Total calculated			
ΔH^{calc}	19	16	82
ΔG^{calc}	8	(8)	(73)
Experimental			
ΔH_d (f)	22.6	-	-
ΔG_d (f,g)	14.5	13.1	19.0

(a) van der Waals and electrostatic energies of interaction were calculated with X-PLOR after 300 cycles of conjugated gradient minimization; the dielectric constant was ε=3. (b) rotational/translational entropy loss estimated as in [22]. (c) side-chain immobilisation entropy estimated as in [23]. (d) for the lysozyme-HyHEL5 complex, the loss of vibrationnal entropy was estimated by comparing the calculated ΔC_d to the calorimetric value and converting the difference into an entropy change with the help of an empirical relation proposed by Sturtevant [24]; in the absence of calorimetric data, the same value of ΔS^{vib} was used for the other two complexes. (e) coefficients from [19] were applied to the buried surface areas. (f) calorimetric data at 25°C [6]. (g) data from [25] and [2].

12

On the other hand, every entropy term listed in Table 2 favours dissociation. This is obvious for ΔS^{rt} (calculated at $c_\emptyset = 1M$) and for the side-chain entropy. This is also true of vibrational entropy, because association usually makes the main chain more rigid, in the interface at least. Nevertheless, there is a major entropy term that favours association: the entropy derived from the hydrophobic effect. It is implicit in our analysis and can be made explicit by comparing the numbers ΔH^{npol} and ΔG^{npol} in Table 2 that refer to the hydration of non-polar groups. They are of opposite sign and their difference, about 30 kcal.mol^{-1} in all three complexes, points to a large stabilizing entropy contribution of the hydrophobic effect. However, the hydrophobic effect is by no means exclusively entropic. Interactions between non-polar groups at the interface are stronger than with the solvent and their enthalpy favour association. Moreover, it should be recalled that the heat capacity of non-polar hydration is large, making these values highly temperature-dependent [11].

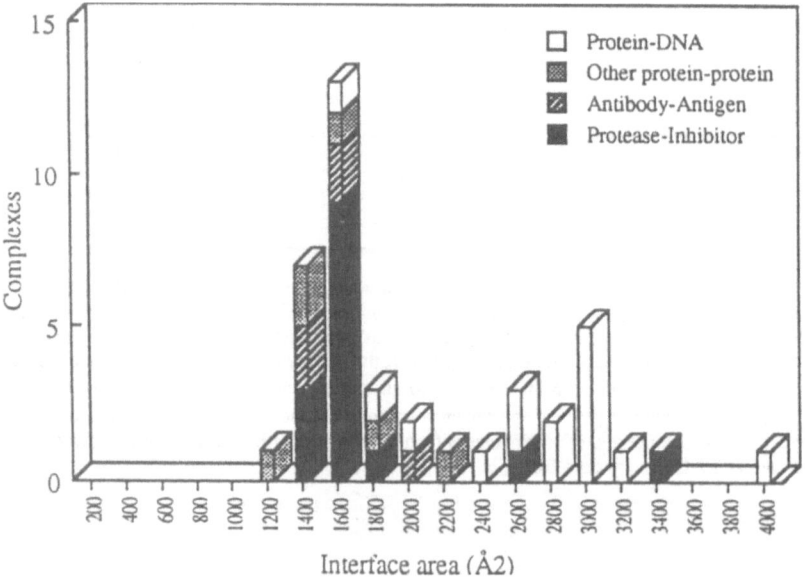

Figure 2: Histogram of interface areas in protein-protein and protein-DNA complexes. A sample of 15 protease-inhibitor complexes including trypsin-PTI [26] and subtilisin-eglin [27]; 5 antigen-antibody complexes including lysozyme-D1.3 [28] and lysozyme-HyHEL5 [29]; 6 other protein-protein complexes including barnase-barstar [30]; and 15 protein-DNA complexes. References and individual *B* values are listed in [19] from which this figure is taken. Two protease-inhibitor complexes have *B*>2000 Å^2, both are thrombin-hirudin complexes [31].

6. A correlation between affinity and interface areas

Since no exact and reliable number can yet be derived from theory, a logical alternative is to turn to an empirical analysis. The simplest hypothesis has affinity be a function of the size of the contact zone, easily estimated as the interface area B. The hydrophobic contribution, the number of van der Waals contacts and hydrogen bonds, are more or less in proportion with B [12-13]. Data on the water solubility of alkanes at 25°C suggest a linear correlation [14]:

$$\Delta G_d \approx \gamma B + \text{constant} \tag{11}$$

where $\gamma \approx 25$ cal.mol^{-1}.Å^{-2}. Fig. 2 shows that a large majority of protease-inhibitor and antigene-antibody complexes have $B = 1500 \pm 250$ Å^2 or $\gamma B = 38 \pm 6$ kcal.mol^{-1}. This being independent of their affinity implies that the two are uncorrelated. Moreover, some much larger interfaces exist, in the thrombin-hirudin complex for instance. Leech hirudin has for thrombin an affinity comparable to that of PTI for trypsin, whereas B is twice larger and the mode of interaction is quite different. In addition to a globular domain that blocks the protease active site, hirudin has a 17 residue C-terminal tail that occupies an external site. In solution, RMN shows the tail to be disordered. Thus, binding involves a transition from disorder to order. This is observed in some protein-protein complexes and in the majority of protein-ADN complexes for which structural data are available. Fig. 2 shows protein-ADN complexes to have interfaces with $B \approx 3000$ Å^2 and above, twice that of protein-protein complexes. In a more general way, any departure from rigid-body association destroys the correlation between B and ΔG_d, and also any departure from complementarity. Thus, mutating a residue at the interface may perfectly well change the affinity but not the interface area. Mutant R59A of barnase in Table 1 is not the example we want, because residue Arg 59 contributes 20% of the barnase surface that is buried in contacts with barstar. A better example is mutant R69K of hen lysozyme. The mutation drops the affinity for antibody HyHEL5 by a factor of 10^3, yet it has essentially no effect on the structure of the complex and the value of B [15].

Nevertheless, Horton & Lewis [16] showed that a correlation exists between B and ΔG_d. Albeit poor in Eq. 11 (R≈10%), it becomes significant (R=80%) after removing complexes that depart from rigid-body approximation, and excellent (R=96%) if we take the chemical nature of the buried groups into account to write:

$$\Delta G_d \approx \sum \sigma_i B_i + \text{constant} \tag{12}$$

14

The σ coefficients in the weighted sum are derived from those of Eisenberg & MacLachlan [10]; for non-polar groups, they are very close to the γ of Chothia [14]. The constant noted $-\Delta G_{(0)}$ in Fig. 3 is the expected value of ΔG_d for $B=0$, when no contact is made. It includes among other things the entropic cost of correlating the positions and orientations of the two components of a complex [22].

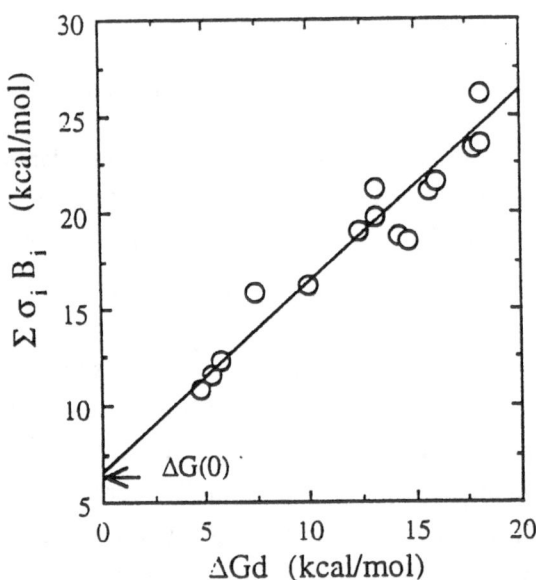

Figure 3: A linear correlation between the interface area and the free enthalpy of dissociation. The weighted sum $\Sigma\, \sigma_i\, B_i$ is plotted against the experimental ΔG_d at 25°C. B_i is the contribution of the five atom types i (neutral C, O, N and S, charged O or N) to the interface area B in 15 protein-protein or protein-peptide complexes. The regression line has unit slope and extrapolates to $\Delta G_{(0)}$ on the vertical axis and to $-\Delta G_{(0)}$ for $B=0$ on the horizontal axis. Adapted from [16].

7. The statistical physics of specificity

Whereas a single number, K_d or ΔG_d, suffices to quantify affinity, specificity is by no means as easy to characterize. This is often done by comparing the affinity of a same protein for two ligands having related structures, for instance the same antibody for two variants of an antigen. This type of measurement poorly describes a situation that prevails *in vivo* and *in vitro* : competition is not between related molecules, but between the specific epitope and an extremely heterogeneous mixture of molecular surfaces, all present in solution. Even though they have only a weak affinity for the antibody, they are in very large number and form transitory interactions with the antibody. At least, they contribute to the background that limits the sensivity of the test for the specific epitope.

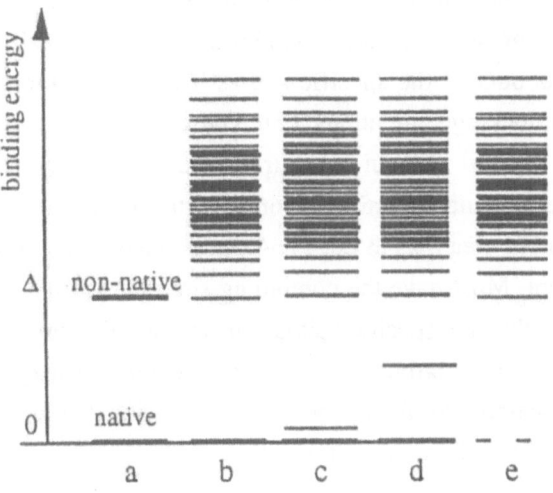

Figure 4: A statistical physics view of specificity. A binding energy is attributed to each possible mode of association of ligand B to the protein. Only the native mode yields a specific complex. It competes with non-native modes where the same ligands binds in 'wrong' orientations at the specific site, or binds at other sites on the protein, or where competing ligands bind. Five situations are considered: (a) a two-state model; (b) the native mode is unique and separated from a quasi-continuous spectrum of non-native modes by a gap Δ; (c) multiple specificity with more than one native mode; (c) non-native modes below the continuous spectrum; (d) no specific binding.

Such a situation can be analyzed with the methods of statistical physics by enumerating all possible modes of association and listing their energies E. To do so, I applied to specificity the so-called Random Energy Model which was originally developed to describe complex physical systems like spin glasses [17]. It is already in use in protein folding studies [18]. If the system has very many states, they form a quasi-continuous energy spectrum described by the spectral density function m(E) or the related entropy function S(E) = R ln m(E). In the presence of competing ligands, m(E) and S(E) depend on their concentrations. At thermodynamic equilibrium, complexes are distributed between modes following Boltzmann law. In Fig. 4 , the native mode natif has zero energy by convention, and also zero entropy, being unique. Non-native binding modes compose the spectrum.

The simplest model that represents two ligands competing for the same site is the two-state model (a). The energy gap Δ between the two states is equivalent to ΔΔG_d and it fully describes the system. The situation in spectrum (b) is more realistic. It has a native mode where ligand B binds protein A by specific recognition. The native mode

16

competes with others which have an energy no less than Δ and form a quasi-continuous spectrum above Δ. These modes may yield complexes with non-specific ligands, or complexes where B binds outside the specific site as it does in computer-generated complexes resulting from docking simulations. Fig. 5 describes a simulation where the docking algorithm places hen lysozyme in an arbitrary orientation at the combining site of antibody HELHy5. We attribute the artificial complex thus formed, an energy that is a linear function of the buried area B as in Eq. 11. All possible orientations are tested to derive the energy spectrum. Most have the combining site interacting with regions of the lysozyme surface other than the specific epitope and structurally unrelated to it. One may therefore assume that non-native modes generated with lysozyme alone are representative of the interaction with the many different proteins that are present under physiological conditions or in an immunochemical test.

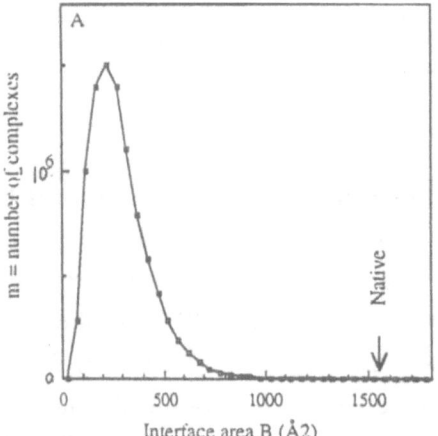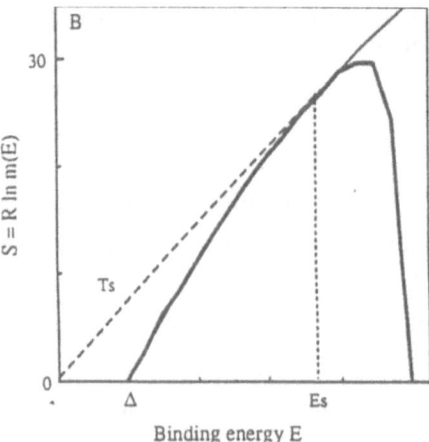

Figure 5: The energy spectrum for hen lysozyme association with antibody HyHEL5. Association is simulated in the computer by docking the antigen molecule onto the combining of the antibody in all possible orientations (32-33). The orientation observed in the crystal structure of the complex yields the native mode, all others yield non-native modes. Each has its interface area B. (A) Histogram of B values. (B) An entropy/energy curve is derived from the histogram by taking E to be a linear function of B and writing S = R ln m. Temperature is given by $\frac{\partial S}{\partial E} = \frac{1}{T}$. Non-native modes are a majority above the specificity transition temperature T_S defined by the dashed tangent line.

A histogram of B values achieved after docking show that the great majority of non-native modes bury much less surface area than the native $B \approx 1500$ Å^2. Their energy is therefore much higher. Still, they are many, and they dominate the equilibium distribution when the temperature is above T_s, the value where E-T_sS≈ 0. T_s is the specificity transition temperature, given by the slope of the tangent from the origin to the S(E) curve. More generally, the non-native-to-native ratio at equilibrium is equal to the partition function:

$$r = \int_\Delta^\infty m(E) \exp{-\frac{E}{RT}} \, dE = \int_\Delta^\infty \exp{-\frac{E\text{-}TS}{RT}} \, dE \qquad (13)$$

r is a measure of specificity in a given environment; its value depends on the shape of the energy spectrum rather than on the gap Δ alone as it does in a two-state model. Indeed, the non-native modes that compete with the native are not the ones of low energy $E \approx \Delta$, which are few, but those of low free energy E-TS, their large number compensating for their higher energy. The composition of the solution, the types and concentrations of all chemical species it contains, all affect m(E). If for instance we consider the ratio r is the noise-to-signal ration in an immuno-detection experiment, we expect r to increase when the solution becomes more complex and heterogeneous even if we keep constant the total concentration of contaminating species, because the m(E) distribution should become wider under such conditions.

8. Conclusion

The forces that stabilize protein-protein complexes and make specific recognition possible are ubiquitous. Therefore, they play very different functions in different physical and biological processes. Two extreme situations are protein crystallization and the assembly of oligomeric proteins. When polypeptide chains assemble to make an oligomer, the component structures often exist only in a transitory way, and their affinity is almost always much to high for a dissociation constant to be measured. There, the rigid-body approximation rarely applies. It does when proteins crystallize, a phenomenon that has much in common with protein-protein recognition even though it lacks specificity [19]. Crystallization can be viewed as a conter-example to specific recognition, the study of one helping in understanding the other by pinpointing the differences between specific and non-specific. The statistical physics model developed above does apply to crystallization - but there is no native mode in this case. We

18

submitted packing interfaces in protein crystals to a statistical analysis [20] and found that interfaces observed in a crystal bury no more surface than random interfaces formed by docking in the computer. Crystal packing contacts are a potential source of experimental data on non-specific protein-protein interaction. In addition to a statistical approach, they can be submitted to site-directed mutagenesis and to the many other theoretical and experimental approaches that were developed for the study of specific recognition, examples of which are the antigen-antibody, protease-inhibitor and the barnase-barstar complexes analyzed here.

9. Acknowledgements

We acknowledge a long standing collaboration with Dr. C. Chothia (Cambridge) and S. Wodak (Brussels) on the subject of protein-protein recognition. Various stages of the studies described here have involved Dr. J. Cherfils, F. Rodier, S. Duquerroy and C. Robert in my laboratory.

10. References

1. Pauling, L. and Delbrück, M. (1940). The nature of the intermolecular forces operative in biological processes. *Science* 92:77-79.
2. Schreiber, G. & Fersht, A. (1995) Energetics of protein-protein interactions: analysis of the barnase-barstar interface by single mutations and double mutants cycles*J. Mol. Biol.* 248:478 .
3. Schreiber, G. & Fersht, A. (1996) Mechanism of rapid, electrostatically assisted, association of proteins. *Nature Struct. Biol.* 3:427-431.
4. Hawkins, R.E., Russell, S.J., Baier, M. & Winter, G. (1993). The contribution of contact and non-contact residues of antibody in the affinity of binding to antigen. The interaction of mutant D1.3 antibodies with lysozyme. *J. Mol. Biol.* 234:958-964.
5. Wiseman, T., Williston, S., Brandts, J.F. and Lin, L.N. (1989). Rapid measurement of binding constants and heats of binding using a new titration calorimetry. *Anal. Biochem.* 179:131-137.
6. Hibbits, K.A., Gill, D.S. and Wilson, R.C. (1994). Isothermal titration calorimetric study of the asociation of hen egg lysozyme and the anti-lysozyme antibody HyHEL5. *Biochemistry* 33:3584-3590.
7. Janin, J. (1995a). Elusive affinities. *Proteins* 21:30-39.
8. Makhatadze, G.I. & Privalov, P.L. (1995). Energetics of protein structure *Adv. Prot. Chem.* 47:307-425.
9. Ooi, T., Oobatake, M., Nemethy, G., Scheraga, H.A (1987). Accessible surface areas as a measure of the thermodynamic parameters of hydration of peptides. Proc. Nat. Acad. Sci. USA 84:3086-3090.
10. Eisenberg, D. and McLachlan, A.D. (1986). Solvation energy in protein folding and binding. *Nature* 319:199-203.
11. Privalov, P.L., Gill, S.J. (1988). Stability of protein structure and hydrophobic interaction. *Adv. Prot. Chem.* 39:193-234.
12. Chothia, C. and Janin, J. (1975). Principles of protein-protein recognition. *Nature* 256:705-708.
13. Janin, J. & Chothia, C. (1990) The structure of protein-protein recognition sites. *J. Biol. Chem.* 265:16027-16030.
14. Chothia, C. (1974). Hydrophobic bonding and accessible surface area in proteins. *Nature* 248:338-339.
15. Chacko, S., Silverton, E., Kam-Morgan, L., Smith-Gill, S., Cohen, G. and Davies, D. (1995). Structure of an antibody-lysozyme complex. Unexpected effect of a conservative mutation. *J. Mol. Biol.* 245:261-274.

16. Horton, N. and Lewis, M. (1992). Calculation of the free energy of association for protein complexes. *Protein Sci.* 1:169-181.
17. Janin, J. (1996). Quantifying biological specificity: the statistical mechanics of molecular recognition. *Proteins* (in press).
18. Bryngelson, J.B., Wolynes, P.G. (1987) Spin glasses and the statistical mechanics of protein folding. *Proc. Natl. Acad. Sc. USA* 84: 7524-7528.
19. Janin, J. (1995).Principles of protein-protein recognition from structure to thermodynamics. *Biochimie* 77: 497-505.
20. Janin, J. & Rodier, F. (1995). Protein-protein interaction at crystal contacts. *Proteins* 22:580-587.
21. Vincent, J.P. & Lazdunski, M. (1972). Trypsin-pancreatic trypsin inhibitor association. Dynamics of the interaction and the role of disulfide bridges. *Biochemistry* 11:2967
22. Finkelstein, A.V. & Janin, J. (1989). The price of lost freedom: entropy of bimolecular complex formation. *Protein Engin.* 3:1-3.
23. Pickett, S.D. and Sternberg, M.J.E. (1993). Empirical scale of side chain conformational entropy in protein folding. *J. Mol. Biol.* 231:825-839.
24. Sturtevant, J.M.(1977). Heat capacity and entropy changes in processes involving proteins. *Proc. Nat. Acad. Sci. USA* 74:2236-2240.
25. Ascenzi, A., Amiconi, G., Menagatti, E., Guarneri, M., Bolognesi, M., Schnaebli, H.P. (1988). Binding of the recombinant proteinase inhibitor Eglin C from leech Hirudo medicinalis to human leukocyte elastase, bovine α-chymotrypsin and subtilisin Carlsberg: Thermodynamic study. J. *Enzyme Inhibition* 2:167-172.
26. Huber, R., Kukla, D., Bode, W., Schwager, P., Bartels, K., Deisenhofer, J., Steigemann, W. (1974). Structure of the Complex formed by Bovine Trypsin and Bovine Pancreatic Trypsin Inhibitor. II Crystallographic Refinement at 1.9 Å Resolution. *J. Mol. Biol.* 89:73-101.
27. Bode, W., Papamokos, E., Musil,D (1987). The High-Resolution X-Ray Crystal Structure of the Complex Formed Between Subtilisin Carlsberg and Eglin C, an Elastase Inhibitor from the Leech Hirudo Medicinalis. Structural Analysis, Subtilisin Structure and Interface Geometry. *Eur. J. Biochem.* 166:673-692.
28. Amit, A.G., Mariuzza, R.A., Phillips, S.E.V., Poljak, R.J. (1986). Three-dimensional structure of an antigen-antibody complex at 2.8 Å resolution. *Science* 233:747-753.
29. Sheriff, S., Silverton, E. W., Padlan, E. A., Cohen, G. H., Smith-Gill,S. J., Finzel, B. C., Davies D. R. (1987) Three-Dimensional Structure of an Antibody-Antigen Complex. *Proc. Nat. Acad. Sci. USA* 84:8075-8079.
30. Guillet, V., Lapthorn, A., Hartley, R.W., Mauguen, Y. (1993). Recognition between a bacterial ribonuclease, barnase, and its natural inhibitor, barstar. *Structure* 1:165-177.
31. Rydel, T.J., Tulinsky, A., Bode, W., Huber, R. (1991). The Refined Structure of the Hirudin-Thrombin Complex. *J. Mol. Biol.* 221:583-601.
32. Cherfils, J., Duquerroy, S., Janin, J. (1991). Protein-protein recognition analyzed by docking simulation. *Proteins* 11:271-280.
33. Janin, J. & Cherfils, J. (1993). Protein Docking Algorithms: Simulating Molecular Recognition. *Current Opinion Struct. Biol.* 3:265-269.

APPROACHES TO PROTEIN-LIGAND BINDING FROM COMPUTER SIMULATIONS

WILLIAM L. JORGENSEN, ERIN M. DUFFY,[‡] JONATHAN W. ESSEX,[&] DANIEL L. SEVERANCE,[§] JAMES F. BLAKE,[‡] DEBORAH K. JONES-HERTZOG,[#] MICHELLE L. LAMB and JULIAN TIRADO-RIVES
Department of Chemistry, Yale University
New Haven, Connecticut 06520-8107, USA

Abstract. Accurate computation of protein-ligand binding affinities is a challenging goal with great potential value in the design of therapeutic agents. Applications of statistical mechanics simulations to the problem are considered that feature full atomic-level descriptions of the protein, ligand and aqueous environment. Basic concepts on the methodology and intermolecular interactions in solution are presented along with results of Monte Carlo simulations for binding of inhibitors by trypsin and thrombin.

1. Introduction

The ability to compute the effects of structural changes in a ligand on its binding affinity with a protein has great potential importance in drug design. Modifications that lead to enhanced binding affinity may be beneficial in increasing specificity and lowering toxicity for inhibitors. Diverse approaches are being taken to the problem ranging from ligand docking and growing procedures with empirical scoring functions to fluid simulations with full atomic detail.[1] We have pursued the latter course owing to the direct connections that can be made between observed and predicted free energies of binding and to the detailed structural information that can be obtained to help understand the variations in binding. In this case, there are several key choices: (a) the sampling procedure, e.g., molecular dynamics or Monte Carlo statistical mechanics, (b) representation of the solvent as a continuum or as discrete molecules, (c) the force field, and (d) the methodology for the free energy calculations.

Though molecular dynamics (MD) has been the dominant choice of sampling procedure, recent results suggest that Monte Carlo (MC) calculations may be particularly efficient for conformational sampling of protein side chains.[2] In view of this and numerous successes of MC simulations for organic host-guest complexation,[1c,3]

Current affiliations: [‡] Pfizer Inc., Groton, CT. [&] Univ. of Southampton, UK. [§] Roche Bioscience, Palo Alto, CA. [#] Neurogen Corp., Branford, CT.

G. Vergoten and T. Theophanides (eds.), Biomolecular Structure and Dynamics, 21–34.
© 1997 *Kluwer Academic Publishers.*

recent work in our laboratory has explored the MC approach for protein-ligand binding. The solvent, water, is represented as discrete molecules with the TIP4P potential;[4] the description of specific interactions such as hydrogen bonding with continuum models is a concern for their application along with the loss of detail on variations in water structure. The remaining interactions in our studies have employed the OPLS force fields, which have been parameterized to give correct conformational energetics and properties for organic liquids.[5] Finally, free energy changes can be computed rigorously with free-energy perturbation (FEP) or thermodynamic integration (TI) methods.[1c,d] In both approaches, a free energy change is computed for converting a system with a molecule A to one with a molecule B over a series of non-physical intermediate states. The two methods are closely related and are comparably effective; they differ primarily by a choice of computing the total energy difference between the reference and perturbed state or the derivative of the total energy with respect to a coordinate that represents the perturbation. The calculations are computationally taxing owing to the need to run the series of calculations for the intermediate states with adequate configurational sampling for each. Consequently, more efficient, approximate methods such as Åqvist's linear interaction energy procedure are being evaluated for facile treatment of larger numbers of structurally diverse ligands.[6-9]

2. Methodology

Results are presented here for free energy profiles (potentials of mean force - pmf) for the association of two prototypical systems, N-methylacetamide (NMA) dimer and propane dimer, in order to illustrate fundamental points about binding in solution. Then, results are presented for complexation of series of benzamidine inhibitors with trypsin and of sulfonamides with thrombin. The free energy calculations for the pmfs and trypsin were performed with the FEP method, while those for thrombin used an approximate approach, linear-response plus surface area (LRSA). Monte Carlo sampling was performed with the BOSS program[10] for the free energy profiles and with MCPRO[11] for the protein-ligand systems.

Force Field and Sampling. A classical force field is used in which the energy expression consists of harmonic terms for bond stretching and angle bending, a Fourier series for each torsional angle, and Coulomb and Lennard-Jones interactions between atoms separated by three or more bonds (eqs 1-4). The latter "non-bonded" interactions are also evaluated between intermolecular atom pairs, and they are reduced by a factor of 2 for intramolecular 1,4-interactions. The calculations for the NMA and propane dimers

$$E_{bond} = \Sigma_i k_{b,i} (r_i - r_{0,i})^2 \tag{1}$$

$$E_{bend} = \Sigma_i k_{\vartheta,i} (\vartheta_i - \vartheta_{0,i})^2 \tag{2}$$

$$E_{torsion} = \Sigma_i \{ V_{1,i}(1 + \cos \varphi_i)/2 + V_{2,i}(1 - \cos 2\varphi_i)/2 + V_{3,i}(1 + \cos 3\varphi_i)/2 \} \tag{3}$$

$$E_{nb} = \Sigma_i \Sigma_j \{ q_i q_j e^2/r_{ij} + 4 \varepsilon_{ij} [(\sigma_{ij}/r_{ij})^{12} - (\sigma_{ij}/r_{ij})^6] \} \tag{4}$$

used the OPLS-AA (all-atom) force field.[5c] The protein inhibitors were also represented in an all-atom format with OPLS Lennard-Jones parameters and partial charges obtained by fitting to electrostatic potential surfaces from *ab initio* RHF/6-31G* calculations for the benzamidines and RHF/6-31+G* calculations for the sulfonamides.[12] The proteins are described by the OPLS force field with all hydrogens explicit except those on aliphatic carbon.[5a,b] The TIP4P water and OPLS chloroform[13] molecules are treated as rigid bodies that only translate and rotate, while the sampling for NMA, propane and the inhibitors included translations, rotations, bond angle variations, and torsional motion in each case. The calculations for NMA and propane also included bond length variations, so these OPLS-AA models are fully flexible. For the proteins, attempted MC moves involved variation of the bond angles and dihedral angles for the side chain of one randomly-picked residue at a time; the protein backbone has been held fixed in the present calculations.

Free Energy Methods. The FEP calculations use the Zwanzig expression (eq 5) to compute the free energy change between the reference system 0 and the perturbed system 1.[1c,d] The average is taken for sampling configurations of the reference sytem.

$$\Delta G(0 \rightarrow 1) = - k_B T \ln <\exp [-(E_1 - E_0)/k_B T]>_0 \qquad (5)$$

For the pmf calculations, the perturbations are for a reaction coordinate that is defined as a CO --- HN distance for the NMA dimer and as the C2 --- C2 distance for the propane dimer. For the trypsin binding calculations, perturbations were made to convert one ligand to another using the thermodynamic cycle shown below. The

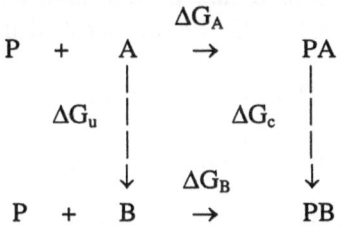

corresponding reaction coordinate involves a coupling parameter, λ, that causes one molecule to be smoothly mutated to the other by changing the force field parameters and geometry. The difference in free energies of binding for the ligands A and B then comes from eq 6. Two series of mutations are performed to convert A to B unbound in water and complexed to the protein, which yield ΔG_u and ΔG_c.

$$\Delta\Delta G_b = \Delta G_A - \Delta G_B = \Delta G_u - \Delta G_c \qquad (6)$$

The approximate approach that is illustrated here for binding with thrombin follows from the work of Åqvist et al.[6] They introduced a procedure based on linear response (LR) theory for estimating free energies of binding. In this model, the free energy of interaction of a solute with its environment is given by one-half the electrostatic (Coulombic) energy plus the van der Waals (Lennard-Jones) energy scaled by an empirical parameter, α. For binding a ligand to a protein, the differences in the interactions between the ligand in the unbound state and bound in the complex then provide an estimate of the free energy of binding, ΔG_b, via eq 7. A value of 0.162 for α

$$\Delta G_b = \beta \, \Delta{<}E_{elec}{>} \quad + \quad \alpha \, \Delta{<}E_{vdw}{>} \tag{7}$$

was determined empirically by fitting to experimental data for a series of inhibitors.[6] The required energy components were obtained from molecular dynamics simulations for the inhibitors in water and for the protein-inhibitor complexes in water. Key advantages over FEP methods are (a) absolute free energies of binding are readily obtained, and (b) only simulations at the endpoints of a mutation are required, i.e., the simulations for the intermediate states are eliminated, which allows much easier application to structurally diverse sets of molecules. In spite of the approximations in eq 7, the approach has yielded promising results for several applications.

Extension of the LR approach to calculate free energies of hydration (ΔG_{hyd}) incorporated a third term proportional to the solute's solvent-accessible surface area (SASA), as an index for cavity formation within the solvent.[7] The latter term is needed for cases with positive ΔG_{hyd} such as alkanes. It was also found that additional improvement occurred when both α and β were allowed to vary. Eq 8 gives the corresponding LRSA expression for ΔG_b. Results of calculations with this relationship

$$\Delta G_b = \beta \, \Delta{<} E_{elec} {>} + \alpha \, \Delta{<} E_{vdw} {>} + \gamma \, \Delta{<}SASA{>} \tag{8}$$

are presented here for a series of sulfonamide inhibitors with human thrombin.

System Setup and MC Details. The pmf calculations were performed in the isothermal, isobaric (NPT) ensemble at 25 °C and 1 atm. The systems consisted of the solutes plus 740 TIP4P water molecules or 189 OPLS chloroform molecules in a rectangular cell, ca. 26 x 26 x 38 Å, with periodic boundary conditions. The reaction coordinate was changed by 0.125 Å for each perturbation, so 0.25 Å was covered in each simulation with double-wide sampling. The intermolecular interactions were spherically truncated at 12 Å separations between molecular centers with quadratic feathering to zero over the last 0.5 Å. Convergence was carefully studied and achieved by very long MC runs. Each simulation for chloroform involved 10M configurations of equilibration followed by 16M configurations of averaging. The corresponding numbers for water were 16M and 24M configurations.

The MC simulations for the proteins and inhibitors were carried out in water caps with 20 Å radius at 37 °C for thrombin and at 15 °C for trypsin to be consistent with the temperatures for the experimental binding studies. For the complexes, amino acid residues more than ca. 18 Å from the binding site were removed and only residues within ca. 16 Å were active (sampled). For trypsin, 153 of the original 223 residues were retained, while 164 of 247 residues were included for thrombin. The number of water molecules in the calculations is ca. 1100 for the unbound inhibitors and ca. 475 for the complexes. An attempt to move a protein residue was made every 10 configurations and the period was 115 configurations for the benzamidines and 24 configurations for the more flexible sulfonamides. The remaining moves were for the water molecules. The protein fragments were neutral, so no couterions were added. Residue-based cutoffs were used with truncation of the non-bonded interactions at 12-15 Å. The FEP calculations for the benzamidine mutations were performed for 11 windows with $\Delta\lambda$ = 0.05. The MC run for each window consisted of 10M configurations of equilibration and 5M configurations of averaging. The LRSA calculations for the sulfonamides covered ca. 10M configurations for equilibration, followed by averaging for ca. 10M configurations with the unbound inhibitors and for at least 3M configurations with the complexes. Initial coordinates for the calculations were derived from the 3ptb entry for the bovine trypsin-benzamidine complex and the 1dwc entry for the human thrombin-MD805 complex in the Brookhaven Protein Data Bank.[14]

3. Results for the NMA and Propane Dimers

The results of gas-phase geometry optimizations and of the pmf calculations in water and chloroform for the NMA and propane dimers are summarized in Figure 1. The results provide important insights into the value of hydrogen bonding and solvophobic interactions for binding in solution. The lowest energy structure for the NMA dimer in the gas phase has the expected hydrogen bond and a net interaction energy of 8.7 kcal/mol with the OPLS-AA force field. Of course, the optimal interaction between non-polar propane molecules is much weaker at 1.9 kcal/mol. For reference, the lowest energy structure for the water dimer in the gas phase has an interaction energy of 6.2 kcal/mol with the TIP4P model. It should be noted that these potential functions give excellent descriptions of the structure and thermodynamic properties of the corresponding pure liquids including only ca. 1% errors for heats of vaporization and densities.[4,5]

The pmfs illustrate the free energy profiles for association of the dimers in solution. This now includes the effects of thermal and orientational averaging and the competition between the association of the solutes with each other and with the solvent. For NMA in chloroform, association still occurs and yields a hydrogen-bonding well with a depth of ca. 2.0 kcal/mol at a CO---HN separation of 1.9 Å. However, in water, the pmf for the NMA dimer is purely repulsive, so there is no association of the amides. Formation of an NMA-NMA hydrogen bond in water is not favorable enough to offset the poorer hydration of the complex than the separated monomers. The small size of water coupled with its four hydrogen-bonding sites makes it the most effective agent for

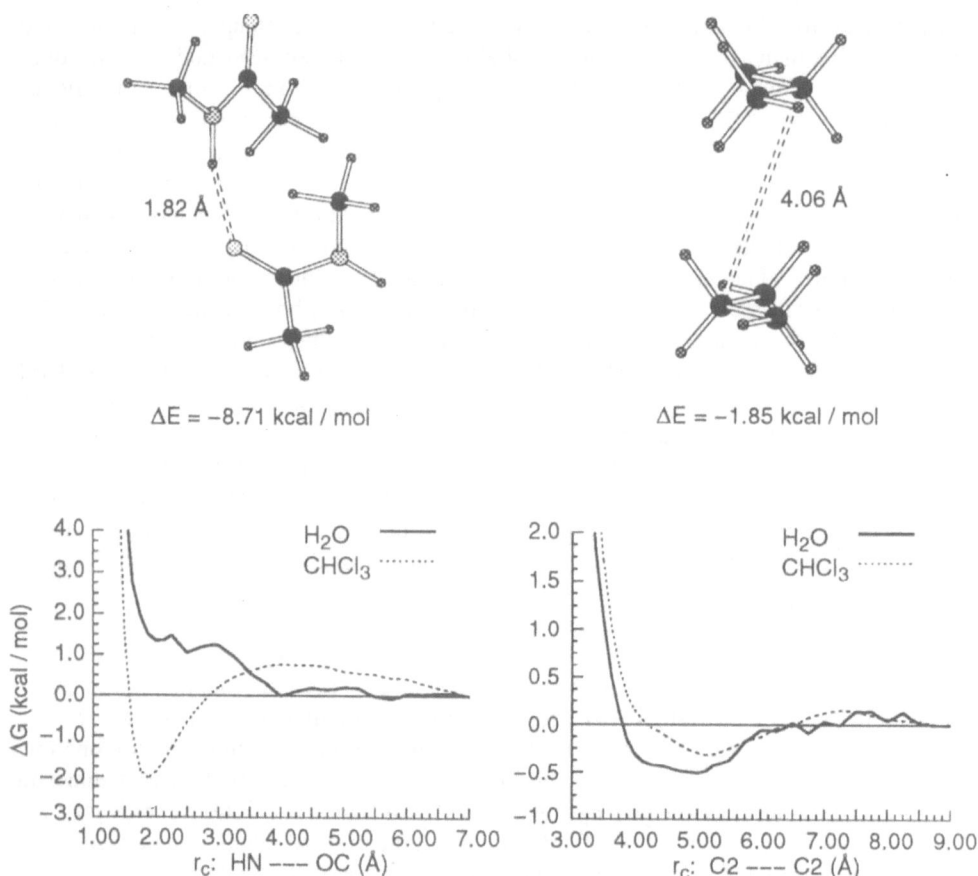

Figure 1. (*Top*) Structures and interaction energies from gas–phase optimizations with the OPLS–AA force field for NMA (*left*) and propane (*right*) dimers.

(*Bottom*) Computed pmfs for the separation of the dimers in water and chloroform at 25 °C. The reaction coordinate is the distance in Å between the amide H of one NM and the carbonyl O of the other (*left*) or the central carbon of one propane monomer to the other (*right*).

disrupting hydrogen bonds. As noted previously, when forced to short separations, the NMA dimer prefers to form stacked structures that keep the CO and NH units exposed for maximal hydrogen bonding with water.[15] The present results are also fully consistent with experimental observations that while NMA dimer has an association constant, K_a, of ca. 3 M^{-1} in chloroform, association is not observed spectroscopically in aqueous solution.[16]

The results for the propane dimer in solution are qualitatively different. The configurational averaging makes the free energy wells in solution quite shallow; however, association of propane dimer in both chloroform and water is favorable. Indeed, the attraction is significantly greater in water, which is consistent with basic notions about hydrophobic association.[17] The attraction between hydrophobic groups is enhanced in water, while that between hydrogen-bonding groups is nullified. The implications for protein-ligand binding and protein folding, which is akin to protein self-binding, are profound. The most straightforward gains in stability come from removing non-polar regions from water and burying them in a non-polar region of the protein. Formation of a hydrogen bond between a ligand and an isolated hydrogen-bonding site of a protein is generally unfavorable in view of the loss of hydrogen bonding to water. Ligand binding to the outer surface of a protein is generally not observed owing to the avoidance of substantial non-polar patches on the surface and the extensive hydration of polar surface groups. Ligand binding in cavities is the norm. Cavities may not be fully hydrated and are capable of supporting larger patches of non-polar surface. It is important that the ligand replaces hydrogen bonds that are lost between cavity water molecules and the protein , while deriving the key stability gains from matching up non-polar regions of ligand and cavity. Exceptions to the dominance of hydropohobic association are possible when there is an exquisite match between a constellation of hydrogen bonding groups in the protein cavity and a complementary rigid array of hydrogen-bonding sites on the ligand. In this case the rigidity of the ligand can lead to entropy and enthalpy gains over the competing cluster of water molecules in the cavity. This situation appears to occur for the hallmark streptavidin-biotin complex, which features five hydrogen bonds between the protein and ureido group in addition to burial of the non-polar surface area of the tetrahydrothiophene ring and attached butyl chain.[18]

4. Results of MC/FEP Calculations for Trypsin-Benzamidine Complexes

The four benzamidines that were considered are the parent and the *para* amino, methyl and chloro derivatives. The structures are shown below along with the experimentally determined inhibition constants, K_i.[19] The results from the MC/FEP calculations are summarized in Table 1 for the three mutations that were performed from the most strongly bound *para*-amino derivative. The computed relative free energies of binding, $\Delta\Delta G_b$, are compared with the observed values derived from the K_i's. Though the differences in $\Delta\Delta G_b$ are not large, the calculations correctly find *p*-aminobenzamidine to be the strongest inhibitor, while the principal error is in predicting benzamidine itself to be too weakly bound by 1.5 kcal/mol. The average error is 0.6 kcal/mol.

	NH$_2$	H	CH$_3$	Cl
K$_i$ (nM)	8.3	16.6	26.5	54.0

TABLE 1. Free energy changes (kcal/mol) for binding benzamidines by trypsin.

Benzamidines	ΔG_u	Calculated ΔG_c	$\Delta\Delta G_b$	Exptl. $\Delta\Delta G_b$
p-NH$_2$ → *p*-H	-0.3 ± 0.2	1.6 ± 0.1	1.9 ± 0.2	0.40
p-NH$_2$ → *p*-CH$_3$	0.8 ± 0.2	1.9 ± 0.1	1.1 ± 0.2	0.67
p-NH$_2$ → *p*-Cl	-3.1 ± 0.1	-2.0 ± 0.1	1.1 ± 0.2	1.08

A key aspect of this study was to test the precision of the calculations. This was done by completing the perturbation cycles, *p*-amino → *p*-chloro → benzamidine → *p*-amino. The net free energy change should be exactly zero. In fact, the hysteresis for the FEP-calculations for the unbound inhibitors was 0.3 kcal/mol and it was 0.4 kcal/mol for the complexes. There was some cancellation of the discrepancies for the two cycles such that the hysteresis for their difference, the $\Delta\Delta G_b$ values, was 0.1 kcal/mol. These are very propitious results for further application of the methodology. The findings are also consistent with the standard deviations (±1σ) that are reported in Table 1 for the computed quantities, which were obtained by the batch means procedure with batch sizes of ca. 3M configurations. The high precision suggests that sampling problems are probably not to blame for the predicted overly-weak binding of the parent benzamidine. A more likely possibility is a deficiency in the partial charges from the RHF/6-31G* calculations.

Much structural information has also been obtained from the simulations. For example, the accommodation of *p*-aminobenzamidine in the binding site is illustrated in Figure 2. Five to six hydrogen bonds are found for both the inhibitor in water and in the complex. In the complex, these involve the salt-bridge with Asp189, and hydrogen bonds with the

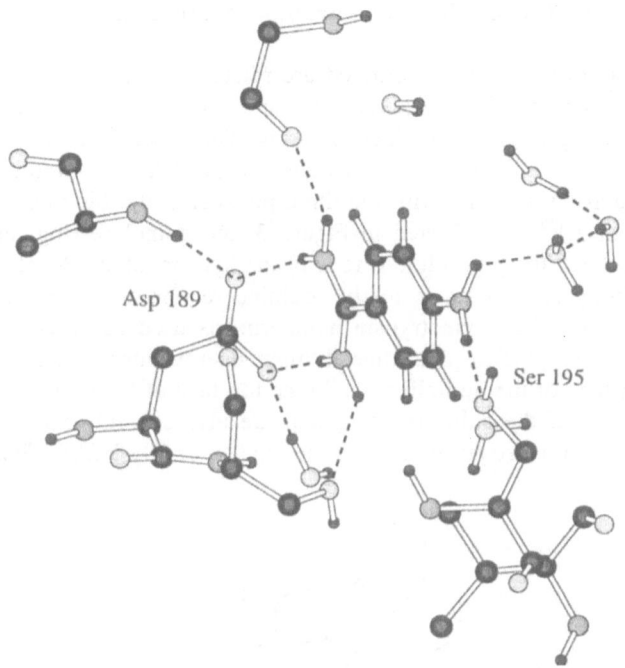

Figure 2. Hydrogen bonding pattern for *p*-aminobenzamidine bound to trypsin. An arbitrary configuration from the Monte Carlo simulation is illustrated.

side chains of Ser195 and sometimes Ser190, the oxygen of Gly219, and 1-2 water molecules. The structural results from the MC simulations of the other complexes also reveal a full complement of hydrogen bonds for the benzamidines, so the variations in binding affinities do not arise from any qualitative changes in hydrogen bonding. In the experimental binding study,[19] a roughly linear relationship was found between pK_i and the Hammett σ_p substituent constant for the *para* substituent with weaker binding for more electron-withdrawing groups. The trend was rationalized with reference only to effects on the energetics for the bound structures. The explanation invoked a new dipole-dipole attraction in the complexes with an electron-donating *para* substituent like amino and a new dipole-dipole repulsion for an electron-withdrawing substituent such as chloro or nitro. The present results indicate that this explanation is faulty. The energy components in Table 1 for the amino to chloro perturbation state that this perturbation is favorable both free in water and in the complex. Electron-withdrawing groups make benzamidine more polarized with greater positive charge on the amidinium group. This strengthens the interactions both with water and in the binding site. Simply, water is more polar than the protein binding site, so a more polar ligand is more stabilized in water and shows weaker binding.

5. Results of MC/LRSA Calculations for Thrombin-Sulfonamide Complexes

The thrombin inhibitors that were considered are related to the sulfonamide MD-805 (**1**). The seven structures differ by the stereochemistry, and presence and absence of the carboxylate and methyl groups in the piperidine ring. The observed inhibition constants cover a greater than 10^4-fold range, as illustrated below.[20] The inhibitors have three subunits, which match up with the with the three pockets in the binding site from the crystal structure with **1**.[14b] As reflected in Figure 3, the quinoline ring resides in the hydrophobic "D-pocket", the piperidine ring is in the hydrophobic "P-pocket", and the quanidinium-containing side chain is in the arginine recognition site forming a salt bridge with Asp-H189. (The chymotrypsin numbering is used here, as in the Protein Data Bank entry, 1dwc.[14b]) The piperidine is more deeply buried in the protein and contacts the bottom face of the quinoline, while the top face of the quinoline is solvent-exposed. The oxygens of the sulfonamide group are also solvent exposed, while the sulfonamide NH forms a hydrogen bond with the oxygen of Gly-H216. The P-pocket is

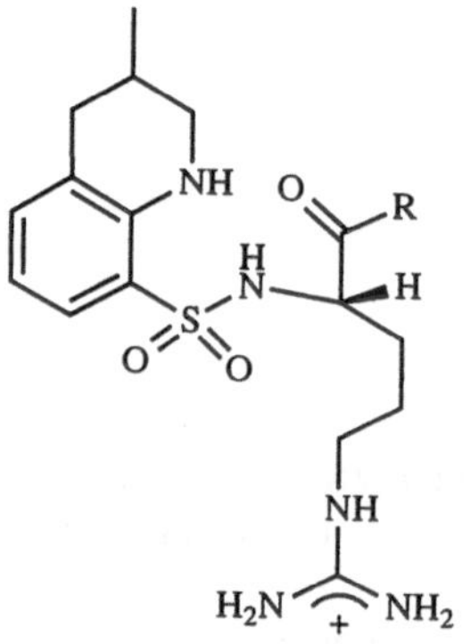

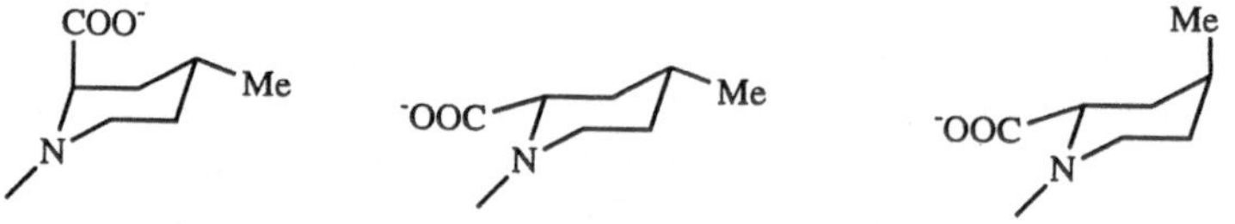

	1	**2**	**3**
R =			
K_i (μM)	0.019	1.9	280

	4	**5**	**6**	**7**
R =				
K_i (μM)	0.31	0.030	0.13	0.24

formed from the side chains of His-H57, Trp-H60D, Tyr-H60A, and Leu-H99. A space-filling representation reveals a tight steric match between the P and D-pockets and their contents, the piperidine and quinoline rings. The carboxylate group attached to the piperidine ring forms a hydrogen bond with the hydroxyl group of Ser-H195 and with two water molecules in the crystal structure. It does not appear to be particularly well-accommodated unless there are more hydrogen bonds with water molecules and it is also close to the carboxylate group of Glu-H192; two of the O-O distances are below 6 Å. In contrast, the other group that is modified for **1-7**, the piperidine C4 methyl group, snuggly fills the end of the hydrophobic P-pocket for the equatorial orientation in **1**.

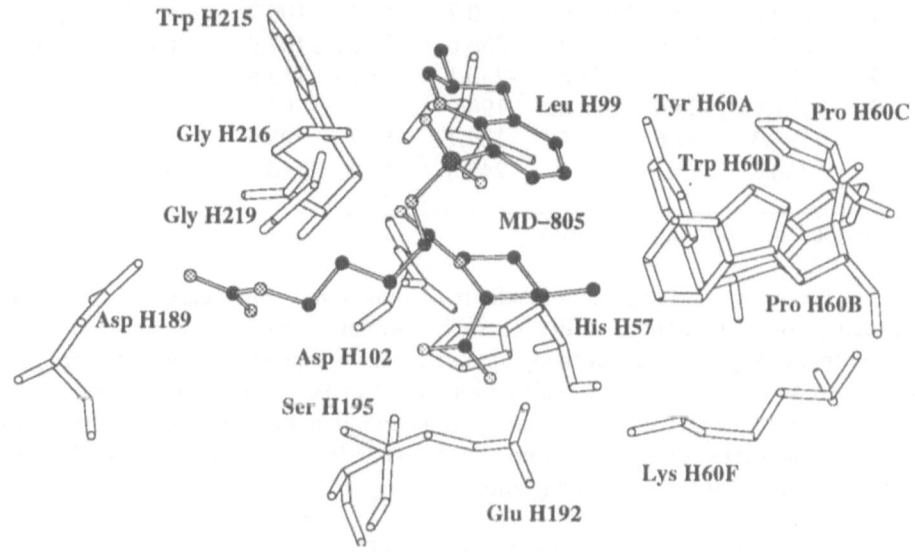

*Figure 3.*Residues of thrombin near the binding site for the sulfonamide inhibitors.

The computed energy components and solvent-accessible surface areas from the Monte Carlo simulations are listed in Table 2. These results were used in several fits to eqs 7 and 8. The best fit with eq 7 yielded an average error of 1.2 kcal/mol for the free energies of binding ($\beta = 0.165$, $\alpha = 0.476$). Addition of the surface area term lowered the average error to 0.8 kcal/mol with $\alpha = \beta = 0.130$ and $\gamma = 0.014$ with still only two variables. The predicted ΔG_b values from this fit are compared with the experimental results[20] in Table 2. The principal discrepancy, 1.9 kcal/mol, is for inhibitor **1**. The weak binding of **2** and **3** with the equatorial carboxylate groups is well-reproduced along with the strong binding for the most hydrophobic inhibitors, **5** and **6**.

The average unsigned error for the fit to eq 8 with seven inhibitors and two parameters of 0.8 kcal/mol can be compared with values of 0.4 and 1.8 kcal/mol reported by Åqvist for the smaller sets of four endothiopepsin and three HIV proteinase inhibitors.[6a,c] Clearly, it is desirable to test the methodology against still larger

TABLE 2. Average interaction energies, areas, and free energies of binding for thrombin inhibitors.[a]

Inhib.	Unbound			Bound			ΔG_b	
	E_{elec}	E_{vdw}	SASA	E_{elec}	E_{vdw}	SASA	Calc.	Exptl.
1	-267.2	-26.0	716.9	-246.6	-51.3	109.3	-9.1	-11.0
2	-259.5	-25.8	688.0	-230.7	-53.8	100.2	-8.1	-8.1
3	-280.9	-24.6	704.2	-236.6	-53.6	107.9	-6.4	-5.0
4	-256.3	-25.3	677.8	-250.1	-48.7	113.5	-10.1	-9.2
5	-191.6	-33.4	686.8	-190.9	-57.5	97.5	-11.1	-10.7
6	-195.3	-32.6	669.4	-184.7	-57.2	98.5	-9.8	-9.8
7	-266.6	-24.1	689.2	-245.8	-47.2	110.4	-8.4	-9.4

[a] Energies in kcal/mol, SASA in Å^2.

databases. Nevertheless, the present results in conjunction with the earlier studies[6-9] are encouraging for the potential utility of such correlative methods as a short-cut to predicted free energy changes for many applications. As expanded upon elsewhere,[9] the convergence and precision of the calculations need to be carefully monitored. In the present case, the total statistical uncertainty in the computed ΔG_b values from the noise in the MC simulations is ca. 0.5 kcal/mol, which sets a lower bound on potential accord between the computed and observed results.

The MC simulations also provided insights into the interactions occurring in the active site and the origins of variations in ΔG_b. Though details are provided elsewhere,[9] some key points are: (1) equatorial placement of the carboxylate group at C2 in the piperidine ring of the inhibitors causes electrostatic destabilization with the side chain of Glu-H192, (2) the number of water molecules hydrogen-bonded to the C2 carboxylate group in the complex also declines from 5-6 for the axial orientation to 3-4 for the equatorial epimers owing to increased shielding by the trimethylene spacer of the guanidinium unit, and (3) axial disposition of the C4-methyl group reduces favorable hydrophobic interactions in the P-pocket of the enzyme. It is also apparent from comparison of the observed ΔG_b's for 1 and 5 or 4 and 6 that the axial carboxylate group makes negligible contribution to the binding. Free in water, it has the usual complement of six hydrogen bonds, which are replaced by the hydrogen bonds to Ser-H195 and five water molecules in the bound structure. On the other hand, the sensitivity of ΔG_b to the match between the non-polar surfaces of the P-pocket and of the piperidine ring is apparent in the results for 2, 3, 5, and 6. These observations are consistent with the insights on binding inferred above from the pmfs for the associations of polar and non-polar molecules.

6. Conclusion

The results summarized here illustrate the progress that has been made with statistical mechanics simulations at understanding intermolecular interactions in solution and at the quantitative calculation of binding affinities for systems as complex as proteins. Methodologies based both on rigorous free-energy perturbation (FEP) calculations and on a more empirical linear-response-surface-area (LRSA) approach in conjunction with Monte Carlo sampling show great potential as tools in ligand design. The structural results from the Monte Carlo simulations further enrich the quantitative predictions by providing additional insights on the origin of variations in binding affinities. Future applications on larger sets of ligands and for diverse proteins will undoubtedly lead to further enhanced methodology and predicitve ability.

7. Acknowledgments

Gratitude is expressed to the National Institutes of Health and the National Science Foundation for support of this research.

8. References

1. (a) Lybrand, T. P. (1995) *Curr. Opin. Struct. Biol.* **5**, 224. (b) Verlinde, C. L. M. J. and Hol, W. G. H. (1994) *Structure* **2**, 577. (c) Jorgensen, W. L. (1991) *Chemtracts - Org. Chem.* **4**, 91. (d) Kollman, P. A. (1993) *Chem. Rev.* **93**, 2395.
2. Jorgensen, W. L. and Tirado-Rives, J. (1996) *J. Phys. Chem.* **100**, 14508.
3. See, for example: Jorgensen, W. L. and Nguyen, T. B. (1993) *Proc. Natl. Acad. Sci. USA* **90**, 1194. Duffy, E. M. and Jorgensen, W. L. (1994) *J. Am. Chem. Soc.* **116**, 6337.
4. Jorgensen, W. L., Chandrasekhar, J., Madura, J. D., Impey, R. W. and Klein, M. L. (1983) *J. Chem. Phys.* **79**, 926.
5. (a) Jorgensen, W. L. and Tirado-Rives, J. (1988) *J. Am. Chem. Soc.* **110**, 1657. (b) Jorgensen, W. L. and Severance, D. L. (1990) *J. Am. Chem. Soc.* **112**, 4768. (c) Jorgensen, W. L., Maxwell, D. S. and Tirado-Rives, J. (1996) *J. Am. Chem. Soc.* **118**, 0000.
6. (a) Åqvist, J., Medina, C. and Sammuelsson, J.-E. (1994) *Protein Eng.* **7**, 385. (b) Åqvist, J. and Mowbray, S. M. (1995) *J. Biol. Chem.* **270**, 9978. (c) Hansson, T. and Åqvist, J. (1995) *Protein Eng.* **8**, 1137.
7. Carlson, H. A. and Jorgensen, W. L. (1995) *J. Phys. Chem.* **99**, 10667.
8. Paulsen, M. D. and Ornstein, R. L. (1996) *Protein Eng.* **9**, 567.
9. Jones-Hertzog, D. K. and Jorgensen, W. L., submitted for publication.
10. Jorgensen, W. L. (1995) *BOSS, Version 3.6*; Yale University; New Haven, CT.
11. Jorgensen, W. L. (1996) *MCPRO, Version 1.4*; Yale University; New Haven, CT.
12. Frisch, M. J., Trucks, G. W., Head-Gordon, M., Gill, P. M. W., Wong, M. W., Foresman, J. B., Johnson, B. G., Schelgal, H. B., Robb, M. A., Replogle, E. S., Gomperts, R., Andres, J. L., Raghavachari, K., Binkley, J. S., Gonzalez, C.,

34

Martin, R. L., Fox, D. J., Defrees, D. J., Baker, J., Stewart, J. J. P. and Pople, J. A. (1994) *Gaussian94, Revision A*; Gaussian, Inc.; Pittsburgh, PA.

13.   Jorgensen, W. L., Briggs, J. M. and M. L. Contreras (1990) *J. Phys. Chem.* **94**, 1683.

14.   (a) Marquart, M., Walter, J., Deisenhofer, J., Bode, W. and Huber, R. (1983) *Acta Cryst. B* **39**, 480. (b) Banner, D. W. and Hadvary, P. (1991) *J. Biol. Chem.* **266,** 20085.

15.   Jorgensen, W. L. (1989) *J. Am. Chem. Soc.* **111**, 3770.

16.   Krikorian, S. E. (1982) *J. Phys. Chem..* **86**, 1875. Klotz, I. M. and Franzen, J. S. (1962) *J. Am. Chem. Soc.* **84**, 3461.

17.   Blokzijl, W. and Engberts, J. B. F. N. (1993) *Angew. Chem. Int. Ed. Engl.* **32**, 1545.

18.   Weber, P. C., Wendoloski, J. J., Pantoliano, M. W. and Salemme, F. R. (1992) *J. Am. Chem. Soc.* **114**, 3197.

19.   Mares-Guia, M., Nelson, D. L. and Rogana, E. (1977) *J. Am. Chem. Soc.* **99**, 2331.

20.   Kikumoto, R., Tamao, Y., Teauka, T., Tonomura, S., Hara, H. and Ninomiya, K. (1984) *Biochemistry* **23**, 85.

DYNAMICS OF BIOMOLECULES : SIMULATION VERSUS X-RAY AND FAR-INFRARED EXPERIMENTS

S. HERY, M. SOUAILLE AND J.C. SMITH

Section de biophysique des protéines et des membranes, DBCM CEA-Saclay, 91191 Gif-sur-Yvette cedex, France

1. Introduction

It is now well established that the biological activity of proteins is related not only to their mean molecular structure but also to their intramolecular mobility [1]. In many areas of molecular biology, such as muscle contraction or conformational changes in allosteric proteins, motion is self-evidently crucial. In facilitating such processes as the access of ligands to protein binding sites or in accomodating the change in protein and substrate geometries associated with an enzyme reaction, molecular motion is required.

Knowledge of dynamic processes in proteins, and biomolecules in general, initially can be obtained from spectroscopic methods. Theoretical works involving simulations using fast computers can also be used and enable a detailed description of the dynamics of molecules in condensed phases to be obtained. The calculated dynamical trajectories can be used to derive experimental quantities such as scattering intensities. These quantities can then be compared with experiment and interpreted in detail using the simulations. Furthermore, quantities which are not directly accessible experimentally can be evaluated from the atomic trajectories.

We discuss here the combination of molecular dynamics (MD) simulations with recent work on X-ray crystallography and far-infrared spectroscopy.

2. X-ray diffraction

2.1. INTRODUCTION

It was widely believed that X-ray crystallography was an inherently static technique, incapable of providing information about the dynamic

G. Vergoten and T. Theophanides (eds.), Biomolecular Structure and Dynamics, 35–46.
© 1997 *Kluwer Academic Publishers.*

properties of molecules. However, although X-ray crystallography has indeed supplied much information on the static organization of proteins, it can also furnish information on atomic displacements. Analysis of isotropic atomic temperature factors, obtained from refinement against the Bragg diffraction intensities have provided an overall view of the amplitudes of fluctuations of atoms throughout complete protein structures. A further step towards the understanding of protein dynamics, especially correlated motions, is the analysis of the diffuse, non-Bragg scattering arising from displacements of atoms from their average positions.

2.2. INFORMATION AVAILABLE FROM X-RAY EXPERIMENTS OR CALCULATIONS

Every atom in a crystal structure interacts with numbers of atoms with forces of various types. Assuming classical mechanics, the geometry position of the atom at $0°K$ is that corresponding to the minimum potential energy. At higher temperatures the atoms vibrate about their mean positions with amplitudes which increase with the temperature of the solid. At physiological temperatures, diffusive motions may also contribute. These vibrations will affect the relative positions of the atoms and hence the X-ray scattering pattern.

A real crystal structure can be considered as an ideal periodic structure with slight perturbations. When exposed to X-rays, a real crystal gives rise to two scattering components: the set of Bragg reflexions, I_B, arising from the periodic structure, and the scattering outside the Bragg spots (diffuse scattering), I_D, which arises from the structural perturbations:

$$I_T(\vec{q}) = I_B(\vec{q}) + I_D(\vec{q}) = |F_{hkl}(\vec{q})|^2 \tag{1}$$

where

$$F_{hkl}(\vec{q}) = \int_{-\infty}^{+\infty} \rho(\vec{r}) e^{2i\pi \vec{q}\cdot\vec{r}} d\vec{r} \tag{2}$$

2.2.1. *Temperature factor*

To interpret the observed electron density distribution in terms of particular types of atomic motion, the application of a physical model is required. That commonly used for macromolecular crystal structure analysis is the simple Debye-Waller model. In the Debye-Waller treatment of atomic motion, the probability of finding an atom at a given distance x from its equilibrium position x_0 ($x = x(t) - x_0$) is considered to be Gaussian. The probability is also considered to be isotropic. In this case, the model states that, in any direction, the motion can be characterized in terms of a mean-square displacement, $\langle x^2 \rangle$. Application of this model is usually done in reciprocal space, the space of the diffraction pattern. It is possible to calculate the expected diffraction from an atom, modified by a Gaussian function that is related to the estimated mean-square displacement of that atom in real space.

The form of the Gaussian is $exp(-B sin^2 \frac{\theta}{\lambda^2})$ where B, the atomic temperature factor or Debye-Waller factor, is related to the mean-square displacement by

$$B = 8\pi^2 \langle x^2 \rangle \tag{3}$$

Therefore, for small amplitude displacements, the Bragg intensity I_B can be derived from the expression for scattering by a perfect crystal by multiplying each atomic scattering factor by the factor $exp(-B sin^2 \frac{\theta}{\lambda^2})$. The effect of temperature on the crystal is hence to reduce the Bragg intensity by the latter factor.

In MD simulations, atomic mean square displacements or fluctuations $\langle x^2 \rangle$ can be directly calculated from the atomic trajectories and thus compared with experimental values.

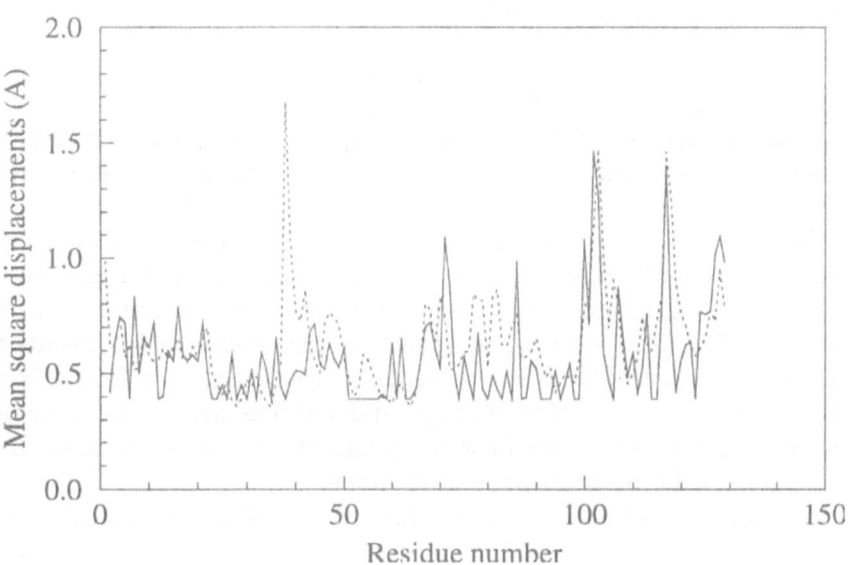

Figure 1. Mean square displacements of C_α atoms obtained experimentally (solid line) and from a simulation of an isolated lysozyme molecyle simulation (dotted line).

Figure 1 shows the mean-square displacements of the C_α atoms of lysozyme obtained experimentally and from a 500-*ps* molecular dynamics simulation of lysozyme. We can see there is reasonable agreement between the two curves except for the regions around residues 40 and 80. These differences can be explained partly by the presence of contacts in the crystal, reducing the atomic motions, which are not present in a simulation in vacuo.

2.2.2. *Diffuse scattering*
Theoretical background

Direct information on correlated displacements is not available from the Bragg scattering nor the spectroscopic techniques hitherto employed to probe motions in proteins. However, diffuse X-ray scattering presents a way of obtaining this information.

The diffuse scattering intensity I_D can be written as:

$$I_D(\vec{q}) = N \sum_m < (F_n - < F >)(F_{n+m} - < F >)^* > e^{-2i\pi\vec{q}.\vec{r}_m} \qquad (4)$$

where F_n is the structure factor of the nth unit cell, $\vec{q}$ is the scattering vector and the sum $\sum_m$ runs over the relative position vectors $\vec{r}_m$ between unit-cells [2].

The diffuse scattering intensity I_D can be written in terms of the structure factor fluctuation correlation function ϕ_m:

$$\phi_m(\vec{q}) = < (F_n - < F >)(F_{n+m} - < F >)^* > \qquad (5)$$

Therefore, the presence of diffuse scattering on an X-ray diffraction pattern indicates the presence of electron density fluctuations from the mean crystal density.

At least three types of diffuse scattering are found in protein crystal scattering pattern: haloes around each Bragg spots (thermal diffuse scattering) due to long range displacements correlated over different unit cells; diffuse scattering located along reciprocal lattice planes which results from anisotropic intermolecular displacements correlated over a few unit cells; and low-intensity, very diffuse background patches arising from displacements that are not correlated between different unit cells and may contain a contribution from internal protein motions.

If we focus on the very diffuse scattering, we can simplify equation (5):

$$I_D(\vec{q}) = < |F(\vec{q})|^2 > - | < F(\vec{q}) > |^2 \qquad (6)$$

In the approximation of independant molecules (intramolecular motions only)

$$I_D(\vec{q}) = \sum_j^M < |F^j(\vec{q})|^2 > - | < F^j(\vec{q}) > |^2 \qquad (7)$$

where M is the number of asymmetric units in the unit cell.

In principle, the information that could be obtained from diffuse scattering would substantially augment that obtainable from crystallographic temperature factors: there is the possibility of describing large displacements of whole subunits or domains of proteins and the presence or absence of cooperativity between different displacements could be established.

There is, however, a major deficiency in the information about dynamics available from both X-ray diffraction and X-ray diffuse scattering: both supply time- and space-averaged information on the distribution of matter within the crystals and give no information about the timescales of the displacements they detect. Indeed, the displacements discovered may not be dynamic at all, but may result from static disorder in the crystal. There have been attempts to discriminate between static and dynamic disorder by doing experiments at different temperatures [3], but because dynamic disorder can 'freeze' into static disorder at low temperatures, such studies are not always conclusive.

Previous results

A number of studies combining simulations with diffuse scattering calculations have been performed to investigate various types of motions in biomolecules.

Doucet *et al.* [4] have investigated the atomic and molecular displacements in hen egg-white lysozyme using X-ray diffuse scattering analysis. Their results proved the existence of intermolecular rigid-body displacements which are correlated within short rows of aligned molecules along $\vec{a}$ et $\vec{c}$.

Several formalisms have been proposed to interpret the very diffuse scattering depending in part on the correlation distances of the displacements involved. Caspar *et al.* [5] have interpreted the very diffuse scattering found in crystals of lysozyme and insulin in terms of "liquid-like" motions, random atomic displacements correlated over distances $< 0.6\ nm$. Faure *et al.* have examined further the origins of the very diffuse scattering in orthorhombic hen egg-white lysozyme [6]. Diffuse scattering patterns calculated from normal mode analysis and a molecular dynamics simulation were compared to the experimental scattering patterns. The scattering pattern obtained from a harmonic description of the protein motions is in reasonable accord with the observed data. As the lowest-frequency modes dominate the mean square displacements in the harmonic approximation, this study indicate that intramolecular displacements correlated over long distances exist in lysozyme. This is in contrast with the previous analysis of insulin and lysozyme in which a phenomenological model involving "liquid-like" motion was fitted to the data.

In figure 2 and 3 is shown an example of the patterns that can be obtained experimentally and from a 400-*ps* simulation of hen-egg white lysozyme analyzed with the program SERENA (Scattering of Ex-Rays Elucidated by Numerical Analysis) [7].

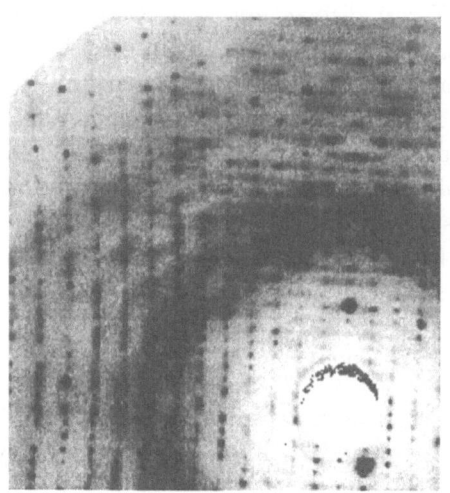

Figure 2. Experimental X-ray scattering pattern from orthorhombic hen egg-white lysozyme crystal from ref. [7]

As we can see on figure 2, the very diffuse scattering is of very low intensity compared to the bragg peaks. The very diffuse scattering is visible as patches all over the pattern. These patches are clearly structured, being located mostly in a ring that has an average intensity maximum around $0.39\ nm^{-1}$.

The simulated scattering pattern on figure 3 agrees reasonably well with the experimental one concerning the radius of the average ring and the coarse-grained description of the intensity variation within the ring. However the finer details are not always in accord.

These recent results suggest that the combination of diffuse scattering and computer simulation may become a powerful approach for both the interpretation of experimental scattering data in terms of atomic correlations and the determination of the dynamical modes accessible to functional proteins. However, recent studies has pointed out the convergence of diffuse scattering calculated from molecular dynamics is slow [8]. It has been suggested that 100 *ns* is an upper bound for the convergence time [9].

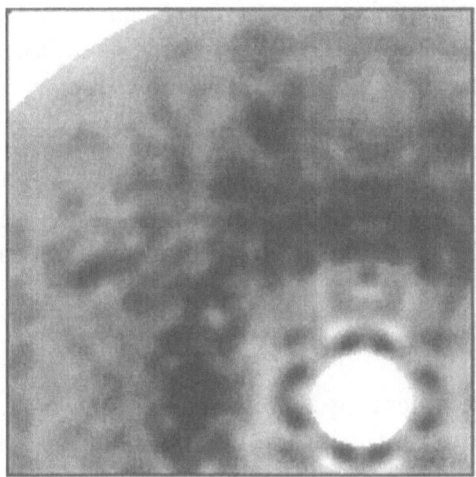

Figure 3. X-ray diffuse scattering pattern calculated from molecular dynamics simulation of orthorhombic hen egg-white lysozyme crystal.

3. Far-infrared spectroscopy

3.1. MOLECULAR DYNAMICS STUDIES

The information obtained from neutron scattering, nuclear magnetic resonance, Raman scattering and infrared absorption have permitted the characterization of a range of inter- and intramolecular motions in condensed phase system. However, the complete interpretation of these experiments is made difficult by the fact that the dynamical properties determining a spectral profile may involve the sampling of a large number of degrees of freedom. Molecular dynamics simulation can be of valuable help in this regard as it provides a detailed representation of atomic dynamics in such systems. The atomic trajectories can be used to calculate spectral intensities for comparison with experiment. Spectral features can then be attributed to the dynamical features and, where appropriate, to the charge fluctuations present in the simulation.

Light absorption in the infrared region involves transitions between vibrational levels of molecules. Atomic charges have two distinguishable effects on far-infrared spectra. The first is indirect: Coulombic interactions between charges play a role in determining atomic dynamics. In molecular dynamics simulation this enters into the potential energy function used for calculating the force between the atoms. The second effect is that, once the nuclear trajectories have been determined, the atomic charge fluctuations associated with them will directly determine the absorption.

3.1.1. *Theoretical background*

Gordon has shown how a system interacting with an electric field of frequency ω is able to absorb or emit quanta of energy $\hbar\omega$.

Consider a system of N interacting molecules in quantum state, i. Let the Hamiltonian of this system be H_0. The system interacts with an electric, monochromatic field [10]:

$$\vec{E}(t) = E_0 \, \vec{e} \, \cos\omega t \qquad (8)$$

where $\vec{e}$ is a vector indicating the direction of the field.

Since the field is uniform, the interaction between the field and the molecules can be written as

$$H_1 = -\vec{M}.\vec{E}(t) \qquad (9)$$

where $\vec{M}$ is the total electric dipole moment operator of the N-body system. According to the Golden Rule of time-dependant quantum-mechanical perturbation theory, the probability per unit time that a transition from the state i to the state j takes place is given by

$$P_{i \rightarrow j}(\omega) = \frac{\pi E_0^2}{2\hbar^2} \, |\langle f|\vec{e}.\vec{M}|i\rangle|^2 \, [\delta(\omega_{fi} - \omega) + \delta(\omega_{fi} + \omega)] \qquad (10)$$

where $\omega_{fi} = \omega_f - \omega_i$.

The infrared absorption coefficient for such a system is given by

$$\alpha(\omega) = \frac{4\pi^2}{3cn(\omega)V} \frac{\omega}{\hbar} \, (1 - e^{-\beta\hbar\omega}) \, C(\omega) \qquad (11)$$

where

$$C(\omega) = \sum_{i,f} \rho_i \, |\langle f|\vec{e}.\vec{M}|i\rangle|^2 \, \delta(\omega - \omega_{fi}) \qquad (12)$$

where c is the velocity of the light, $n(\omega)$ is the refractive index of the medium, $\beta = 1/k_B T$, $\vec{M}$ is the dipole moment of the system and V its volume.

For macromolecules, for which Hamiltonian eigenfunctions cannot yet be accurately calculated, it is possible to calculate the absorption coefficient from the dipole moment autocorrelation function. It can be shown $C(\omega)$ is the Fourier transform of the dipole moment autocorrelation function [10]:

$$C(\omega) = \int \frac{dt}{2\pi} \, e^{-i\omega t} \langle \vec{M}(0).\vec{M}(t)\rangle \qquad (13)$$

The total dipole moment $\vec{M}$ can be expressed as a sum of the microscopic dipoles occuring in the sample. This comprises the contribution of the atomic charges (giving rise to molecular dipoles which can be considered as permanent, p_i, in the case of small molecules simulated over reasonable timescale), and the induced dipole, μ_i (dipole-induced dipole (DID) model).

Consider a system of N atoms in an external electric field $\vec{E}_{ext}$. Each atom possesses a polarizability α_i and is polarized by $\vec{E}_{ext}$ giving rise to an induced dipole moment μ_i which itself contributes to the total electric field $\vec{E}_{tot}$. This can be written in the following way:

$$\vec{\mu}_i = \alpha_i \vec{E}_{tot} \tag{14}$$

where the total electric field is the sum of the field created by the induced dipoles and the external field:

$$\vec{E}_{tot} = \vec{E}_{ext} + \sum_{i \neq j} T_{ij} \vec{\mu}_j \tag{15}$$

where T_{ij} is the dipolar field tensor.

To calculate the induced part of the dipole moment, the above procedure must be adapted. The external field $\vec{E}_{ext}$ is replaced by $\vec{E}_i$, the field acting on the ith atom due to the permanent charges q_i. Eq. 14 then becomes

$$\vec{\mu}_i = \alpha_i \left(\vec{E}_i + \sum_{i \neq j} T_{ij} \vec{\mu}_j \right) \tag{16}$$

To calculate the induced dipoles on the molecules, Eq. 16 can be solved for each molecular dynamics trajectory frame generated. This can be done using an iterative procedure in which the $\vec{\mu}_i$ vectors calculated in the nth step are used as $\vec{\mu}_j$ in the $(n+1)$th step. The iteration is repeated until convergence of $\vec{\mu}_i$. The iteration of Eq. 16 leads to a self-consistent representation of the local field and the induced dipoles.

3.1.2. Recent work

Many molecular dynamics simulations associated with far-infrared (FIR) absorption have been realized on liquid water systems. The experimental 300 K far-infrared spectrum of water contains a wide absorption band at $\sim 600\ cm^{-1}$, attributed to the librations of water molecules in their local hydrogen-bond networks, and a band at $\sim 200\ cm^{-1}$, attributed to hydrogen-bond stretching [12] [13]. A low frequency mode assigned to the flexibility of O-H...O units is detected but its presence is more questionable [12] [14] [15]. Attempts have been made to reproduce these features using molecular dynamics simulations [16] [17] [18].

Madden *et al.* [16] have combined a molecular dynamics simulation of liquid water with far-infrared absorption coefficient calculations using a simple dipole-induced dipole model. This model was able to reproduce experimentally observed bands at 60 and 200 cm^{-1}. Although this study emphasized the importance of the dipole-induced dipole mechanism in making spectrally active the translational oscillations of the water molecules, some important questions were not adressed, in particular the relative importance of the spectral intensities coming from the permanent dipoles, the induced dipoles and the electronic overlap dipoles as well as the importance of quantum corrections in the evaluation of $C(\omega)$.

Guillot [17] examined the influence of electronic overlap dipole in the calculation of far-infrared spectrum of water (in the range 0.5-1000 cm^{-1}) from a molecular dynamics simulation. He has shown that the intermolecular motions generated permit the satisfactory reproduction of the overall shape of the FIR absorption spectrum of liquid water. In particular, the introduction of the DID mechanism was crucial to obtain the correct magnitude of the absorption intensity over a large domain of frequency (0.5 → 1000 cm^{-1}) and the appearance of the shoulder at 200 cm^{-1}, which is the signature of the O-H...O stretching mode. If the DID mechanism was neglected, the translational band disappeared and the high frequency libration band had an absolute intensity approximately two times greater than the experimental value. This work also indicated that quantum corrections are important for frequencies greater than $\sim$ 200 cm^{-1}. Their effect is to desymmetrize the absorption profile and to enhance the intensity of the libration band by roughly a factor 2 compared to the classical profile. Guillot also showed that the modulation of the overlap dipoles, occurring between O and H atoms, by the oscillations of the hydrogen bond network is not sufficiently large to affect significantly the spectral intensity.

In these previous simulation studies, induced molecular dipole moments were calculated by multiplying the molecular polarizability matrix by the local electric field; the induced dipole contribution was calculated in a non-iterative and implicit manner. In recent work by Souaille and Smith [19] the approach was refined in two ways. First, the induced dipole moments were calculated not on the molecules but on the individual atoms in the system using a modified dipolar field tensor introduced by Thole. The use of atomic point dipoles increases the level of detail of the description of the induced charge distribution. Second, the induced dipole moments were evaluated by a procedure in which their contribution to the local electric field acting on them is calculated in an iterative, self-consistent way. A further question studied in this paper concerns the role of long range electrostatic interactions in determining the induced dipoles. The results of the calculations indicate that the inclusion of explicit polarization terms in the molecular dynamics potential is not needed to produce the nuclear dynamics responsible for the 200 cm^{-1} band. However, the peak is not seen in the spectrum calculated with the permanent charges (whereas the 600 cm^{-1} peak is present) and explicit dipole-induced dipole absorption is required to make the band spectroscopically active (figure 4).

Improved agreement with experiment is seen when the induced dipole are calculated using the self-consistent, iterative method. Long-range Coulombic interactions are important in determining the hydrogen-bond vibrations involved in the far-infrared absorption via an effect of cross-correlations with the permanent molecular dipoles. The inclusion of the long-range interactions in the potential function by Ewald summation is also necessary for agreement with experiment. However, the agreement with experiment is still not perfect: the intensity of libration band is $\sim$ 20 % lower than experimentally. One possible reason for this is the inaccuracy in quantum corrections, which have been shown to have a significant effect on calculated spectra [17] [23].

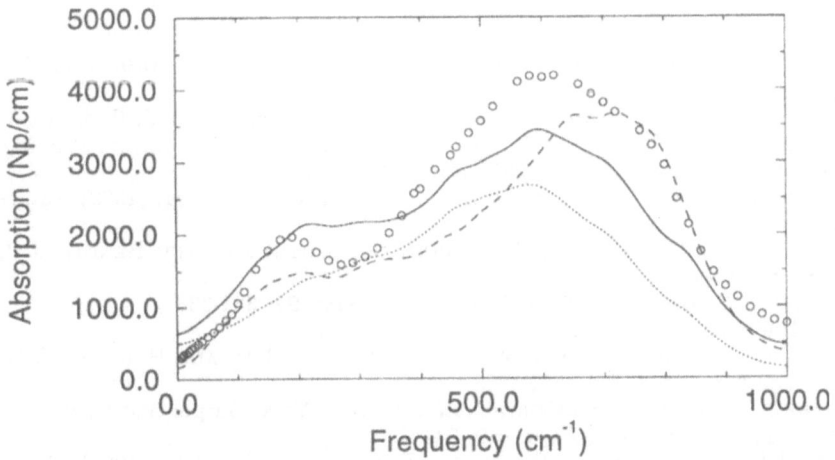

Figure 4. Total dipole far-infrared spectra. The experimental spectrum is from [20], [21], [22] and has been multiplied by the refractive index $n(\omega)$. Experiment (o); simulation EWALD using induced-dipole methode EWRF (Eq. 16 is solved iteratively to convergence) (—); simulation EWALD using induced-dipole method NONIT (non-iterative method) (...); simulation RC (cut-off) using induced-dipole method EWRF (- - -).

We have seen good simulation models for calculating polarization effects on far-infrared spectra in liquid systems are now available and therefore can be extended to investigations of greater systems such as biomolecules. Using the national Synchrotron Light Source at Brookhaven, far-infrared absorption in the frequency range 15-45 cm^{-1} was detected in samples of lysozyme at different hydratations [24]. The form of the absorption profile was found to be temperature independent but varied significantly with the hydratation of the protein. At higher hydratations, the profile resembles closely that of water in the region 20-45 cm^{-1}. At a low hydratation, marked differences were seen with, in particular, the appearence of an absorption maximum at 19 cm^{-1}. A parallel theoretical investigation has been undertaken [25]. Preliminary results suggest that far-infrared absorption from lysozyme contains a significant component from induced dipole absorption.

4. Conclusion

Scattering and spectroscopic techniques can be combined with molecular dynamics simulation to probe motions in biological molecules. The steady improvement in available computer power, coupled with the recent improvements in synchrotron radiation sources and instrumentation, will lead to ever clearer pictures of functionally-important dynamics in proteins and nucleic acids.

References

1. FERRAND, M., DIANOUX, A.J., PETRY, W. AND ZACCAÏ, G. Proc. Natl. Acad. Sci. USA (1992) **90** 9668-9672
2. GUINIER, A. Théorie et Technique de la radiocristallographie. (Dunod, Paris, 1956)
3. FRAUENFELDER, H., PETSKO, G.A. AND TSERNOGLOU, D. (1979) Nature **280**, 558
4. DOUCET, J. AND BENOIT, J.P. (1987) Nature **325**, 643
5. CASPAR, D.L.D., CLARAGE, J., SALUNKE, D.M. AND CLARAGE, M. (1988) Nature **332**, 659
6. FAURE, P., MICU, A., PERAHIA, D., DOUCET, J., SMITH, J.C. AND BENOIT, J.P. (1994) Struct. Biol. **1**, 124
7. MICU, A, AND SMITH, J. (1995) Comp. Phys. Comm. **91**, 331-338
8. HÉRY S. Rapport de DEA. 1994
9. CLARAGE, J.B., ROMO, T., ANDREWS, B.K., PETTITT, B.M. AND PHILLIPS, G.N. (1995) Proc. Natl. Acad. Sci. **92**, 3288
10. MCQUARRIE, D.A. (1976) Statistical mechanics New-York: Harper and Row
11. THOLE, B.T. (1981) Chem. Phys. **59**, 341
12. HASTED, J.B., HUSAIN, S.K., FRESCURA, F.A. AND BIRCH, J.R. (1985) Chem. Phys. Lett. **118**, 622
13. ROBERTSON, C.W. AND WILLIAMS, D. (1971) J. opt. Soc. Amer. **61**, 1316
14. SIMPSON, O.A., BEAN, B.L. AND PERKOWITZ, S. (1980) J. Opt. Soc. Amer. **69**, 1723
15. VIJ, J.K. AND HUFNAGEL, F. (1989) Chem. Phys. Lett. **155**, 153
16. MADDEN, P.A. AND IMPEY, R.W. (1986) Chem. Phys. Lett. **123**, 502
17. GUILLOT, B. (1991) J. Chem. Phys. **95**, 1543
18. NIESAR, U., CORONGIU, G., CLEMENTI, E., KNELLER, G.R. AND BHATTACHARYA, D.K. (1990) J. Phys. Chem. **94**, 7949
19. SOUAILLE, M. AND SMITH, J.C. (1996) Molec. Phys. , **87**, 1333 (1996)
20. ROBERTSON, C.W. AND WILLIAMS, D. (1971) J. Opt. Soc. Amer. **61**, 1316
21. RUSK, A.N., WILLIAMS, D. AND QUERRY, M.R. (1971) J. Opt. Soc. Amer. **61**, 895
22. AFSAR, M.N. AND HASTED, J.B. (1977) J. Opt. Soc. Amer. **67**, 902
23. BORISOW, J., MORALDI, M. AND HARRISON, M.C. (1985) Molec. Phys. **56**, 913
24. MOELLER, K.D., WILLIAMS, G.P., STEINHAUSER, S. HIRSCHMUGL, C. AND SMITH, J.C. (1992) Biophys. J. **61**, 276
25. SOUAILLE, M. AND SMITH, J.C. in preparation

SEMIEMPIRICAL AND AB INITIO MODELING OF CHEMICAL PROCESSES

From aqueous solution to enzymes.

Richard P. Muller, Jan Florián, Arieh Warshel*

Department of Chemistry

University of Southern California

Los Angeles, California, 90089-1062, USA

1. Introduction

The origin of the catalytic power of enzymes is one of the most fundamental problems in molecular biophysics. Our opinion is that enzyme catalysis can be accounted for by the available physical concepts and does not require a new paradigm in thermodynamics or quantum mechanics. The difficulty encountered in studying these systems computationally is associated with the fact that enzymes are complex systems and the science of modeling complex systems is, in many respects, in its infancy. Nonetheless, computational and conceptual studies of enzymes have yielded two important points. First, even the most rigorous modeling of the isolated solute reaction in the gas phase cannot provide meaningful information about the corresponding reaction in an enzyme [1-3]. Second, including the enzyme but neglecting the solvent surrounding it similarly provides unreliable results [4]. In our opinion, it is necessary to model the entire system, the solute, enzyme and surrounding solvent, in order to obtain reliable results for a reaction in an enzyme active site or even a reaction in bulk solvent. Consequently, one is often forced to represent the complete system in an approximate way rather than examining a smaller part of it in a rigorous way. Only when this complete solute-protein-solvent model produces meaningful results (e.g. reproduces asymptotic energetics or pKa's in solution) should a gradual improvement of the components of that model, such as improving the level of theory describing the solute, be

47

G. Vergoten and T. Theophanides (eds.), Biomolecular Structure and Dynamics, 47–77.
© 1997 Kluwer Academic Publishers.

attempted. We emphasize this point since the tendency in science is to progress in an incremental way, building parts of the puzzle step by step; we argue here is that in dealing with complex systems one cannot use this incremental approach in much the same way that in designing an airplane one cannot start with the design of a perfect passenger seat.

After this rather philosophical introduction, we will describe in the upcoming sections the requirements for modeling enzymatic reactions. We focus on what is possible with current technology, recognizing that this technology will progress significantly over the next few years. We first consider approaches for modeling reactions in water. Here, after a short review, we will examine the calculation of hydration free energies of neutral and ionic solutes by Langevin dipole models of the solvent. Also, we will illustrate the performance of this approach in untangling a part of the mechanistic puzzle of phosphoryl transfer reaction. In the second part, we will move to enzyme modeling to provide an analysis of the current state of the art of the field, examining the effect of a mutation in subtilisin on that protein's ability to catalyze the nucleophilic attack step of a proteolysis reaction.

2. Approaches for modeling chemical reactions in solution

Enzyme catalysis is best defined by considering the given reaction in the enzyme relative to the corresponding reaction in solution. Typically enzymes provide rate accelerations of 10^7 or more over the rates in aqueous solution [5]. By comparing reaction profiles in aqueous solvent to those in the protein we may identify the essential elements of the protein responsible for the rate acceleration. At the same time, the similarity of protein and aqueous solvent energetics and charge distribution (as opposed to the corresponding values in the gas phase, which are qualitatively different) insures that our comparison is meaningful. Furthermore, it is frequently easier to obtain experimental information about solution reactions than about gas phase processes.

In the following section we consider briefly different methods for modeling chemical reactions in solution. At present the most effective methods for modeling chemical reactions in solution are the so-called hybrid quantum mechanical/classical methods, which include hybrid quantum mechanical / molecular mechanics (QM/MM) and quantum mechanical / continuum dielectric (QM/CD) techniques. The basic idea of these models is to represent the reacting region quantum mechanically and the surroundings classically. Table 1 gives a concise overview of early contributions in the field, and below we consider the main developments.

TABLE 1. Early Hybrid QM/Classical Reactions in Solution[a]

Reaction	Solvent	Solute	Averaging	Ref.
Proton Trans./General Acid	dipolar	MINDO	Average over EM	[6-8]
Proton Trans.	reaction field	INDO/ab initio	—	[9,10]
Proton Trans./General Acid	all-atom	EVB	FEP/US/MD	[11,12]
Sn2	all-atom	ab initio[b]	FEP/MC	[13]
Nucleophilic Attack	all-atom	ab initio	EM	[14]
Proton Transfer	all-atom	ab initio	EM/MD	[15]
Sn2	dipolar	ab initio	EM	[16]
Sn2	all-atom	AM1	FEP/MD[c]	[17]
Sn2	all-atom	EVB	FEP/MD	[18]
Sn2	reaction field	AM1/MNDO	—	[19]

a: FEP/US denotes a free energy perturbation/umbrella sampling approach. EM denotes energy minimization.
b: No consistent solute-solvent coupling.
c: Used gas phase structures in the mapping procedure.

The earliest treatment of actual chemical reactions by such models dates back to the work of Warshel and Levitt [6], who introduced QM/MM approaches to reactions in condensed phases; similar ideas were implemented in subsequent studies [7,8]. The strategies of using hybrid QM/CD models remained rather qualitative until the emergence of discretized continuum models [10,20,21] and the recognition that the van der Waals or effective Born radius should be parameterized to reproduced observed solvation energies [8]. Recent years saw a renaissance in hybrid QM/MM approaches [17,22-37], with attempts to implement such approaches with more rigorous microscopic solvation models and to use them in free energy perturbation (FEP) calculations. The earliest attempts to progress in this direction were done using combined valence bond/MM methods within the empirical valence bond (EVB) formulations, which have been reviewed extensively elsewhere [38,39]. At present, the implementation of molecular orbital (MO) approaches in FEP calculations of reaction profiles in solution is problematic (see discussion in [40]) and only sophisticated techniques of controlling the progress of the reaction from reactants to products, such as using the EVB mapping surface as a mapping potential, allow one to rigorously determine activation free energies. Note in this respect that the interesting work of Bash et al. [17] simulated an S_n2 reaction using a mapping potential based on the ground and transition state structures in the gas phase. Such an approach does not provide a general method for FEP studies since it is not expected to work in heterolytic bond cleavage reactions, when the gas phase transition state structure is drastically different than the corresponding solution structures.

Density functional theory (DFT) treatments of chemical reactions in solution are now starting to emerge. These include combined DFT/MM approaches [41], and attempts to treat the entire solvent quantum mechanically, either with the direct approach of Parrinello and coworkers [42], who try to treat the entire solute-solvent systems on the same level of DFT, or our frozen DFT (FDFT) [43-45] or constrained DFT (CDFT) [46] approaches where the density of the solvent molecules is constrained, achieving significant savings in computer time. The DFT based approaches were reviewed recently [47] and will not be considered here any further.

After this brief review it is useful to establish the validity of our current models. The accuracy of our FEP approach and our Langevin dipoles (LD) approach, which uses empirically adjusted van der Waals radii, was discussed in [48] and [49]. We have also developed more rigorous and more expensive approaches [40,50]. Here we report the recent version of what is our simplest and, in fact, oldest strategy of combined QM/Langevin dipole (LD) solvent models [6,8,28]. In its current version, this model uses electrostatic potential-derived atomic charges that are evaluated at the Hartree-Fock (HF) level using the 6-31G* basis set. We denote this method QM(ai)/LD to denote that the Langevin dipoles solvate charges obtained from ab initio quantum mechanics. These charges are used to evaluate electrostatic part of the hydration free energy, ΔG_{lgvn}, as well as the potential-dependent hydrophobic surface of the solute, ΔG_{phob}, a term approximating the relaxation energy of the solute, ΔG_{relax}, and the Born correction, ΔG_{Born}, that accounts for the bulk contribution. Thus, the total solvation free energy is given by

$$\Delta G_{sol} = \Delta G_{lgvn} + \Delta G_{relax} + \Delta G_{vdw} + \Delta G_{Born} + \Delta G_{phob}. \tag{1}$$

In order to obtain a quantitative agreement with experimental ΔG_{sol} of *ionic* solutes, ab initio-calculated parameters must be combined with a set of empirically adjusted vdW radii of solute atoms. In general, the vdW radii need to be hybridization dependent. To illustrate this point and the accuracy obtainable with the LD solvation model, we compare in Table 2 the ΔG_{sol} values calculated by using the LD model and by using the polarized continuum model (PCM) of Tomasi [10], with the experimental results for a wide range of neutral and ionic solutes. The vdW radii used in these calculations are presented in the legend to Table 2, other details of our current LD implementation will be published elsewhere [51].

The results of both the iterative and noniterative LD calculations are presented. In the iterative LD method (ILD), the electrostatic field at a given Langevin dipole is given as the sum of contributions from the solute and other Langevin dipoles, evaluated iteratively. In the noniterative LD method (NLD), this electrostatic field is generated by the solute charge distribution by using a distance-dependent dielectric constant.

Among the computational approaches considered here, the ILD method provided the most accurate results. However, even with this method, the errors in calculated hydration energies of ionic solutes may reach as much as 5 kcal/mol. The larger disagreement obtained for HSO_4^- ion probably reflects the uncertainty in the value of the experimental dissociation constant that must be extrapolated from concentrated sulfuric acid to 1M aqueous solution.

The accuracy of the NLD method is only slightly inferior to the ILD one. However, one should realize that the NLD method is about two orders of magnitude faster than the ILD method. Therefore, the good performance of the noniterative approach is promising for calculations of biochemical reactions. The NLD method performs especially well for neutral and monoanionic solutes, whereas for solutes with larger charges NLD hydration energies are significantly underestimated.

TABLE 2. Calculated and Observed Solvation Energies

Solute	PD[a]		PCM[b]	experiment[c]
	non-iter	iter		
Hydrocarbons				
methane	1.8	1.8	0.0	1.9
ethane	2.0	2.0	0.0	1.8
propane	2.0	2.0	0.0	2.0
butane	2.0	2.0	0.0	2.2
pentane	2.0	2.0	0.0	2.3
hexane	2.1	2.1	0.0	2.6
cyclohexane	1.8	1.8	0.0	1.2
cyclopentene	1.6	1.5	0.0	0.6
benzene	0.2	-0.1	0.0	-0.9
naphthalene	-1.9	-2.3	0.0	-2.4
cyclopentadiene	0.7	0.5	0.0	
$C_5H_5^-$	-59.3	-62.3	-64.0	-65±3
C,H,N Compounds				
ammonia	-3.5	-4.1	-4.8	-4.3
methylamine	-2.5	-3.0	-3.9	-4.6
ethylamine	-2.2	-2.8	-3.9	-4.5
dimethylamine	-0.8	-1.3	-3.1	-4.3
trimethylamine	0.2	-0.1	-2.1	-3.2
NH_4^+	-86.8	-81.0	-86.7	-80±2
$MeNH_3^+$	-78.0	-74.2	-75.8	-72±2
$Me_2NH_2^+$	-71.1	-68.2	-67.7	-65±2
Me_3NH^+	-65.9	-63.5	-63.6	-60±2
aniline	-4.2	-5.0	-2.2	-4.9
anilineH$^+$	-65	-65.2	-62	-69±4
pyridine	-3.1	-3.6	-3.4	-4.7
pyridineH$^+$	-59.0	-58.4	-61.4	-58±2
imidazole	-6.5	-7.6	-5.6	
imidazoleH$^+$	-63.2	-61	-66.5	-61±3
HCN	-4.5	-4.9	-3.5	
CN$^-$	-77.6	-74.8	-89.1	-75±3
acetonitrile	-6.6	-6.9	-4.1	-3.9
CH_3CNH^+	-71.6	-67.4	-73.6	-69±3

TABLE 2. Calculated and Observed Solvation Energies (cont.)

		C, H, O Compounds		
H_2O	-8.2	-9.3	-6.0	-6.4
H_3O^+	-108.3	-101.5	-94.1	-105±3
OH^-	-119.5	-115.9	-120.7	-109±3
methanol	-5.4	-6.4	-4.7	-5.1
$MeOH_2^+$	-87.5	-82.6	-79.8	-85±2
MeO^-	-98.0	-95.9	-104.5	-98±3
ethanol	-4.5	-5.5	-4.6	-5.0
$EtOH_2^+$	-80.3	-77.2	-73.3	-79±2
EtO^-	-94.2	-93.5	-102	-94±3
propanol	-4.3	-4.2	-5.9	-4.8
butanol	4.3	-4.1	-6.0	-4.7
ethanediol	-8.2	-9.9	-7.9	-9.6
phenol	-4.4	-5.3	-2.9	-6.6
$C_6H_6O^{-,d}$	-74.2	-77.8	-82.5	-75±3
dimethylether	-2.0	-2.5	-3.2	-1.9
diethylether	-0.6	-1.0	-2.8	-1.6
1,4-dioxane	-4.2	-5.3	-6.2	-5.1
tetrahydrofuran	-3.3	-4.0	-3.9	-3.5
acetaldehyde	-4.8	-5.3	-4.8	-3.5
formic acid	-5.4	-6.3	-6.5	
formate	-85.2	-83	-87.8	-80±3
acetic acid	-6.0	-6.9	-7.0	-6.7
acetate	-85.4	-83.6	-88.5	-82±3

		C, H, S Compounds		
H_2S	-0.3	-0.4	-0.2	-0.7
HS^-	-80.2	-78.7	-87.3	-75±3
methanethiol	-1.5	-1.7	-0.5	-1.2
MeS^-	-77.4	-76.9	-84.4	-74±3
ethanethiol	-1.5	-1.7	-0.6	-1.2
$EtSH2^+$	-64.2	-63.1	-68.7	-68±3
EtS-	-74.6	-75.0	-81.4	-72±3
thiophenol	-1.4	-1.7	-0.3	-2.6
$C_6H_6S^{-,d}$	-67.1	-71.1	-68.0	-65±5

TABLE 2. Calculated and Observed Solvation Energies (cont.)

	Inorganic compounds			
HNO$_2$	-4.0	-4.5	-2.5	
NO$_2^-$	-75.9	-73.2	-84.9	-73±5
HNO$_3$	-5.1	-5.8		
NO$_3^-$	-69.7	-66.8		-66±3
PH$_3$	1.3	1.2	-0.1	0.6
H$_3$PO$_4$	-8.8	-10.6	-13.1	
H$_2$PO$_4^-$	-67.7	-68.2	-80.6	-67±6
HPO$_4^{2-}$	-218.6	-240.6	-274.9	-244±10
PO$_4^{3-}$	-429.3	-525.4		-535±14
H$_2$SO$_4$	-8.9	-10.3	-8.9	
HSO$_4^-$	-66.8	-66.5	-73.9	-57±6
SO$_4^{2-}$	-218.7	-240.0	-269.6	-232±10

	Four atom types			
nitromethane	-6.3	-6.8	-5.2	
CH$_2$NO$_2^-$	-85.5	-83	-84.2	-81±5
formamide	-7.9	-8.6	-7.9	
formamideH$^+$	-80.5	-76.6	-77.1	-77±3
acetamide	-8.3	-9.0	-9.4	-9.7
acetamideH$^+$	-73.1	-70.1	-69.9	-70±2
cytosine	-15.7	-17.1	-14.5	
cytosineH$^+$	-65.6	-65.5	-67.2	-65±3
methylphosphoric acid	-9.2	-10.5	-12.1	
Me-phosphate$^-$	-68.4	-69.1	-79.8	-66±6
Me-phosphate^{2-}	-214.8	-238.8	-272.4	-241±10

a: LD VdW radii (Å): 2.60 (C_sp^3), 3.20 (C_sp^2), 2.65 (N), 2.20 (O_sp^3), 2.90 (O_inorganic), 2.00 (H_inorganic), 3.10 (S), 3.10 (P). VdW radii of hydrogen atoms are equal to 0.88 times VdW radius of the adjacent heavy atom, with the exception of H atoms attached to the inorganic oxygen. The VdW radii as well as other parameters of the LD model are identical for both the iterative and noniterative methods. Note that our VdW radii implicitly include VdW radii of the solvent atoms. In PCM and other continuum based methods, a constant, usually 1.4Å, is added to the atomic VdW radii to mimic the VdW radii of solvent molecules.
b: The ab initio calculations by using the PCM method with the HF/6-31G* geometry and wave function, by using the Gaussian94 program [52]. Merz-Kollman VdW radii (Å) of 1.2 (H), 1.50 (C), 1.50 (N), 1.40 (O), 1.75 (S), 1.80 (P) scaled by a factor of 1.2 , were used in this calculation.
c: Experimental hydration free energies of neutral molecules were taken from Cabani et al [53]. For ionic solutes, they were determined from experimental pKa values [54], gas phase proton affinities/acidities [55], and hydration energies of related neutral solutes, using the hydration enthalpy of H+ of 267 kcal/mol. In cases where experimental gas phase data or hydration energies of neutral solutes were not available, they were substituted by MP2/6-31+G**//HF/6-31G* results and/or iterative LD results, respectively. The given uncertainties reflect the accuracy of data used in the derivation of 'experimental' ΔGsol.
d: phenolate/thiophenolate ion

The highly charged systems present a problematic case also for the polarized continuum method (PCM), which has larger CPU demands than ILD or NLD methods.

Too large PCM hydration energies of dianionic and trianionic ions can be rationalized by the lack of saturation in continuum methods. In contrast, the nonlinear dependence of induced dipoles upon the magnitude of the solute electrostatic field is expressed in terms of the Langevin function in LD-based approaches. The continuum method is also deficient for nonpolar neutral solutes. Here, a simple remedy can be provided by augmenting the electrostatic hydration energy by the van der Waals and field dependent hydrophobic terms as in the LD approach or field independent as in the implementations by Cramer and Truhlar [56], Marten et al. [57], and Stefanovich and Truong [58]. Except for the mentioned extremes, PCM and LD methods exhibit similar behavior. They tend to overestimate hydration energies of ammonium cations, and acetate, formate, phenolate, hydroxide and sulfur-containing anions. On the other hand, they underestimate hydration energies of solutes protonated at the sp^3 oxygen, including the hydronium cation. These systematic errors can be expected to compensate themselves for larger systems with multiple functional groups. Indeed, this seems to be the case for cytosine or acetamide. We believe that calculation of large systems will benefit from more generic parameterization and from the consistency brought by the use of ab initio charges. In those instances when the whole system cannot be treated quantum mechanically, the charge distribution can be built from individual group constituents, e.g. amino acids. In fact, the use of potential-derived HF/6-31G* charges and the build-up strategy of the LD approach is similar to that used by Kollman and coworkers in the AMBER program [59], wherein hydration energies are calculated using an all-atom solvent model and FEP calculations. Thus, one may use current libraries of atomic charges to calculate hydration energies with the more economic LD approach. For neutral systems, such as amines [60] or nucleic acid bases [61], LD and AMBER calculations provide similar results, whereas for ionic solutes the LD approach is more accurate and efficient.

Before moving to enzymes, let us consider the use of the QM(ai)/LD approach to study potential energy surfaces of chemical reactions in aqueous solution. More specifically, the reaction mechanisms of the phosphoester monoanion bond cleavage, which can be formally expressed as

$$R\text{-}OH + HO\text{-}P(O2)\text{-}OR'^{(-)} \rightarrow RO\text{-}P(O2)\text{-}OH^{(-)} + R'OH, \qquad (2)$$

$$R = -CH3, R' = -CH3$$

This reaction is an important component of several biochemical processes. For example, in RNA hydrolysis, R and R' correspond to the C2' and C5' atoms of riboses forming phosphodiester linkage. In DNA hydrolysis, ROH stands for the attacking water molecule (R = H), and R'OH is the deoxyribose-C5'OH group. In modeling both DNA and RNA, it is convenient to model the C3' atom of the ribose ring with hydrogen. This is a reasonable simplification since the mentioned hydrogen is not transferred to other groups during the reaction, nor does it form hydrogen bonds. In addition, hydrogen is present at this position in the third type of biochemical reaction that we intend to study - the transfer of terminal phosphoryl group of nucleoside mono-, di-, and tri-phosphates. This class of reactions includes ATP $\rightarrow$ ADP + P_i and GTP $\rightarrow$ GDP + P_i hydrolyses. Here, ROH and R'OH represent the attacking H2O molecule and the leaving β-phosphoryl group, respectively. Because pKa of the terminal phosphoryl group is about 6, this group is present in the cell as either dianion or monoanion. In this account, we will limit ourselves to the study of the monoanion hydrolysis.

Clearly, the nucleophilic attack of the oxygen atom of ROH at phosphorus is needed for reaction (2) to proceed, but different opinions exist as to its detailed mechanism. Some studies favor a dissociative mechanism, in which the metaphosphate anion PO_3^- is formed in the first reaction step [62-64], while others favor associative mechanism involving the formation of the pentacoordinated phosphorane intermediate [65,66]. Concerted mechanism, in which the metaphosphate like preassociated structure is considered to be transition state rather then intermediate, has also been suggested [67]. The controversy about the actual mechanism is disconcerting considering the importance of this reaction and the enormous experimental effort made to elucidate this problem.

Our strategy for resolving the mechanistic puzzle in phosphate hydrolysis is based on comparing the calculated activation barriers for different assumed mechanisms to the observed barriers (that can be deduced from the observed reaction rates). Provided our QM(ai)/LD or QM(ai)/FEP methods are sufficiently reliable, we can eliminate mechanisms whose activation energy is significantly higher than the one with the lowest barrier. Such elimination strategy has not been applied before to phosphate hydrolysis.

Furthermore, although the great importance of relative solvation energies in phosphate hydrolysis reactions was pointed out as early as 1978 [68], these effects were neglected in numerous subsequent theoretical studies (see e.g. [69-71] and references therein). Only recently, continuum solvation models were used to investigate hydrolysis of inorganic pyrophosphate [72] and strain effects in the base-catalyzed hydrolysis of cyclic phosphates [73]. However, none of these studies explored the crucial question about relative importance of alternative mechanisms for phosphate hydrolysis.

Our study of phosphate hydrolysis is based on calculating the relative free energies for the individual reaction steps in solution. This is done using MP2/6-31+G(d,p)//HF/6-31G(d) gas phase energies and ILD hydration free energies by using Gaussian 94 program [52] and generic LD parameterization described above. The use of ab initio methods such as MP2 that recover a significant part of electron correlation is necessary for the evaluation of relative energies of transition states for bond breaking/forming processes. Still, the expected error bounds for the relative free energies of individual reaction steps are generally large. Given the estimated 5 kcal/mol error for the solvation part and 4 kcal/mol uncertainty of MP2 energies the total error for 2 component system can be significant. Fortunately, most of these errors are of systematic character and as such they cancel each other when relative energies are evaluated. In our case, solvation energies of individual reactants and products are determined with accuracy of 3 kcal/mol (Table 2). Similarly, for other systems it is often possible to adjust solvation parameters for a given class of compounds using relevant experimental data measured for related compounds and increase thus the reliability of the AM(ai)/LD method to the 5 kcal/mol limit. Compared to this uncertainty, the errors originating from the neglected zero point and thermal contributions to the reaction enthalpies, or from gas phase entropies are negligible. Moreover, these contributions are likely to be unimportant in solution since their changes are usually compensated by changes in solvation free energies.

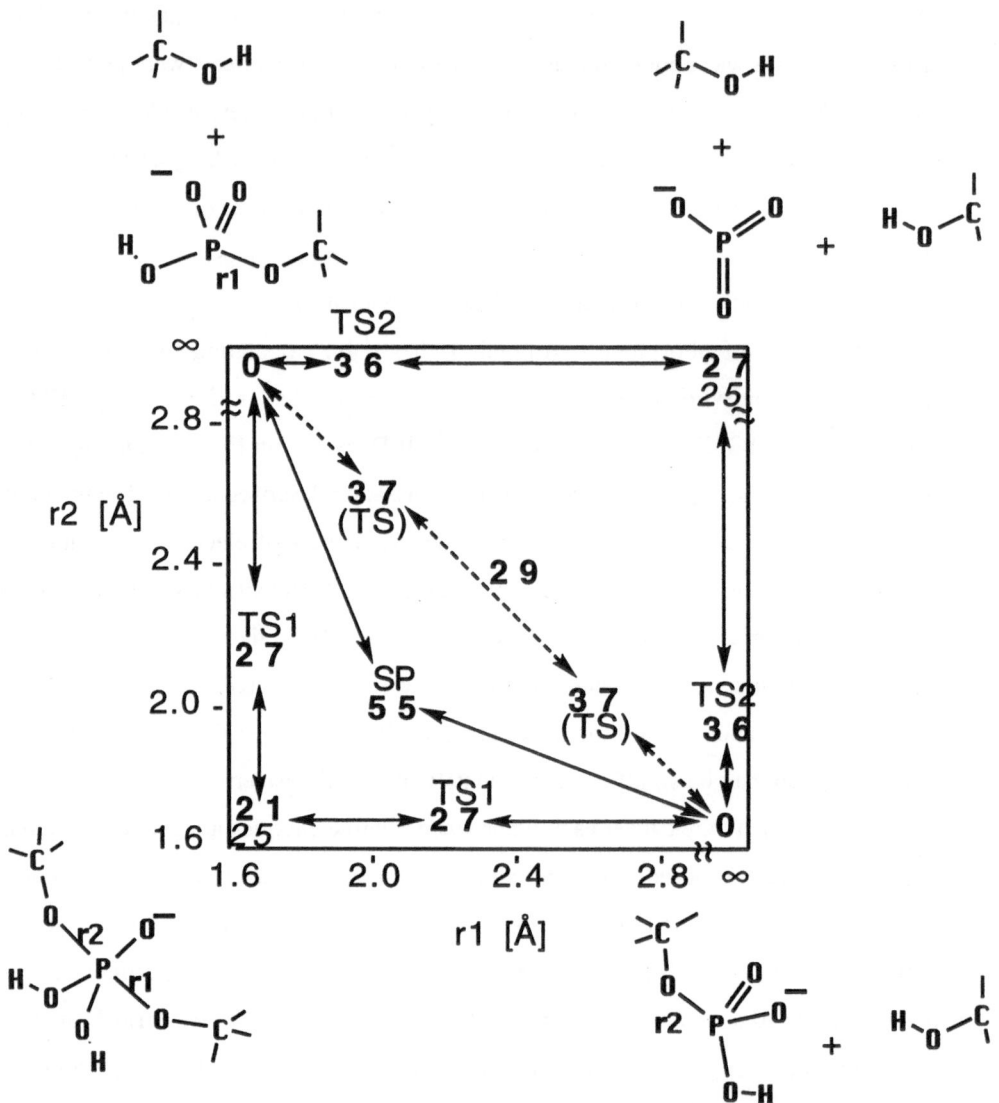

FIGURE 1. MP2/LD free energy surface for reaction (2). The calculated free energies [kcal/mol] are given as numbers printed in bold face. The experimental values (where available) are given as Latin numbers. TS and SP denote transition states and the saddle point of the second order, respectively. Reaction mechanisms discussed in the text are indicated by arrows.

In Figure 1, the calculated free energy surface is presented as a function of the bond length between phosphorus and oxygens of the incoming and leaving groups. The corners of this diagram are occupied by the reactants and products and by the pentacovalent and metaphosphate intermediates. Upon going from the reactants (upper left corner) to either intermediate, system must pass through the transition states, denoted TS1 and TS2. Energies of these points determine activation barriers for the associative and dissociative reaction pathways, respectively. According to our calculations, the associative stepwise mechanism is favored by 9 kcal/mol. By assuming that reaching TS1 is the rate determining step and by using the exponential prefactor of $6*10^{12}$ s^{-1}, the calculated activation barrier of 27 kcal/mol can be used to predict the rate constant for the hydrolysis by the associative mechanism as k = 10^{-7} s^{-1}. This rate constant agrees well with the rates measured for the uncatalyzed reaction of phosphate methyl and ethyl esters in neutral aqueous solution [65,74].

Also the relative energies of both intermediates agree well with the experimental estimates [74] (Figure 1). The dissociative and associative regions are separated by the high energy region located in the vicinity of the saddle point of the second order (SP). The reaction path intersecting this saddle point corresponds to the concerted mechanism, in which two protons simultaneously transfer between methanol and phosphate oxygens. In addition, relative energies of several points that lie near the center of the diagram have been determined by the partial HF/6-31G(d) geometry optimization followed by MP2 and ILD single-point calculations. Positions of these points were chosen to lie on a path that corresponds to the pre-association-assisted dissociation mechanism. The highest points on this path denoted "(TS)" were obtained by the partial transition state optimization, in which PO bond distances of incoming and leaving groups were fixed. The calculated gradient and energy reveal that the simple dissociation mechanism is more favorable then this simulated reaction path (denoted in Figure 1 by a dashed line) going through the metaphosphate-like pre-associated structure .

In general, the activation barrier for the nucleophilic attack by ROH molecules will always have a substantial proton transfer contribution. However, the barrier for proton transfer increases with increasing separation of atoms between which this transfer occurs. Consequently, any attempts to locate transition state for concerted dissociative reaction mechanism on the right side of SP (Figure 1) will result in high energy structures.

Components of the relative free energies are analyzed in Table 3. The largest solvation related destabilization of 11 kcal/mol (compared to reactants) was predicted for pentacovalent phosphorane anion, whereas metaphosphate intermediate was found to be more stable in solution than in gas phase. However, the solvation contributions for the transition states are nearly the same. This result is probably related to the similarities in their structures, in which proton is only half-way transferred from the methanol oxygen to the equatorial phosphate oxygen. This proton transfer opposes the charge redistribution caused by the formation/disruption of the PO bond so that only relatively minor variations in atomic charges occur in the course of reactions. Consequently, only small changes in the predicted energetics for the competing reaction mechanisms can be expected when the computational model is further improved by optimizing geometries of stationary points in the presence of the solvent. Indeed, we found that the single-point solvation and MP2 corrections along the intrinsic reaction coordinate (IRC) for the both stepwise mechanisms did not change the position of transition states on IRC. However, caution should be taken when using gas-phase geometries for studies of reactions involving larger changes of charges of reacting fragments. Unfortunately, quantitatively correct computational methods that enable to optimize the geometry of solute in the presence of solvent [75,76] are still rather computationally demanding.

TABLE 3. Relative energies of stationary points on the PES of phosphoryl monoanion transfer reaction[a].

System Description[b]	rl [Å]	r2 [Å]	HF	MP2	LD
Reactants	1.64	∞	0	0	0
TS1	1.66	2.31	32.5	22.5	4.2
Phosphorane Anion	1.68	1.68	20.1	10.2	11.0
TS2	2.00	∞	47.9	34.0	2.5
Metaphosphate Anion	∞	∞	33.9	29.7	-2.4
SP (C_2 Symmetry)	2.09	2.09	72.2	45.6	9.2

a: Reaction 2. The total HF/6-31G(d), MP2/6-31+G(d,p)//HF/6-31G(d) and ILD energies of reactants (methanol + methylphosphate monoanion) amount to -795.532708 and -796.930425 a.u., and -75.5.kcal/mol, respectively.
b: See Figure 1

The most important conclusion from these results is the fact that simplified solvation models like the combined QM(ai)/LD method are starting to give sufficiently reliable information on chemical reactions in solution, and can be used in calibrating approaches for studies of enzymatic reactions.

3. How to obtain reliable results in studies of enzymatic reactions.

The challenge of understanding enzymatic reactions has led to significant theoretical activity in recent years [1,2,22,50,77]. However, it appears that despite this impressive progress many studies are still overlooking the major factors and focusing on quantum mechanical features of the reacting fragments that essentially cancel out in the evaluation of catalytic effects. In this section we shall briefly review two techniques that have been used to study reactions in proteins, that of using a gas phase representation of the solute with a few charged residues representing the protein, and that of the EVB/MM technique. We will then discuss our new ab initio QM/MM method and present results for the nucleophilic attack of deprotonated serine on a substrate in a serine protease.

We begin our consideration of techniques for simulating chemical reactions in enzymes by noting that attempts to consider the substrate and several enzyme residues in the gas phase (e.g. [78,79]) are potentially problematic and may even be deceiving because the missing solvation effects can be enormous. For example, recent calculations that consider ionized residues in the reaction of ribonuclease [71] or related studies [79] can be considered as a part of an incremental attempt to systematically understand the reacting system. However, such calculations overlook enormous solvation contributions. These contributions generally change drastically between the ground and transition states. Even inclusion of the entire enzyme and some solvent molecules (in a QM/MM fashion) does not guarantee correct results owing to variations in how the surrounding environment is treated. This is true even in the case of recent QM/MM methods that formally include the entire system except the bulk around the protein region (e.g. [23,80,81]). Most of these methods are not yet able to provide accurate free energies and generally only involve energy minimization. Furthermore, they are ineffective in providing the proper solvent reorganization in response to formation of charges. Finally, the approaches used do not involve proper electrostatic boundary conditions (e.g. surface polarization and bulk effects which are discussed elsewhere [4,48] and cannot reproduce the correct compensation for charge-charge interaction). The importance of these points may be seen from the examples given below. These examples are not meant to detract from significant progress in the field but to point out major traps

One apparent problem with modeling enzyme reactions using a solute and several residues in the gas phase is associated with the relative contribution of ionized residues [23,81]. According to these calculations, ionized groups 15 Å from the reacting system can provide a major contribution to catalysis (up to 10 kcal/mol). However, observations from mutation experiments [82] and more consistent calculations [48] is that charge-charge interactions more than 8 Å away from the reacting region can never contribute more than 1-2 kcal/mol. This issue is discussed in great length in many of our papers (e.g. see [4]). Apparently properly handling electrostatic effects presents a major problem in QM/MM calculations. It is not enough to simply include the residual charges of the solvent and the protein in the calculation; the proper reorientation upon charging must be consistently reproduced.

An additional problem with most current QM/MM approaches is the fact that the quantum mechanical method used is not accurate enough. In many cases, this oversight may contribute very large errors. This is demonstrated by the findings of Lyne et al. [81], where the difference between the AM1 and ab initio barrier for the reaction of chorismate mutase is shown to be around 20 kcal/mol. Since ab initio approaches are often to expensive to use with current methods of describing an enzyme active site, one is forced to use AM1-type methods that are not expected to give accurate results even when the environment is treated properly. To further illustrate our point, we refer the reader to two sets of calculations of the catalytic reaction of triosephosphate isomerase, the combined AM1/MM calculation of Bash et al. [23] and the recent EVB calculation of Åqvist and Fothergill [83]. The AM1/MM calculation does not reproduce quantitative results since the observed rate determining step for the reaction is the first step with an observed barrier of about 12 kcal/mol, whereas this calculation gave a rate determining barrier of 20 kcal/mol at the third step. On the other hand the calculations of Åqvist and Fothergill reproduce the experimental trend with impressive accuracy and without adjusting any parameters in moving from water to the enzyme active site.

The reader might question how the seemingly simple EVB method achieves correct results where other methods fail. The answer is largely in the physical insight of this method. Many theoreticians assume that one can simply take gas phase calculations, add the enzyme, and eventually reproduce the correct results. Although this technique may be feasible in the future with much faster computers, present computational limitations preclude the entire system to be studies at a sufficient level of theory to yield accurate results. On the other hand, the EVB technique reflects the realization that enzyme catalysis involves the difference between the reaction in the enzyme and the corresponding reference reaction in water. Enzyme catalysis is due to environmental effects and is not due to the solute potential surface. The EVB method avoids the need to evaluate the exact free energy surface in the enzyme site by calibrating it on experimental information in solution or fitting it to high level ab initio gas phase calculations [84]. The advance of the QM(ai)/LD techniques described in Section 2 makes it possible to calibrate EVB parameters using ab initio calculations in solution. The essence of the EVB method, however, is the focus on the difference between the potential surfaces in enzyme and solution, ensuring that calculations reproduce experimental results for the reference reaction in solution.

Even with the scope and utility of the EVB method, it is still very important to go beyond this level to try to obtain results for enzymatic reactions by "first principle" approaches. This section concludes with mentioning our progress in this direction.

We have modified our ab initio hybrid QM/MM technique [40] to model chemical reactions in proteins. This method uses a molecular mechanics force field to represent the solvent, and an MP2/6-31G** ab initio quantum mechanical technique that incorporates the electrostatic field from the solvent and/or protein in its Hamiltonian to represent the solute (although a smaller basis set is used for the preliminary results reported here). The EVB mapping potential is used as a reference potential for free energy perturbation (FEP) calculations of the potential surface; this involves running molecular dynamics trajectories with EVB/MM potentials at a variety of different values for the EVB coefficients [1]. The reactions considered in [40] were simple enough to fit by hand EVB parameters that reproduced the ab initio surface. This fitting process becomes more involved for larger and more varied substrates. We consequently have developed an automated procedure that constructs an EVB gas phase potential surface whose minima are as close to the corresponding minima obtained by gas phase ab initio calculations as possible. This automated procedure results in an EVB surface that accurately approximates the ab initio surfaces of the reactant, product, and crucial intermediates, and this potential surface may be further refined by adjusting the appropriate off-diagonal elements and by applying weak Cartesian constraints. Next we use a perturbation approach to "mutate" the system from the EVB mapping potential to the ab initio free energy surface. The automated procedure that forces the EVB surface to reproduce the ab initio minima significantly reduces the amount of sampling needed to obtain the ab initio free energy surface.

One important point that emerges from our studies is that the ab initio and EVB charge distributions along the reaction path are quite similar. That is, our most advanced treatment starts by using the ab initio fragment charges in solution as the EVB fragment charges. Thus, the "solvation" free energy of the ab initio and EVB fragments is quite similar. As the system moves toward the transition state, the EVB charges have a value between that of the reactant and the products. Because the EVB captures correctly the physics of solute polarization (the contribution of the charges of each resonance structure depends on its relative stabilization) it correctly interpolates the charges along the reaction coordinate. The agreement between the EVB and ab initio charges is evident from the fact that the solvation energies obtained by each method along the reaction path are similar. This guarantees stable results, and allows us to use various tricks which will be described briefly below.

As in previous hybrid QM/MM procedures (e.g. [6,23,30]) we have to properly connect the quantum and classical regions. Warshel and Levitt introduced an excellent hybrid orbital treatment [6], which was followed recently by the approach of Rivail and coworkers [30]. In the present approach we use a much simpler approach. We note that errors introduced by the connecting atoms are likely the same in aqueous solvent as they are in the protein. We parameterize our technique so that the overall free energy curve reproduces experimental results in solution, and use the same parameters for the connection procedure in the enzyme. We begin with rather small reacting fragments, where the connecting classical atoms are replaced by hydrogens in the quantum treatment (the hydrogens are shifted along the bond to the proper distance). We subsequently generate results using slightly larger fragments and determine the size needed to obtain converging results for the *difference* between the free energy in the protein and solution, $\Delta\Delta G^{W\rightarrow P}$. This approach is remarkably reliable because we focus on the difference rather than the absolute free energies of the reacting fragments.

Even with the care taken in designing the above ab initio QM/MM methodology, it is still extremely difficult to obtain converged free energies in a reasonable amount of CPU time. Part of this effect is due to the inevitably large amount of time required to simulate a large protein. A significant part, however, is due to fluctuations in the intramolecular energy (i.e. the part of the energy expression that depends solely on the solute) which contributes little to the change in free energy $\Delta\Delta G^{W\rightarrow P}$, and requires a large amount of time to converge properly. In order to avoid these problems, we focused first on the difference in electrostatic ("solvation") energy associated with transforming the reacting fragments from water to protein. The same philosophy was used very effectively in EVB calculations (e.g. [18]) where it was demonstrated that the catalytic effect of the enzyme can be estimated much faster by this approach then by using the full EVB potential. What is new in the current treatment is the fact that we are evaluating ab initio solvation energies in the enzyme active site. As an example of the power of our technique, we report in Figure 2 results on the effect of the Asn155Ala mutation for the nucleophilic attack of deprotonated serine on the substrate protein in subtilisin. Figure 2a shows the free energy curve for the S_n2 attack of deprotonated serine on the protein backbone carbonyl group in the serine protease subtilisin. Figure 2b shows the corresponding results for the Asn155Ala mutation of subtilisin. The (+) symbols in the figures are the EVB results from classical simulation, the details of which are given in the figure caption. At the reactant, transition, and product states, the free energy difference between the EVB potential surface and the ab initio quantum mechanical potential surface is calculated; these points are shown in the figures as (o) symbols. We realize that much longer trajectories are required to obtain truly converged free energies, and we present our results here to show that our hybrid QM/MM technique is capable of studying real problems in enzyme catalysis.

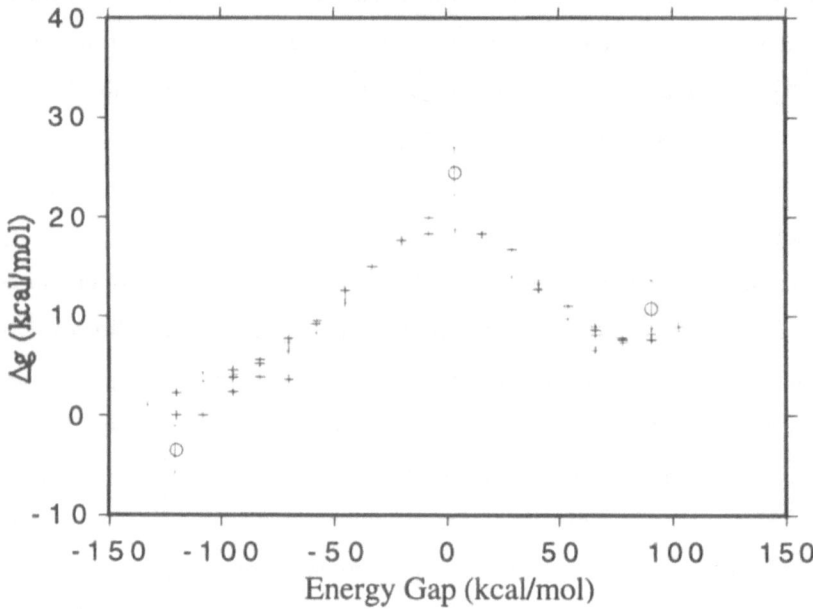

FIGURE 2a. Free energy curve for nucleophilic attack of deprotonated Ser221 on the carbonyl carbon of an amide bond of a tyrosine-glycine substrate in native subtilisin. The calculated curves were obtained via EVB (+) and ab initio (o) techniques. To simulate the EVB potential surface we gradually change the EVB mapping potential [1] from one corresponding to the reagents to one corresponding to the products in a series of 11 steps; each step consists of a 2ps trajectory at 300 K using the ENZYMIX program [48]. The ab initio results are computed by running a shorter 0.5 ps at mapping steps corresponding to the reagent, transition, and product states, and using the GAMESS program suite [85] to compute the difference between the EVB and the MP2/STO-3G potential surfaces.

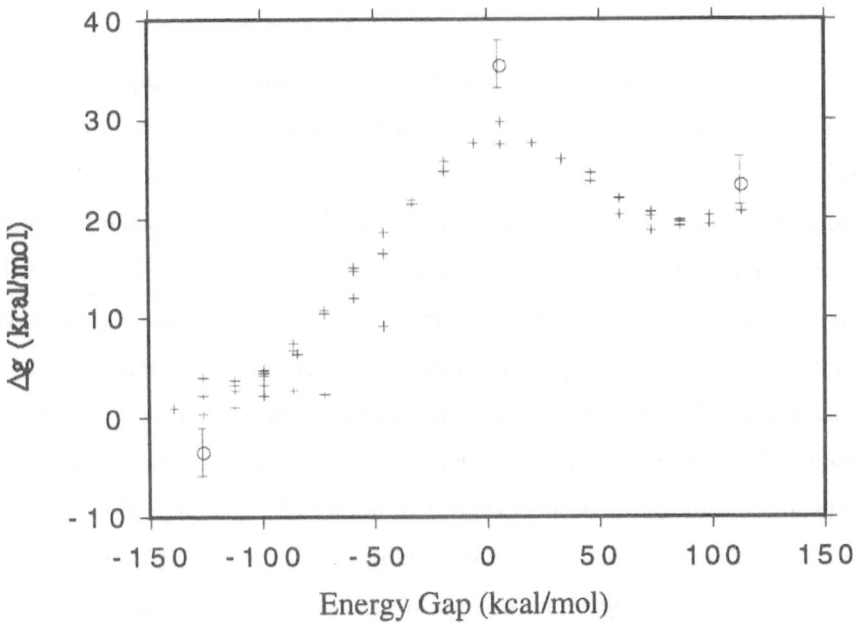

FIGURE 2b. Free energy curve for nucleophilic attack of deprotonated Ser221 on the carbonyl carbon of an amide bond of a tyrosine-glycine substrate in the Asn155Ala mutation of subtilisin. The calculated curves were obtained via EVB (+) and ab initio (o) techniques. The computational details are the same as those described in the caption to Figure 2a.

We note several important points about these curves. First, the EVB method calculates a significantly larger barrier for the mutated protein than for the native protein. Second, the electrostatic free energy change between the EVB and ab initio potential surface is small, which is in accordance with our earlier contention that the EVB mapping potential accurately interpolates the ab initio charges throughout the reaction path. We are confident that with longer simulation times and a more accurate basis set, this QM(ai)/MM technique will provide a powerful tool to computationally probe enzyme catalysis.

4. Conclusions

In this report we have introduced our philosophy of modeling enzymatic reactions and discussed problems associated with some approaches used in this field. We have also summarized our recent progress in attempts to use hybrid ab initio QM/MM techniques to simulate enzymatic reactions. We have discussed two methods in particular. The first, a hybrid QM(ai)/LD approach, reproduces the accuracy of dielectric continuum models, is more efficient, and provides a more realistic physical picture. Moreover, the QM(ai)/LD method is also quite promising because its speed and accuracy allow calibration of EVB parameters in solution rather than the gas phase. Furthermore, this approach allows one to elucidate the mechanism of solution reactions whose understanding is crucial for further progress in studies of key biological systems. One of the best examples is our ability to start probing the nature of phosphate hydrolysis in solution and the relative importance of associative and dissociative mechanisms.

Our earlier studies [40] have shown that it is essential to use the EVB mapping potential as a reference for hybrid QM/MM computations. Until now we have used the results of gas phase ab initio calculations to augment experimental results in deriving parameters for the EVB mapping potential. Clearly the availability of quality ab initio results in solution will greatly simplify this process.

We have also detailed the latest version of our hybrid QM/MM method, which is used to calculate free energy curves for enzymatic reactions. We have introduced the use of an automated procedure to derive EVB parameters that reproduce ab initio geometries. Subsequently, we have used the EVB mapping potential to simulate the nucleophilic attack in a serine protease. Our results for the catalytic effect of the Asn155Ala mutation show reasonable agreement with experimental estimates. To the best of our knowledge they are probably the first ab initio FEP results that reasonably describe the energetics of an enzymatic reaction.

5. Acknowledgments

This work was supported by Grant GM24492 from the National Institutes of Health and by Grant DE-FG03-94ER61945 from the Department of Energy.

6. References

1. Warshel, A. (1991) *Computer Modeling of Chemical Reactions in Enzymes and Solutions*, John Wiley & Sons, New York.

2. Naray-Szabo, G.; Fuxreiter, M.; Warshel, A. (In Press) Electrostatic Basis of Enzyme Catalysis, in Naray-Szabo, G.; Warshel, A. (ed.), *Computational Approaches to Biochemical Reactivity*, Kluwer Academic Publishers,

3. Warshel, A.; Åqvist, J.; Creighton, S. (1989) Enzymes Work by Solvation Substitution Rather Than by Desolvation, *Proc. Natl. Acad. Sci. USA* **86**, 5820.

4. Alden, R. G.; Parson, W. W.; Chu, Z.-T.; Warshel, A. (1995) Calculations of Electrostatic Energies in Photosynthetic Reaction Centers, *J. Am. Chem. Soc.* **117**, 12284.

5. Fersht, A. (1985) *Enzyme Structure and Mechanism*, W. H. Freeman & Co., New York.

6. Warshel, A.; Levitt, M. (1976) Theoretical Studies of Enzymatic Reactions: Dielectric, Electrostatic, and Steric Stabilization of the Carbonium Ion in the Reaction of Lysozyme, *J. Mol. Biol.* **103**, 227.

7. Warshel, A. (1978) A Microscopic Model for Calculations of Chemical Processes in Aqueous Solutions, *Chem. Phys. Lets.* **55**, 454.

8. Warshel, A. (1979) Calculations of Chemical Processes in Solution, *J. Phys. Chem.* **83**, 1640.

9. Tapia, O.; Johannin, G. (1981) An Inhomogeneous Self-Consistent Reaction Field Theory of Protein Core Effects. Towards a Quantum Scheme for Describing Enzyme Reactions, *J. Chem. Phys.* **75**, 3624.

10. Miertus, S.; Scrocco, E.; Tomasi, J. (1981) Electrostatic Interaction of a Solute with a Continuum. A Direct Utilization of ab Initio Molecular Potentials for the Prevision of Solvent Effects., *Chem. Phys.* **55**, 117.

11. Warshel, A. (1982) Dynamics of Reactions in Polar Solvents. Semiclassical Trajectory Studies of Electron-Transfer and Proton Transfer Reactions, *J. Phys. Chem.* **86**, 2218.

12. Warshel, A.; Russell, S. T.; Churg, A. K. (1984) Macroscopic Models for Studies of Electrostatic Interactions in Proteins: Limitations and Applicability, *Proc. Natl. Acad. Sci. USA* **81**, 4785.

13. Jorgensen, W.L.; Ravimohan, C. (1985) Monte Carlo Simulation of Differences in Free Energies of Hydration, *J. Chem. Phys.* **83**, 3050.

14. Weiner, S. J.; Singh, U. C.; Kollman, P. A. (1985) Simulation of Formamide Hydrolysis by Hydroxide Ion in the Gas Phase and in Aqueous Solution, *J. Am. Chem. Soc.* **107**, 2219.

15. Singh, U.C.; Kollman, P.A. (1986) A Combined Ab Initio Quantum Mechanical and Molecular Mechanical Method for Carrying Out Simulations on Complex Molecular Systems, *J. Comp. Chem.* **7**, 718.

16. Kong, Y.S.; Jhon, M.S. (1986) Solvent Effects on Sn2 Reactions, *Theor. Chim. Acta.* **70**, 123.

17. Bash, P.A.; Field, M.J.; Karplus, M. (1987) Free Energy Perturbation Method For Chemical Reactions in the Condensed Phase: A Dynamical Approach Based on a Combined Quantum and Molecular Mechanics Potential, *J. Am. Chem. Soc.* **109**, 8092.

18. Warshel, A., Sussman, F., and Hwang, J.-K. (1988) Evaluation of Catalytic Free Energies in Genetically Modified Proteins, *J. Mol. Biol.* **201**, 139.

19. Ford, G.P.; Wang, B. (1992) Incorporation of Hydration Effects within the Semiempirical Molecular Orbital Framework. AM1 and MNDO Results for Neutral Molecules, Cations, Anions, and Reacting Systems, *J. Am. Chem. Soc.* **114**, 10563.

20. Honig, G.; Sharp, K.; Yang, A.-S. (1993) Macroscopic Models of Aqueous Solutions: Biological and Chemical Applications, *J. Phys. Chem.* **97**, 1101.

21. Rashin, A.A.; Bukatin, M.A.; Andzelm, J.; Hagler, A.T. (1994) Incorporation of Reaction Field Effects into Density Functional Calculations for Molecules of Arbitrary Shape in Solution, *Biophys. Chem.* **51**, 375.

22. Bash, P.A.; Singh, U.C.; Langridge, R.; Kollman, P.A. (1987) Free Energy Calculations by Computer Simulation., *Science* **236**, 564.

23. Bash, P. A.; Field, M. J.; Davenport, R. C.; Petsko, G. A.; Ringe, D.; Karplus, M. (1991) Computer Simulation and Analysis of the Reaction Pathway of Triosephosphate Isomerase, *Biochemistry* **30**, 5826.

24. Field, M.J.; Bash, P.A.; Karplus, M. (1990) A Combined Quantum Mechanical and Molecular Mechanical Potential for Molecular Dynamics Simulations., *J. Comp. Chem.* **11**, 700.

25. Gao, J. (1992) Absolute Free Energy of Solvation From Monte Carlo Simulations Using Combined Quantum and Molecular Mechanical Potential, *J. Phys. Chem.* **96**, 537.

26. Gao, J.; Xia, X. (1992) A Priori Evaluation of Aqueous Polarization Effects through Monte Carlo QM-MM Simulations, *Science* **258**, 631.

27. Gao, J. (1994) Combined QM/MM Simulation Study of the Claisen Rearrangement of Allyl Vinyl Ether in Aqueous Solution, *J. Am. Chem. Soc.* **116**, 1563.

28. Luzhkov, V.; Warshel, A. (1992) Microscopic Models for Quantum Mechanical Calculations of Chemical Processes in Solutions: LD/AMPAC and SCAAS/AMPAC Calculations of Solvation Energies, *J. Comp. Chem.* **13**, 199.

29. Thery, V.; Rinaldi, D.; Rivail, J.-L.; Maigret, B.; Ferenczy, G.G. (1994) Quantum Mechanical Computations on Very Large Molecular Systems: The Local Self-Consistent Field Method, *J. Comp. Chem.* **15**, 269.

30. Ferenczy, G.G.; Rivail, J.-L.; Surjan, P.R.; Naray-Szabo, G. (1992) NDDO Fragment Self-Consistent Field Approximation for Large Electronic Systems, *J. Comp. Chem.* **13**, 830.

31. Stanton, R.V.; Dixon, S.L.; Merz, K.M., Jr. (1995) General Formulation for a Quantum Free Energy Perturbation Study, *J. Phys. Chem.* **99**, 10701.

32. Stanton, R.V.; Hartsough, D.S.; Merz, K.M. Jr. (1993) Calculation of Solvation Free Energies Using a Density Functional/Molecular Dynamics Coupled Potential, *J. Phys. Chem.* **97**, 11868.

33. Stanton, R.B.; Little, L.R.; Merz, K.M. Jr. (1995) An Examination of a Hartree-Fock/Molecular Mechanical Coupled Potential, *J. Phys. Chem.* **99**, 17344.

34. Stanton, R.V.; Hartsough, D.S.; Merz, K.M. Jr. (1995) An Examination of a Density Functional/Molecular Mechanical Coupled Potential, *J. Comp. Chem.* **16**, 113.

35. Hwang, J.-K.; Warshel, A. (1993) A Quantized Classical Path Approach for Calculations of Quantum Mechanical Rate Constants, *J. Phys. Chem.* **97**, 10053.

36. Thompson, M.A. (1995) Hybrid Quantum Mechanical/Molecular Mechanical Force Field Development for Large Flexible Molecules: a Molecular Dynamics Study of 18-Crown-6, *J. Phys. Chem.* **99**, 4794.

37. Thompson, M.A.; Glendening, E.D.; Feller, D. (1994) The Nature of K+/Crown Ether Interactions: a Hybrid Quantum Mechanical Molecular Mechanical Study, *J. Phys. Chem.* **98**, 10465.

38. Warshel, A.; Weiss, R.M. (1980) An Empirical Valence Bond Approach For Comparing Reactions in Solutions and in Enzymes, *J. Am. Chem. Soc.* **102**, 6218.

39. Warshel, A.; Parson, W.W. (1991) Computer Simulations of Electron Transfer Reactions in Solution and Photosynthetic Reaction Centers, *Ann. Rev. Phys. Chem.* **42**, 279.

40. Muller, R.P.; Warshel, A. (1995) Ab Initio Calculations of Free Energy Barriers for Chemical Reactions in Solution, *J. Phys. Chem.* **99**, 17516.

41. Wei, D.; Salahub, D.R. (1994) A Combined Density Functional and Molecular Dynamics Simulation of a Quantum Water Molecule in Aqueous Solution, *Chem. Phys. Lets.* **24**, 291.

42. Tuckerman, M.; Laasonen, K.; Sprik, M.; Parrinello, M. (1995) Ab Initio Molecular Dynamics Simulation of the Solvation and Transport of H3O+ and OH- Ions in Water, *J. Phys. Chem.* **99**, 5749.

43. Wesolowski, T.A.; Warshel, A. (1993) Frozen density functional approach for ab initio calculations of solvated molecules, *J. Phys. Chem.* **97**, 8050.

44. Wesolowski, T.; Warshel, A. (1994) Ab Initio Free Energy Perturbation Calculations of Solvation Free Energy Using the Frozen Density Functional Approach, *J. Phys. Chem.* **98**, 5183.

45. Wesolowski, T.; Muller, R.P.; Warshel, A. (In Press) Ab Initio Frozen Density Functional Calculations of Proton Transfer Reactions in Solution., *J. Phys. Chem.*

46. Wesolowski, T.A.; Weber, J. (1996) Kohn-Sham Equations with Constrained Electron Density: an Iterative Evaluation of the Ground-State Electron Density of Interacting Molecules, *Chem. Phys. Let.* **248**, 71.

47. Muller, R.P.; Wesolowski, T.; Warshel, A. (1996) Calculations of Chemical Processes in Solution by Density Functional and Other Quantum Mechanical Techniques, in Springbourg, M. (ed.), *Density functional methods: Applications in chemistry and materials science.*, John Wiley & Sons, Ltd., London.

48. Lee, F.S.; Chu, Z.T.; Warshel, A. (1993) Microscopic and Semimicroscopic Calculations of Electrostatic Energies in Proteins by the POLARIS and ENZYMIX Programs, *J. Comp. Chem.* **14**, 161.

49. Warshel, A.; Chu, Z. T. (1994) Calculations of Solvation Free Energies in Chemistry and Biology, in Cramer, C. J.; Truhlar, D. G. (ed.), *A C S Symposium Series: Structure and Reactivity in Aqueous Solution. Characterization of Chemical and Biological Systems,*

50. Åqvist, J.; Warshel, A. (1993) Simulation of Enzyme Reactions Using Valence Bond Force Fields and Other Hybrid Quantum/Classical Approaches, *Chem. Rev.* **93**, 2523.

51. Florian, J.; Warshel, A. (1996) Ab Initio-Langevin Dipoles Solvation, *Manuscript in preparation*

52. Frisch, M.J.; Trucks, G.W.; Schlegel, H.B.; Gill, P.M.W.; Johnson, B.G.; Robb, M.A.; Cheeseman, J.R.; Keith, T.R.; Petersson, G.A.; Montgomery, J.A.; Raghavachari, K.; Al-Laham, M.A.; Zakrzewski, V.G.; Ortiz, J.V.; Foresman, J.B.; Cioslowski, J.; Stefanov *Gaussian 94, Revision D.2*, 1995.

53. Cabani, S.; Gianni, P.; Mollica, V.; Lepori, L. (1981) Group Contributions to the Thermodynamic Properties of Non-Ionic Organic Solutes in Dilute Aqueous Solution, *J. Solution Chem.* **10**, 563.

54. Serjeant, E. P.; Dempsey, B. (1979) *Ionization Constants of Organic Acids in Aqueous Solution*, Pergamon Press, Oxford.

55. Lias, S.G.; Bartmess, J.E.; Liebman, J.F.; Holmes, J.L; Levin, R.D.; Mallard, J.G. (1988) Gas-Phase Ion and Neutral Thermochemistry, *J. Phys. Chem. Ref. Data* **17**, Supplement #1.

56. Cramer, C. J.; Truhlar, D. G. (1991) General Parameterized SCF Model for Free Energies of Solvation in Aqueous Solution, *J. Am. Chem. Soc.* **113**, 8305.

57. Marten, B.; Kim, K. Cortis, C. Friesner, R.A.; Murphy, R.B.; Ringnalda, M.N.; Sitkoff, D.; Honig, B. (1996) New Model for Calculation of Solvation Free Energies: Correction of Self-Consistent Reaction Field Continuum Dielectric Theory for Short-Range Hydrogen Bonding Effects, *J. Phys. Chem.* **100**, 11775.

58. Stefanovich, E.V.; Truong, T.N. (1995) Optimized atomic radii for quantum dielectric continuum solvation models, *Chem. Phys. Let.* **244**, 65.

59. Weiner, S.J.; Kollman, P.A.; Nguyen, D.T.; Case, D.A. (1986) An All Atom Force Field for Simulations of Proteins and Nucleic Acids, *J. Comp. Chem.* **7**, 230.

60. Morgantini, P.Y.; Kollman, P.A. (1995) Solvation Free Energies of Amides and Amines - Disagreement between Free Energy Calculations and Experiment, *J. Am. Chem. Soc.* **117**, 6057.

61. Miller, J.L.; Kollman, P.A. (1996) Solvation Free Energies of the Nucleic Acid Bases, *J. Phys. Chem.* **100**, 8587.

62. Westheimer, F. H. (1987) Why Nature Chose Phosphates, *Science* **235**, 1173.

63. Butcher, W.W.; Westheimer, F.H. (1955) Metaphosphate Intermediate, *J. Am. Chem. Soc.* **77**, 2420.

64. Admiraal, S.J.; Herschlag, D. (1995) Mapping the Transition State for ATP Hydrolysis: Implications for Enzymatic Catalysis, *Chem. Biol.* **1995**, 729.

65. Cox Jr., J. R.; Ramsay, O. B. (1964) Mechanisms of Nucleophilic Substitution in Phosphate Esters, *Chem. Rev.* **64**, 317.

66. Breslow, R.; Labelle, M. (1986) Sequential General Base-Acid Catalysis in the Hydrolysis of RNA by Imidazole, *J. Am. Chem. Soc.* **108**, 2655.

67. Williams, A. (1989) Concerted Mechanisms of Acyl Group Transfer Reactions in Solution, *Acc. Chem. Res.* **22**, 387.

68. Hayes, D.M.; Kenyon, G.L.; Kollman, P.A. (1978) Theoretical Calculations of the Hydrolysis Energies of some 'High Energy' Molecules. 2. A Survey of some Biologically Important Hydrolytic Reactions, *J. Am. Chem. Soc.* **100**, 4331.

69. Lim, C.; Karplus, M. (1990) Nonexistence of Dianionic Pentacovalent Intermediates in an Ab Initio Study of the Base-Catalyzed Hydrolysis of Ethylene Phosphate, *J. Am. Chem. Soc.* **112**, 5872.

70. Taira, K.; Uchimaru, T.; Tanabe, K.; Uebayasi, M. (1991) Rate Limiting P-O(5') Bond Cleavage of RNA Fragment: Ab Initio Molecular Orbital Calculations on the Base-Catalyzed Hydrolysis of Phosphate, *Nucl. Acid. Res.* **19**, 2747.

71. Wladkowski, B.D.; Krauss, M.; Stevens, W. (1995) Ribonuclease A Catalyzed Transphosphorylation: An Ab Initio Theoretical Study, *J. Phys. Chem.* **99**, 6273.

72. Colvin, M.E.; Evleth, E.V.; Akacem, Y. (1995) Quantum Chemical Studies of Pyrophosphate Hydrolysis, *J. Am. Chem. Soc.* **117**, 4357.

73. Dejaegere, A.; Liang, X.; Karplus, M. (1994) Phosphate Ester Hydrolysis: Calculation of Gas-Phase Reaction Paths and Solvation Effects, *J. Chem. Soc. Faraday Trans.* **90**, 1763.

74. Guthrie, J.P. (1977) Hydration and Dehydration of Phosphoric Acid Derivatives: Free Energies of Formation of the Pentacoordinate Intermediates for Phosphate Ester Hydrolysis, *J. Am. Chem. Soc.* **99**, 3991.

75. Rivail, J.-L.; Rinaldi, D. (1996) Liquid-State Quantum Chemistry: Computational Applications of the Polarizable Continuum Models, in Leszczynski, J. (ed.), *Computational Chemistry: Reviews of Current Trends*, World Scientific, Singapore.

76. Cossi, M.; Tomasi, J.; Cammi, R. (1995) Analytical Expressions of the Free-Energy Derivatives for Molecules in Solution - Application to the Geometry Optimization, *Int. J. Quant. Chem., Quant. Chem. Symp.* **29**, 695.

77. Kollman, P. (1993) Free Energy Calculations: Applications to Chemical and Biochemical Phenomena, *Chem. Rev.* **93**, 2395.

78. Krauss, M.; Garmer, D.R. (1993) Assignment of the Spectra of Protein Radicals in Cytochrome C Peroxidase, *J. Phys. Chem.* **97**, 831.

79. Nakagawa, S.; Umeyama, H. (1982) *J. Theor. Biol.* **96**, 473.

80. Wazkowycz, B.; Hillier, I.H.; Gensmantel, N.; Payling, D.W. (1991) A Combined Quantum Mechanical-Molecular Mechanical Model of the Potential-Energy Surface of Ester Hydrolysis by the Enzyme Phospholipase-A2, *J. Chem. Soc. Perkin. Trans.* **2**, 225.

81. Lyne, P.; Mulholland, A.J.; Richards, W.G. (1995) Insights into Chorismate Mutase Catalysis from a Combined QM/MM Simulation of the Enzyme Reaction, *J. Am. Chem. Soc.* **117**, 11345.

82. Russell, A.J.; Fersht, A.R. (1987) Rational Modification of Enzyme Catalysis by Engineering Surface Charge, *Nature* **328**, 496.

83. Åqvist, J.; Fothergill, M. (1996) Computer Simulation of the Triosephosphate Isomerase Catalyzed Reaction, *J. Biol. Chem.* **271**, 10010.

84. Hwang, J.-K.; King, G.; Creighton, S.; Warshel, A. (1988) Simulation of Free Energy Relationships and Dynamics of Sn2 Reactions in Aqueous Solution., *J. Am. Chem. Soc.* **110**, 5297.

85. Schmidt, M.W.; Baldridge, K.K.; Boatz, J.A.; Jensen, J.H.; Doseki, S.; Gordon, M.S.; Nguyen, K.A.; Windus, T.L.; Elbert, S.T. *GAMESS Program Suite,* 1995.

PROFESSIONAL GAMBLING.

Rolando Rodriguez[1] and Gerrit Vriend[2]

[1]Center for Genetic Engineering & Biotechnology, Havana, Cuba,
[2]EMBL, 69117 Heidelberg, Germany.

Introduction

Knowledge of the three-dimensional structure is a prerequisite for the rational design of site-directed mutations in a protein and can be of great importance for the design of drugs. Structural information often greatly enhances our understanding of how proteins function and how they interact with each other or it can, for example, explain antigenic behaviour, DNA binding specificity, etc. X-ray crystallography and NMR spectroscopy are the only ways to obtain detailed structural information. Unfortunately, these techniques involve elaborate technical procedures and many proteins fail to crystallize at all and/or cannot be obtained or dissolved in large enough quantities for NMR measurements. The size of the protein is also a limiting factor for NMR.

In the absence of experimental data, model-building on the basis of the known three dimensional structure of a homologous protein is at present the only reliable method to obtain structural information [1-26]. Comparisons of the tertiary structures of homologous proteins have shown that three-dimensional structures have been better conserved during evolution than protein primary structures [27-45,65], and massive analysis of databases holding results of these three dimensional comparison methods [46-49], as well as a large number of well studied examples [e.g. 50-58] indicate the feasibility of model-building by homology.

All structure prediction techniques depend one way or another on experimental data. This is most easily seen for model building by homology, but also secondary structure prediction programs are trained on proteins with a known three dimensional structure, and even molecular dynamics force fields are mainly derived from protein and peptide data. Unfortunately, all protein structures contain errors [61]. Hence, verification of the data used in modeling procedures is a prerequisit for good results [59-64,80,81]. Many of the same verification techniques can of course also be used to get an impression of the quality of the model.

G. Vergoten and T. Theophanides (eds.), Biomolecular Structure and Dynamics, 79–119.

Model building by homology is a multi step process. At almost all steps choices have to be made. The modeller can virtually never be sure that she makes the best choices, and thus a large part of the modelling process consists of serious thought about how to gamble between multiple seemingly similar choices. As this process resembles very strongly what goes on in the mind of a professional gambler who visits the Loutrakis casino, some introduction in game theory seems in place.

Professional gambling, the game of 21.
It is commonly believed that the development of the theory of probabilities started when Fermat and Pascal in the 17-th century worked out the optimal strategy for certain popular gambling games. One does not need much imagination to see a (still) rich nobleman loosing game after game at the card or crap table coming up with the idea to pay somebody really smart, i.e. Fermat or Pascal, to work out a winning strategy for him.

Upon playing the game of 21 in a casino the objective of both the player and the house is to get cards the face values of which add up to as closes as possible to 21, but not over 21. If this were all the rules, both players would have a 50% chance of winning, and the casino would go bankrupt rather soon. Casinos therefore add many rules, the worst one for the player being that upon equal points the house wins. A player who can remember all the cards that were taken of the pile till the moment the decission wether to ask for one more card or not has to be made has much better chances of winning than somebody who assumes that every of the 52 possible cards always has a chance of 1/52 of being the next one on top of the pile. Normally around eight packs of 52 cards are combined and randomly shuffled to make it more difficult to remember all the already used cards, and to ensure that the chances don't deviate too much from 1/52 shortly after a new pile of cards is introduced. Although many casinos use different house-rules for the game 21, on average the chance of the house beating a player who remembers all the cards and plays optimally is better than 50% till about 1/2 till 3/4 of all cards of the pile have been used. And when that fraction of the cards has been used, the not yet used cards normally are discarded and a new pile of cards is introduced.

So, how should one proceed to have the best chances of winning?
Formula 1 describes how the dollars earned, or the profit (Δ$), depend on the external parameters:

$$\Delta\$ = E(\mathbf{Rounds}) * F(\mathbf{Table}) * N(card_1, card_2, ... card_K) \qquad /1/$$

The function E includes the number of rounds played, which is obviously important as that determines the number of cards removed from the pile, and thus the degree of certainty with which the next card can be predicted. The function F includes the table at which the game takes place which obviously is important as that determines the maximal amount of money one can bet per game. The function N includes the cards i (i=1 till K) one gets dealt during the game, but actually only the last card, K, is important because that card determines the difference between winning and going bust. Using only the last card, and taking the natural logarithm of both sides of formula 1 (using some professional gamblers mathematics and the definitions ln(E(Rounds))=**R** and ln(F(Table))=**T**), we obtain:

$$\ln(\Delta\$) = R * T * \ln(N(card_K)) \qquad /2/$$

The logarithm of one dollar is about one Dutch Guilder (G), and the function N of the last card, K, can for simplicity be called K, which converts formula 2 into:

$$\Delta G = RT\ln(K) \qquad /3/$$

a formula that gives a remarkable feeling of 'deja vu'.

What makes the game of 21 so important for homology modelling? Well, upon building a model by homology very often decissions have to be made which essentially boil down to gambling. If the side chain of a residue has to be placed one often has to select between several seemingly good rotamers. At such moments, one should not assume that if there are N rotamers to choose from, each with a 1/N chance of being correct, but rather one should try to obtain as much information as possible to obtain the real probabilities, and select the most probable rotamer. The 21 player can use `theoretical` information, such as the cards already taken from the pile, the card on the bottom of the pile if he were smart enough to look at that one when the opportunity was there, and more experimental data such as the shape of the coffee stain left on the card at a previous occasion. The homology modeller can use theoretical information such as rules obtained from from database searches, multiple sequence alignments, multiple homologous structures, quantum chemical calculations, and more experimental information such as the results of mutagenesis or ligand binding studies, etc. What they have in common is that all sources of information, if treated with proper statistical rigour, will improve the chance that the final outcome is optimal. In the next

82

paragraph the subsequent steps to be taken when modelling a protein by homology will be described. In subsequent paragraphs these steps will be elaborated upon, and the hints that the professional gambler can give the modeller will be listed.

Modeling overview

Differences between three-dimensional structures increase with decreasing sequence identity and accordingly the accuracy of models built by homology decreases [1,2].

range of sequence   key limiting factor
similarity in %   in model building
identical residues   by homology

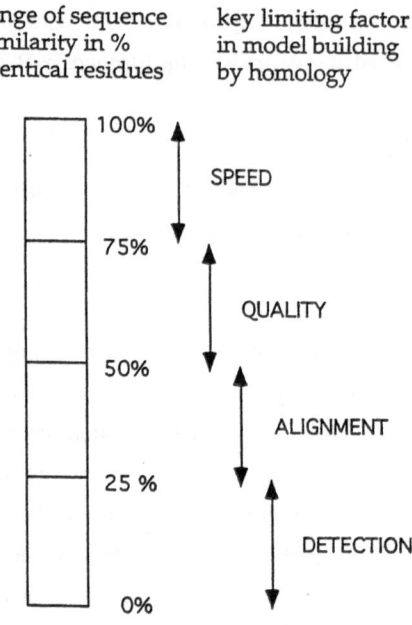

Figure 1. The main limiting steps for model building by homology as function of the percentage sequence identity between the structure and the model.

The errors in a model built on the basis of a structure with >90% sequence identity may be as low as the errors in crystallographically determined structures, except for a few individual side chains [1,62]. If, as a test case, a known structure is built from another known structure, then in case of 50% sequence identity the RMS error in the modeled coordinates can be as large as 1.5 Å, with considerably larger local errors. If the sequence identity is only around 25% the alignment is the main bottleneck for model building by homology, and large errors are often observed. With less than 25% sequence identity the homology often remains undetected. In figure 1 the key limiting factors in modeling as a function of sequence identity are shown.

At present most model building by homology protocols start from the assumption that, except for the insertions and deletions, the backbone of the model is identical to the backbone of the structure. In practice, however, domain motions and 'bending' of parts of molecules with respect to each other is often seen. Even in case of significant bending short range interactions will not differ very much and the model will be perfectly adequate for rational protein engineering, etc. However, the prediction of local differences in the backbone between structures that are homologous in sequence still requires much research, some aspects of which will be described below.

In recent years automatic model-building by homology has become a routine technique that is implemented in most molecular graphics software packages. Currently the emphasis in literature is on a few topics:

• Improvement of the way side chains are positioned in the modelled structure, and improvement of the energetics or database techniques needed to do this fast and accurate.
• Development of techniques to check the quality of models.
• Improvement of the sequence alignment that is at the basis of the model-building procedure, and evaluation of the significance of alignments.
• Development of secondary structure prediction and ab initio 3D modeling techniques.

The modelling process

The modelling process can be subdivided into 9 stages:

• template recognition;
• alignment;
• alignment correction;
• backbone generation;
• generation of canonical loops (data based);

• side chain generation plus optimisation;

• ab initio loop building (energy based);

• overall model optimisation (energy minimisation);

• model verification with optional repeat of previous steps.

What can be modelled?

As will be described described below, the transfer of structural information to a potentially homologous protein is straightforward if the sequence similarity is high, but the assessment of the structural significance of sequence similarity can be difficult when sequence similarity is low or restricted to a short region.

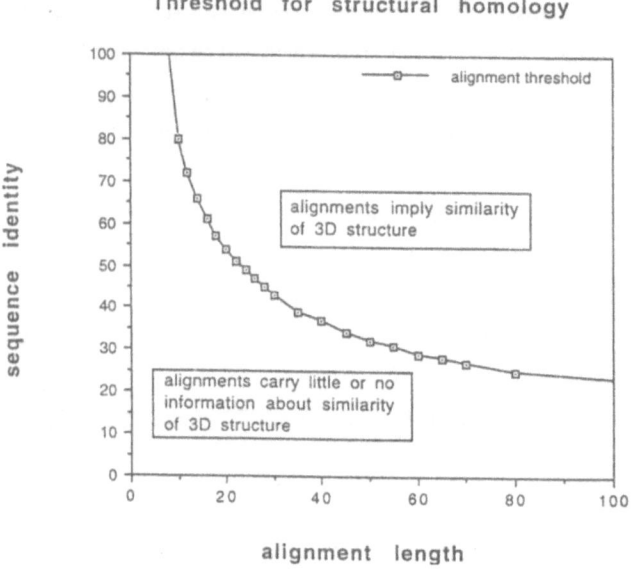

Figure 2. Homology threshold for structurally reliable alignments as a function of alignment length [free after 2]. The homology threshold (curved line) divides the graph into a region of safe structural homology where essentially all fragment pairs are observed to have good structural similarity and a region of homology unknown or unlikely where fragment pairs can be structurally similar but often are not, without a chance of predicting what it will be. At present 15% of the known protein sequences fall in the safe area, which implies that 15% of all sequences can be modelled and thus are open to structure function relation studies.

This indicates a key problem: the shorter the length of the alignment, the higher the level of similarity required for structural significance. Chothia and Lesk [1] have studied the relation between the similarity in sequence and three-dimensional structure for the cores of globular proteins. To quantify this problem, Schneider and Sander [2] calibrated the length dependence of structural significance of sequence similarity. This was done by deriving from the database of known structures a quantitative description of the relationship between sequence similarity, structural similarity and alignment length. The resulting definition of a length-dependent homology threshold (see figure 2) provides the basis for reliably deducing the likely structure of globular proteins down to the size of domains and fragments.

Template recognition.

If the percentage sequence identity between the sequence of interest and a protein with known structure is high enough (more than 25 or 30 %) simple database search programs like FASTA [78] or BLAST [79] are clearly adequate to detect the homology. If, however, the percentage identity falls below 25% detection by straigthforward sequence alignment becomes problematic, and more advanced techniques are required. One technique that is currently being worked on is threading. This topic will be discussed further down in this review.

Alignment

In a 'normal' sequence alignment one puts both sequences along the axes of a big matrix. The matrix elements are filled with numbers indicating the likelyhood that the two corresponding amino acids aught to be aligned. These probabilities are normally obtained from a scorings, or exchange matrix which holds values for the 380 possible mutations and the 20 identities; these matrices normally are symmetric. Figure 3 shows an example of an exchange matrix. It is clearly seen that the values on the diagonal, representing conserved residues, are highest, but one can also observe that exchanges between residue types with similar physico chemical properties get a better score than exchanges between residue types that wildely differ in their properties.

86

The alignment process simply consists of finding the best path through this matrix. The best path is found by taking the little yellow eater from the famous video game 'pac-man' and letting it start near the top left corner and eat its way to somewhere near the lower right corner.

```
     A   C   D   E   F   G   H   I   K   L   M   N   P   Q   R   S   T   V   W   Y
A    5  -2   0   1  -2   0   0  -1   0  -1   0   0   1   0  -1   1   0   0  -2  -2
C   -2   8  -2  -3  -3  -2   0  -2  -3  -3   0  -2  -3  -3  -2  -1  -1  -2  -1  -2
D    0  -2   5   2  -2   0   1  -3   0  -2  -1   2   0   1  -2   0   1  -2  -3  -2
E    1  -3   2   5  -3   0  -1  -2   1  -2  -2   1   1   2   0   1   1  -1  -2  -1
F   -2  -3  -2  -3   6  -3   1   0  -3   2   2  -3  -2  -3  -2  -1  -2   0   3   3
G    0  -2   0   0  -3   5  -1  -2   0  -2  -2   0   0  -1   0   0  -1  -1  -2  -3
H    0   0   1  -1   1  -1   5  -1   1  -1   0   1   0   1   2   0   1  -1   0   1
I   -1  -2  -3  -2   0  -2  -1   5  -2   2   2   2  -2  -3  -2  -1   0   2   0   0
K    0  -3   0   1  -3   0   1  -2   5  -1  -2   1   0   1   2   0   0  -1  -2  -2
L   -1  -3  -2  -2   2  -2  -1   2  -1   5   3  -2  -2   0  -1  -1   0   2   0   0
M    0   0  -1  -2   2  -2   0   2  -2   3   5  -1  -2   0  -1   0   1  -2  -1
N    0  -2   2   1  -3   0   1  -2   1  -2  -1   5  -2   1   0   2   0  -2  -3  -1
P    1  -3   0   1  -2   0   0  -2   0  -2  -2  -2   8   0   0   0   0  -1  -3  -3
Q    0  -3   1   2  -3  -1   1  -3   1   0   0   1   0   5   2   1   0  -1  -1  -2
R   -1  -2  -2   0  -2   0   2  -2   2  -1  -2   0   0   2   5   1   0  -1   0  -1
S    1  -1   0   1  -1   0  -1   0  -1   0  -1   2   0   1   1   5   2  -1   0   0
T    0  -1   0   1  -2  -1   1   0   0   0   0   0   0   0   0   2   5   0  -1  -2
V    0  -2  -2  -1   0  -1  -1   2  -1   2   1  -2  -1  -1  -1  -1   0   5  -1   0
W   -2  -1  -3  -2   3  -2   0   0  -2   0  -2  -3  -3  -1   0   0  -1  -1   6   3
Y   -2  -2  -2  -1   3  -3   1   0  -2   0  -1  -1  -3  -2  -1   0  -2   0   3   6
```

Figure 3.
Example of an exchange or scorings matrix.

Figure 4 describes this process of the alignment of the sequences :

ASTPERASWLGTA and VATTPDKSWLTV.

```
     V    A    T    T    P    D    K    S    W    L    T    V
A    0    5    0    0    1    0    0    1   -2   -1    0    0
S   -1    1    2    2    0    0    0    5    0   -1    2   -1
T    0    0    5    5    0    0    0    2   -1    0    5    0
P   -1    1    0    0    8    0    0    0   -3   -2    0   -1
E   -2    1    1    1    1    2    1    1   -2   -2    1   -1
R   -1   -1    0    0    0   -2    2    1    0   -1    0   -1
A    0    5    0    0    1    0    0    1   -2   -1    0    0
S   -1    1    2    2    0    0    0    5    0   -1    2   -1
W   -1   -2   -1   -1   -3   -3   -2    0    6    0   -1   -1
L    2   -1    0    0   -2   -2   -1   -1    0    5    0    2
G   -1    0   -1   -1    0    0    0    0   -2   -2   -1   -1
T    0    0    5    5    0    0    0    2   -1    0    5    0
A    0    5    0    0    1    0    0    1   -2   -1    0    0
```

Figure 4.
The alignment matrix for two sequences. Scores obtained from figure 3.

Obviously, the path should at every step go at least one line down, and at least one column to the right. The best path is indicated by the characters in bold, but why is the underlined alternative near the lower right corner not better? After all, it 'eats' more points? Well, alignment of sequences is not entirely a video game, but an attempt to capture evolutionary events in numbers. Comparing thousands of sequences and sequence families it became clear that exchanges have a chance of ocurring inversly proportional to the numbers in figure 3, but deletions, like the one indicated by underlining in figure 4 are about as unlikely as at least a couple of non-identical residues in a row. The jump roughly in the middle in the bold face character on the other hand is genuine because after the jump we earn lots of points (5,6,5) which without the jump would have been (1,0,0). Making the right decission about gaps is a whole independent line of research, and this short summary definitely does not give sufficient credits to the scientists working on this problem. But, as shown in the next paragraph, for the homology modeller it is not so very much important to get the gaps right immediately because she can often improve the location of deletions or insertions after building the first model. Just like the 21 player improves his chances a lot if he were allowed to try again with another last card after he went bust, the modeller improves her changes a lot by analysing the original structure and the model, and iteratively adapting the alignment based on the evaluation of the model.

Improving the alignment using multiple sequences

The alignment process described above is clearly open for improvements. The professional gambler will definitely not like the idea that every similar exchange, for example from alanine to glutamic acid, always has the same score, or in other words, the same chance of occurring. In the three dimensional structure of the protein it is surely not likely to see an Ala->Glu mutation in the core, but at the surface this mutation is as likely as every other one. There are two ways to address this problem. One is the use of multiple sequence alignments, often based on profiles [82-90], and the other is called threading [66-77] (unfortunately too often still pronounced: shredding).

The use of multiple sequence alignment methods for the alignment of only two sequences is based on an idea that our professional gambler would not like very much, namely that if something is observed very often, it is probably more likely to occur. Although there is no physical reason to base this assumption on, common sense and practical experience support it strongly. So, rather than aligning just the sequence of interest and the sequence of the template structure, all sequences that are member of this same family are aligned. In such a multi sequence alignment one normally puts higher weights on residue positions where little variation is observed, and tries to get deletions only in areas where the sequences are strongly divergent.

One of the major advantages of the use of multiple sequences can be explained by the following hypothetical (pathological) sequence alignment problem:

Suppose you want to align the sequence LTLTLTLT with YAYAYAYAY. There are two equally poor possibilities:

```
LTLTLTLT      or      LTLTLTLT          Figure 5. Pathological
YAYAYAYAY             YAYAYAYAY         alignment problem.
```

And you can not decide between the two. However, if you additionally align the sequence

```
LTLTLTLT                                Figure 6. Solution for
TYTYTYTYT                               the pathological
YAYAYAYAY                               alignment problem.
```

which has two times 50% sequence identity. This is of course a pathological case, but the same principle also works, although not this beautiful, in the real world.

Improving the alignment using the structure

Threading methods in general incorporate knowledge extracted from the structure of the template, and try to improve the sequence alignment using this information. The aforementioned Ala->Glu mutation would, for example, at the surface get a score of 2 on the scale of figure 3, but in the core of the protein the score would become -5. A good threading program incorporates such, and many more similar considerations. The idea is really simple, and therefore must be brilliant. In practice however, most threading methods are reasonably successful in detecting templates, but strangely enough, when it boils down to the nitty gritty details of the actual alignment, they fail hopelessly [72]. Nevertheless, most modellers use this method manually. If, in the model, a buried glutamic acid is observed the first reaction is to see which alternative alignment would put this negatively charged residue at the surface of the model.

Alignment correction

After the alignment is made, it normally still is not optimal for modelling purposes. Sequence alignment programs are optimised for the detection of evolutionary relations. This is not always the same as a structural relation. The best example is the deletion of a loop as shown in figure 7.

```
  1   2   3   4   5   6   7   8   9   10  11  12  13  14
PHE ASP ILE CYS ARG LEU PRO GLY SER ALA GLU ALA VAL CYS
PHE ASN VAL CYS ARG THR PRO --- --- --- GLU ALA ILE CYS
PHE ASN VAL CYS ARG --- --- --- THR PRO GLU ALA ILE CYS
```

Figure 7. Example of sequence alignment in an area where a deletion needs to be modelled.

The top line in figure 7 is the template sequence, the bottom two lines are two alternative alignments of the model sequence. Everyone would expect the middle line to be correct, because the alignment of the PROs at position 7 give such a strong signal. In figure 8, you see a C_a trace of three loops. The template is given as A. The middle alignment, corresponding to the red C_a trace, leaves a gap of about 10 Å. However, if we use the bottom alignment, we get situation C which has a gap of only about 5 Å. The normal C_a - C_a distance is 3.8 Å. A 5 Å C_a - C_a distance therefor is actually only a 1.2 Å gap, and that can be closed easily with only minimal atomic displacements.

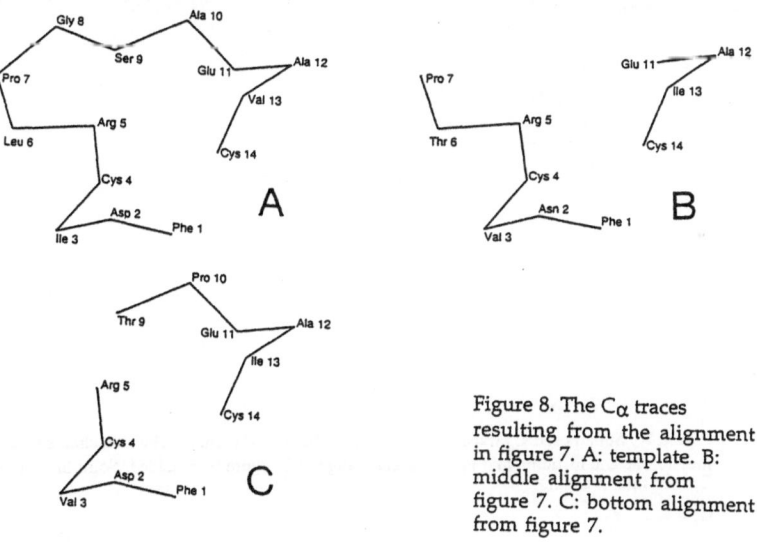

Figure 8. The C_α traces resulting from the alignment in figure 7. A: template. B: middle alignment from figure 7. C: bottom alignment from figure 7.

Another example of alignment problems is the alignment of the haemoglobin alpha and beta chains. This alignment will (almost independent of the alignment program used) look like (figure 9):

```
1        5          10         15         20         25         30
- V L S P A D K T N V K A A W G K V G A H A G E Y G A E A L
V H L T P E E K S A V T A L W G K V - - N V D E V G G E A L

31       35         40         45         50         55         60
E R M F L S F P T T K T Y F P H F - D L S H - - - - G S A
G R L L V V Y P W T Q R F F E S F G D L S T P D A V M G N P

61       65         70         75         80         85         90
Q V K G H G K K V A D A L T N A V A H V D D M P N A L S A L
K V K A H G K K V L G A F S D G L A H L D N L K G T F A T L

91       95         100        105        110        115        120
S D L H A H K L R V D P V N F K L L S H C L L V T L A A H L
S E L H C D K L H V D P E N F R L L G N V L V C V L A H H F

121      125        130        135        140        145
P A E F T P A V H A S L D K F L A S V S T V L T S K Y R
G K E F T P P V Q A A Y Q K V V A G V A N A L A H K Y H
```

Figure 9. Alignment of the α (top) and β (bottom) chain of haemoglobin.

Around residue 50 we see a strong identical triplet: DLS. However, if you study the superposed structures of these two chains, you will see that Asp 49 of the α-chain should be aligned with Gly 48 from the β-chain.

Figure 10. Superposed C_a traces of the haemoglobin α and β chain. The side chains of the methionine and histidine that are incorrectly aligned in figure 9 are added. Solid line: α chain. Dashed line: β chain.

Figure 10 shows the C_a traces of the α and the β chain superposed. The side-chains of His 52 from the α-chain and Met 57 from the β-chain are shown too. Not only do these two sidechains occupy approximately the same space, but an anthroposophical analysis makes clear that they both have the same function: they glue the otherwise rather detached loop around residue 50 to the core of the protein. The histidine in the A subunit does this via a saltbridge with a glutamic acid (which is also indicated in figure 10); the methionine in the B subunit does it via hydrophobic contacts, mainly with the backbone of a nearby helix.

It should be stressed that the alignment errors shown here are not errors in the alignment programs, but errors made by our gambler and our modeller. The alignment programs are optimised to detect evolutionary relationships between proteins, and for that purpose they try to determine what actually happened at the nucleic acid level. Although this can not be seen separately from what happened at the protein level, it is clearly a different thing. Thus, sequence alignments that are perfect in an evolutionary sense, are not always perfect for homology modelling purposes.

Placing new side chains in the structure

When we study the rotamers of residues that are conserved in different proteins with known (similar) three dimensional structure we observe in more than 90% of all cases similar c1 angles. It will be shown below that modellers do not reach that percentage upon modelling, and therefore the advise given by our gambling friend "Don't touch the conserved ones, your odds are better that way" is followed by most modellers. The problem of placing side chains is thus reduced to concentrating on those residues that are not conserved in the sequence.

In case of high sequence homology between the structure and the model the main problem lies just in the replacement of a few side chains. Often this can be done by hand [11,91,92,103], using an interactive molecular graphics package. Alternatively, one tries to find examples of residues in a similar environment [93] in the database of known structures. The problems here are how to define similar environments, and how to extract the information quickly from the database of known structures. In practice, similar environments for residues have most often been defined as having the same secondary structure (α-helix, β-strand or loop). In this review we will emphasize the power of position specific rotamer distributions.

The problem of side chain modelling can be divided in two sub-problems: 1) finding potentially good rotamers, and 2) determining the best one among the candidates.

Until recently side chain conformations were normally selected using standard rotamer libraries [94-96], and a variety of procedures, ranging from manual adaptation to complicated schemes involving energy calculations or Monte Carlo procedures combined with energy calculations, have been used to get rid of Van der Waals clashes, and to find the global optimum among all rotamer combinations [97-100]. Recently, however, several studies indicated that the use of standard rotamer libraries is far from optimal because a large fraction of the information needed to determine the side chain rotamer is contained in the local backbone [100,103-106]. The study by De Filippis et al., [103] has the most direct applicability to statistics based homology modelling and will therefore briefly be summarised.

Position specific rotamers

A single mutation in a protein can dramatically influence its stability or function. However, large structural changes as a result of one single mutation are rare. Generally, the overall structure of proteins does not change upon introduction of a point mutation. If a mutated amino acid does not fit well into its environment, conformational adaptations are made mainly by the mutated residue, to a lesser extent by its entire secondary structure element, and rarely by its three dimensional contact partners. Calculation or even estimates of free energy changes are insufficiently accurate to be used as a routine tool for predicting structural changes. On the other hand, empirical methods based on a variety of approaches work quite well for some purposes [e.g. 108-111]. A reliable and general theory of side chain rotamers, however, does not yet exist.

De Filippis et al., therefore consulted our 21 player and developed a statistical method for predicting structural changes that is based on searches in the database of 3D structure. Given a residue to be mutated, they search the database for residues with a similar environment and then assess which rotamers in the database are statistically preferred and sterically admissible in the current structure. Using simple rules based on this scheme the local structural environment of a point mutation can be predicted correctly in the majority of all cases. They analysed almost 100 point mutated proteins with known wild type and mutant structure.

For every mutant, a position-specific rotamer analysis was performed. This technique has been described in detail elsewhere [112-114]. Briefly, a rotamer distribution at a certain position is determined by extracting from a database of non-redundant protein structures [115,116] all suitable fragments of length 5 or 7 residues (7 in helix and strand, 5 in case of irregular local backbone). Suitable fragments are those that have a local backbone conformation similar to the one around the evaluated position, and have the same residue type at the central position. In these analyses, the RMS deviation of the alpha carbon positions between the structure and the database fragment was kept below 0.5 Ångstrom. The rotamer distributions were then used to answer the following questions:

• What are the differences between the rotamer of the residue mutated by protein engineering and the position specific rotamer distribution obtained from the database of natural proteins?

• Which set of rules would have allowed us to predict the structure of the mutated residue correctly?

In the analysis of all mutations they observed six classes:

1) There is only one way of placing the residue without atomic clashes. In all 4 examples in this class they found just one rotamer that leaves the rest of the molecule unaltered and this is the observed rotamer.

2) The conformation of the mutated residue corresponds to the most populated position-specific rotamer. This occurs in approximately 50% of all cases.

3) The most populated rotamer would lead to atomic clashes and the conformation of the mutated residue corresponds to the second most populated rotamer (4 examples).

4) Neither the most populated, nor the second most populated rotamer fits, and the conformation of the mutated residue is very close to the most populated rotamer (2 examples found, both deviate about 20° from the optimal c_1). Never did they observe that the third, i.e., least populated rotamer was selected.

5) The c_1 of the mutated residue corresponds to the most populated rotamer, but the c_2 or c_3 angle is rotated away from the optimal position-specific rotamer distribution to optimise hydrogen bonding. This is observed in almost 15% of all cases (10 examples).

6) There is no preferred position-specific rotamer, or there are insufficient examples in the database, but analysis of potential hydrogen bonds leads to one clear possibility for the structure. This occurs in 7% of all cases (4 examples). Although in these cases it is possible to fix the charged end of the side-chain in the proper position, it is often difficult to accurately predict the correct conformation of the hydrophobic part of the side-chain.

These six observed classes were converted into a set of rules (See figure 11). Assuming that their dataset is representative of the universe of mutated residues, then this scheme allows to correctly predict the local structures of at least 85% of all mutated residues. The actual predictive ability on proteins not studied here may, of course, be lower.

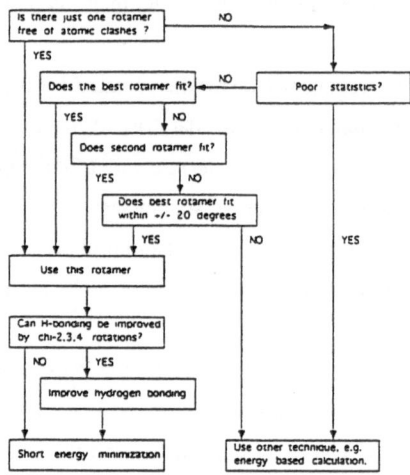

Figure 11. Decision scheme for the prediction of point mutant structures.

Of the six rules, the first three are the most clear-cut, i.e. have a high prediction accuracy. These rules cover about 60% of all cases. In all other cases, the prediction accuracy is lower on average. Reassuringly, however, there is a qualitative correlation between correct prediction and the quality of fit of database fragments on the actual backbone, the number of observed database fragments and the sharpness of the c_1 distribution of the central residues in these fragments.

They conclude with the notion that not all conformational changes resulting from point mutations can be predicted correctly with the techniques available to date. Examples of changes difficult to predict are large domain motions or local structure adaptations induced by co-factor binding.

Combining multiple side chain conformations

In the modelling practice it is not just needed to correctly place one rotamer in an otherwise unmodified protein, but often around half of all side chains need to be altered. That means that in many cases the environment is either incorrect, or which causes slightly less of a problem, not yet complete. This makes the rotamer choice one step more complicated because even the best force field or scorings function can not overcome the problem of the chicken and the egg. In order to correctly position side chain A, all its neighbours need to be in the perfect position already, but these perfect positions can not be determined before side chain A is correctly modelled.

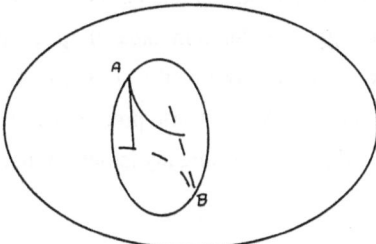

Figure 12. The chicken and the egg problem for side chain modelling. A cavity in the protein is available to accomodate the side chains of residues A (solid) and B (dashed). If A is modelled as a straight line, we can only model B as a straight line too. But the same holds for the bend lines. But what now, if residue A does not really know what she likes, but residue B likes to be straight? If B were modelled only after A made up her mind, he has a 50% chance only of being happy....

Figure 12 indicates in a simplified fashion what is the problem. If there is limited space to model two side chains, there could be, for example, two seemingly equally good solutions. In such a simple case the quality of the outcome is determined solely by the quality of the force field, or scorings function. Our professional gambler likes this situation, because some understanding of statistics will help. If the number of choices is limited the modeller can simply enumerate them and select the one with the highest probability of being right. If there are more than two side chains whose positions influence each other, it seems that we rapidly run into a combinatorial explosion. The dead-end elimination method [101,102] was the first attempt to solve this combinatorial problem without the use of Monte Carlo methods. The rationale of the authors was: "Why gamble if everything can be calculated"? Chinea *et al.*, [107] actually studied this problem, and concluded that it is not a problem at all, or at least the problem is not what previously thoughtlessly was assumed that it was.

Rotametric entropy

If many new side chains have to be placed in the model, multiple side chains can potentially occupy the same space. A big problem for most modeling methods is that they are based on an energy function that includes contacts between residues distant in the sequence, but close in space. This implies that the whole molecule needs to be build before any selected rotamer can be evaluated. This leads to a "chicken and the egg" problem. In order to place the first residue correctly, all other residues should already have been placed correctly. So, where to start? Several techniques have been described to overcome this problem. Monte Carlo procedures [97,100] seem the most apropriate for this purpose, but Desmet *et al.*, [101,102] already indicated that other solutions might exist.

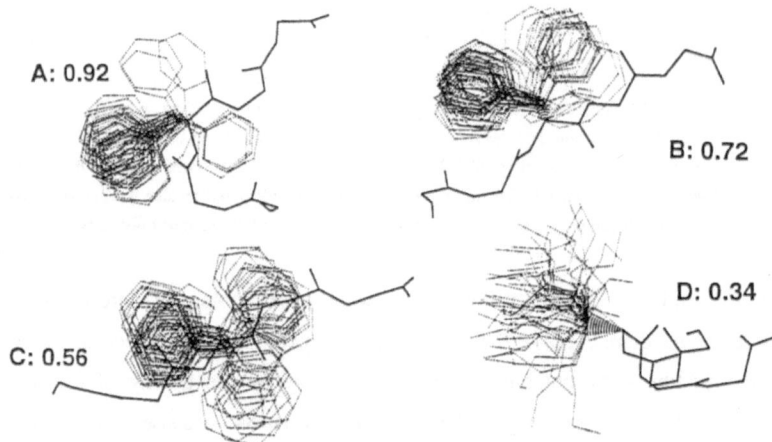

Figure 13. Examples of position specific rotamer distributions. The numbers indicate rotameric entropies as explained in the text.

Figure 13 shows several examples of position specific rotamer distributions. Figure 13A shows an example where the position specific rotamer distribution is extremely narrow. If such a case would ocurr in a modeling study, this residue should be modeled immediately, and never looked at again. Figure 13D shows an example of a very wide rotamer distribution. Such a residue should obviously be modeled late in the modeling procedure, because it can much more easily adapt to the space left to it after all other side chains have been placed. The equivalent of figure 13D for our professional gambler is a set of cards that requires a lot of knowledge about the the distribution of cards still to come in order to make the best decision. His chances of earning some money would in such a case greatly enhance if the completion of this game could be postponed till a later stage. And that is what our modeler is allowed to do. If it is not (yet) clear how to model a side chain, she just waits, builds other side chains first and hopes that thereby the number of choices for the problematic side chains are reduced.

Chinea et al., [107] based their modeling strategy on simple probability principles. The narrower the rotamer distribution, the higher the probability that this is the rotamer needed in the structure to be modeled. To quantify rotamer distributions a rotameric entropy was defined. In figure 13 some examples of rotamer distributions are shown and the derived rotameric entropies are given. The rotameric entropy is defined by $E=(P_{tot}/P)*(F_{tot}/F)$ in which P is the population with c1 within 45° of the most populated of the three standard c1 values (60°, 180°, 300°), P_{tot} is the sum of all rotamers that fall within 45° of any of these three standard c1 values, F is the total number of rotamers in this distribution, and F_{tot} is the maximal number of rotamers obtainable for any distribution. The normalisation with respect to F_{tot} avoids that residues for which only very few position specific rotamers are found get a large value.

In the modeling process first a sparse model is generated. In this sparse model all conserved residues were left untouched, but other residues were mutated into alanine, unless they had to become glycine or proline. For all alanines that subsequently had to be mutated the rotameric entropy in this sparse model was determined. Side chains were then placed in order of decreasing rotameric entropy.

The rotameric entropy is defined by $E = (P_{tot}/P)*(F_{tot}/F)$ in which P is the population with c1 within 45° of the most populated of the three standard c1 values (60°, 180°, 300°), P_{tot} is the sum of all rotamers that fall within 45° of any of these three standard c1 values, F is the total number of rotamers in this distribution, and F_{tot} is the maximal number of rotamers obtainable for any distribution. The normalisation with respect to F_{tot} avoids that residues for which only very few position specific rotamers are found get a large value.

In the modeling process first a sparse model is generated. In this sparse model all conserved residues were left untouched, but other residues were mutated into alanine, unless they had to become glycine or proline. For all alanines that subsequently had to be mutated the rotameric entropy in this sparse model was determined. Side chains were then placed in order of decreasing rotameric entropy.

The sorting of residues as function of the rotameric entropy has an obvious advantage. Early in this modeling process the residues are built that have a very narrow rotamer distribution, which indicates that the conformation is mainly determined by the local backbone, and the absence of many not yet modeled residues is not a disadvantage. Residues with wider rotamer distributions which therefore are more influenzed by the rest of the molecule are built later when more residues are already completed. The advantage of this process is best seen from figure 14.

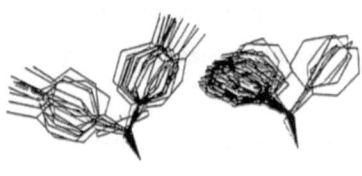

Figure 14. Two overlapping rotamer distributions. If the tyrosine gets modeled first, then the right hand rotamer would be selected. The phenylalanine, however, has a much lower rotameric entropy, and therefore is modelled first, which leads to the more favourable situation with both rotamers pointing to the left.

In practice, cases like the one described in figure 14 occur very often. Normally the number of rotamers left for side chains with many seemingly similar rotamer possibilities reduces strongly if the 'simple' residues are modelled first. For once, proteins are well behaved, and Murphy's laws don't apply.

Database retrieval.

It is obvious that database systems that allow for fast, easy and flexible retrieval of specific information are crucial for model building procedures. Several general [118-122] and single purpose [2,126] data storage or retrieval systems have been developed to extract information about protein sequences and structures from databases. Some of them hardly (re)organize data, but merely combine a database of three-dimensional protein structures with a set of algorithms for pattern recognition, data analysis and graphics. In general, these systems provide very flexible tools, but this flexibility is paid for by a rather low speed when the algorithms are applied to large amounts of data. PKB [118] and to some extent the parameter correlation method [119] are good examples of such systems. They are well suited for prototyping queries or searches in small subsets of the database, but less suitable for practical use if fast data extraction is required.

If retrieval times must be reduced to a minimum, one resorts to systems that pre-process and reorganize data to speed up the process of extracting information. Two important classes of such systems are object-oriented database systems (OODBS) and relational database systems (RDBS) [117]. An OODBS can easily search for many related objects, but the organization of the data makes it slow at doing sequential scans [121]. P/FDM [121] is a good example of an OODBS. Its high level query language Daplex is very concise and approaches the power of a programming language for complex queries.

In a protein RDBS many structural properties such as accessibility, torsion angles and secondary structure are stored in tables and queries are performed by logical combination of these tables. BIPED [120] and SESAM [122] are examples of such systems. SESAM does not fit the relational model exactly as it also provides some algorithms on top of the RDBS to allow for otherwise impossible or prohibitively slow queries. Advantages of a generalized RDBS [117] are the generally high speed of searches and the intuitive way in which queries are constructed.

When one wants to use a standard RDBS to aid with model building by homology one major problem is encountered: Entries in the same database table are assumed to be unrelated; or in other words, the database does not know which residues sit next to each other in the sequence. So, a query like 'buried - accessible - buried - accessible' to find surface β-strands is inherently beyond the capabilities of a standard relational system [120].

The SCAN3D database system [113] was specifically designed to bypass this problem. SCAN3D exploits the sorted character of protein structures in that it stores the residues or their characteristics in the database tables in the sequential order in which they occur in the protein. This allows to easily search for stretches of consecutive residues with specified characteristics, a feature that is especially important for the modeller because she is never interested in just one residue, but always in residues in their environment.

The Brookhaven Protein Databank (PDB) [123] contains atomic coordinates and some related information for more than 4000 macromolecular structures in plain text files. SCAN3D uses a representative set of slightly more than 300 protein structures [115,116]. These proteins are carefully selected to avoid bias towards a small number of abundantly present protein families. Therefore, results obtained from the use of SCAN3D queries are representative for the whole universe of presently known protein structures. Which allows this module of the WHAT IF program to also be used for basic studies of protein sequence-structure relations that later can lead to improved rules for homology modelling.

Modelling loops

Rodriguez *et al.*, [131] studied the structure of surface accessible loops because they realised that such loops are the major source of errors during modelling experiments. They studied 34 pairs of known structures and analysed what happened if they pretended not to know one of the two and then model it based on the other partner of the pair. Of course this is not fair because in the real world the structure of the model is never known, but from this study we can learn a lot about what can all go wrong upon modelling proteins. They concentrated on loops with equal length in the two structures to avoid having to model insertions and deletions and found three reasons why one often observes different conformations in similar structures:

- Difference in the symetry contacts in the crystals of the template and the real structure to be modelled.

- The introduction of bulky for small side chains underneath a loop thereby pushing the loop a bit aside.
- The mutation of a residue to proline in a location where the backbone of the template structure can not accomodate a proline because it has incompatible dihedral angles. Table 1 summarises the results for 116 loops that had an RMS deviation (RMSd) between the model and the real structure that was two times larger than total RMSd.

Symmetry contact in model or in template	46
Symmetry contact in model and in template	43
No symmetry contacts in either of the two	27
Mutations involving proline	25
Mutations involving glycine	26
Mutations involving proline and glycine	4
Cases without any obvious reason for model problems	17
Symmetry contacts combined with proline or glycine	38

Table 1. Most probable reason for the conformational differences between loops in homologous structures. In 75% of all cases crystal symmetry contacts are involved. Not all numbers add up correctly because multiple problems can occur at the same time. Only 17 out of 116 cases were not trivially put in any of the three major problem categories.

The largest fraction of all problems is clearly different symmetry contacts. This problem provides a principal limit to the accuracy of the model and to the posibilities of estimating the reliability of models. Figure 15 and 16 show two examples of symmetry induced conformational differences.

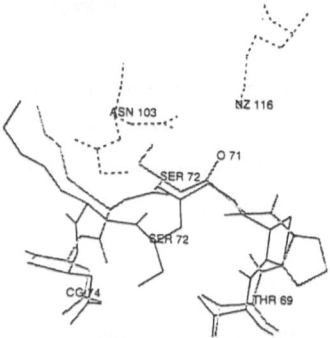

Figure 15. Symmetry induced structural differences between two 'identical' lysozyme molecules. In the gray crystal form the dashed lysine from a partner molecule in the crystal attracts the local backbone. In the black crystal form the dashed asparagine from a differently situated crystal partner would bump with the same local backbone, and thus pushes the loop away, 'downwards'.

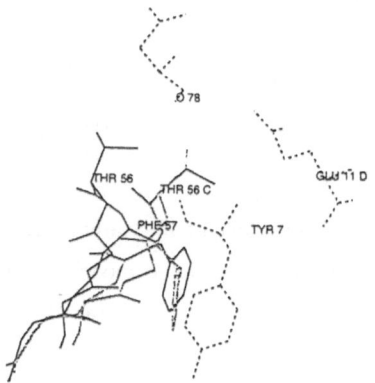

Figure 16. Symmetry induced structural differences between two highly similar molecules. In the gray crystal a hydrogen bond is formed between the threonine and a glutamic acid from a crystal partner. In the black crystal form this threonine forms a hydrogen bond with a backbone oxygen of a crystal partner (in this case there are several more contact differences). The consequence is a rather unexpected but highly significant difference between the gray and the black loop.

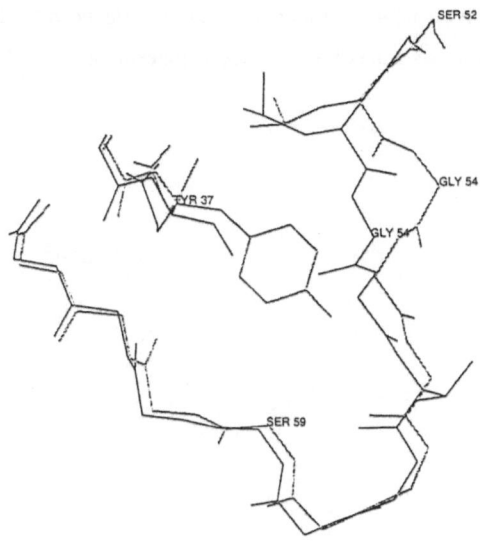

Figure 17. Structural differences induced by a small <--> large residue mutation. The serine at position 37 in the black molecule is mutated into a tyrosine in the gray molecule. The big gray loop needs to be displaced to make space for the big tyrosine. The glycine at position 54 probably facilitates this displacement.

A more predictable scenario ocurrs when the backbone has to move to make space for a bulky residue as can be seen in figure 17 in which the 2.99 Å displacement of the loop from residues 54 to 59 in 1POH [130] relative to 1PTF [132] is probably caused by the mutation of Ser37 in 1POH to Tyr in 1PTF.

Modelling a proline is another big problem. When a proline replaces another residue in many cases the existing backbone has torsion angles that are very unfavourable for proline, and the proline insertion leads to local backbone adaptations. The worst cases are often found for the Gly->Pro mutation, because glycine can have almost every conformation without restrictions. Figure 18 shows the superposition of 5HPV [133] and 1IVP [134]. The loop from residues 34 to 42 is shown to illustrate the change in the backbone conformation due to the mutation of Gly to Pro at position 39 in 5HPV. The backbone at this position can not accomodate a proline (f=115.2, y=131.0, w=177.9). Proline 39 sits in 1IVP in a favourable conformation (for a Pro) (f=-87.5, y=137.9,

104

w=179.6). The residues in the neighbourhood are are also influenced by these backbone torsion angle differences and the loop shows a maximum C_a-C_a displacement of 3.2 Å for the C_a of residue 40.

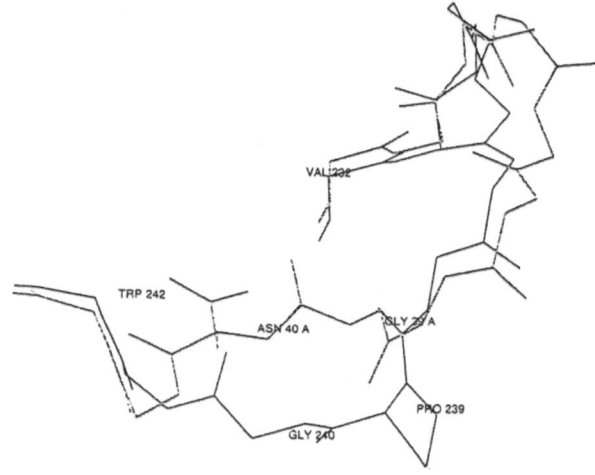

Figure 18. Structural differences induced by a Gly <--> Pro mutation. The gray glycine 39 is mutated into a proline (239) in the black molecule. The backbone torsion angles are different for these two residues. Leading to a severe local distortion. The black glycine (240) probably helps facilitate this distortion.

Building loops if there is an insertion in the model

In case of 75% or more sequence identity between the structure and the model one seldomly encounters insertions or deletions, and when they are encountered, they normally are short.

One of the major problems in model building with intermediate homology is the insertion of loops. If an insertion in the sequence occurs relative to the structure, there is no template to model on, and other techniques have to be applied. The techniques used to model loops are:

• Searching databases for loops with endpoints that match the residues in the structure between which the loop has to be inserted. This technique is based on the same distance geometry loop searching algorithm as described above for the retrieval of position specific rotamers from the structure database [114,135]. All major molecular graphics and modeling programs use this technique (e.g., WHAT IF [135], QUANTA, Hydra [136], FRODO [137], Insight [138], O [139], BRAGI [140]).

• Ab initio building can be done in many ways, but a few points are common to all ab initio techniques: the fact that endpoints have to match is a very strong constraint. Further, Van der Waals clashes have to be prevented, and all rules about proper stereochemistry and energetics should be obeyed. Many articles have been published about these topics, [141-147], but loop building is still a wide open field as is perhaps best indicated by the fact that for more than five years already no significantly new methods have appeared in literature. The world-wide modelling competition held in 1994 [148] made clear that correctly ab-initio modelling of loops is at present not yet possible.

Verification of the quality of the model(s)

All models built by homology will have errors. Sidechains can be placed incorrectly, or whole loops can be misplaced. As with most errors, they become less of a problem when they can get localized. For example, upon modeling a protease it is probably not important that a loop far away from the active site is placed incorrectly.

The most important step in the process of model building by homology is therefore undoubtedly the verification of the model, and the estimation of the likelihood and magnitude of errors.

There are two principally different ways to estimate errors in a structure.

• Determination of the energy of the structure using energy minimization and molecular dynamics programs.

• Determination of normality indices that indicate how well a given characteristic of the model resembles the same characteristic in real structures.

The key aspect is the development of criteria with sufficient discriminatory power to distinguish a good model from a bad one. An example is provided by deliberately misfolded proteins in which the sequence of a protein known to have an all-helical 3D structure is placed into a known structure of a completely different type, an antiparallel b-barrel, and *vice versa*. For the evaluation of the quality of these clearly incorrect hypothetical structures, intramolecular energy, calculated in vacuum using standard empirical potentials, is not a sensitive criterion [Novotny et al 84, 88]. The free energy difference between the folded and unfolded states would be an ideal criterion, but present theories are not capable of calculating free energy differences to sufficient accuracy.

Faced with the lack of an accurate theory of protein folding, empirical observations of regularities gleaned from the database of solved structures can be very useful. A variety of statistical criteria, which measure the preferential distribution of hydrophobic side chains in the interior of proteins, have been used successfully to discriminate between deliberately misfolded and native structures [64,149-151].

Normality indices for structures have already proven their power in structure verification. Many characteristics of protein structures lend themselves for normality analysis. Most of them are directly or indirectly based on the analysis of contacts, either inter residue contacts, or contacts with water. Some published examples are:

• General checks for the normality of torsion angles, bond angles and bond lengths etc. [135,152] are good checks for the quality of experimentally determined structures, but are less suitable for the evaluation of models because the better model building programs simply do not make this kind of errors.

• Inside outside distributions of polar and apolar residues can be used to detect completely misfolded models [149].

• Solvation potentials can detect local errors as well as complete misfolds [81].

• Residue transfer energy is another way of looking at inside/outside distributions [153].

• Packing rules have been implemented for structure evaluation [154].

• Atomic contacts that are not abundant in the protein structure database are good indicators of local model building problems [155].

Atomic contacts are observed because they are energetically favored. Real structures cannot tolerate too many unfavorable interactions. Thus for a model to be correct only a few infrequently observed atomic contacts are allowed. We made a detailed analysis of atom atom contacts [155]. WHAT IF [135] holds a module that compares the local contact patterns with the average contact patterns for similar residue-residue contacts found in the database. This method can be summarized as follows: If a residue-residue contact has the same contact patterns and the same spatial orientation as a contact that occurs often in the database then a high score is given. If a contact in the modeled molecule seems rather unique, either from a point of view of which residues make the contact, or from a point of view of directionality of the contact, a low score is given. This 'quality control' of local packing has proven to be a powerful tool for the detection of abnormal structures. Most methods used for the verification of protein structures can also be used for the verification of models. Not all methods will be useful because certain experimental errors simply are not made by the better modelling programs. In general, however, a verification report is very helpful for the modeller and her friends when they are using the model for the analysis of experimental results or prediction of new experiments.

How good are the models actually?

The quality of protein models built by homology to a template structure is normally determined by the RMS errors in models of proteins of which the structure is known. Rodriguez *et al.*, selected from the PDB [ref] 34 pairs of protein structures that superpose well, have 35% to 98% sequence identity, and have no insertions or deletions. They created this test-set to analyze what could potentially be the major sources for errors in protein modelling and in the assessment of the model quality.

The dataset was carefully selected to be representative for the universe of proteins, but they made sure that they would not encounter big surprises. The models are thus representative for the best scenario one can expect in practical cases, and not for a typical scenario. The selection of 34 pairs of proteins was done using the following criteria:

- No insertions or deletion upon sequence alignment. N- or C-terminal insertions or deletions were allowed as long as they were shorter than five amino acids. This was done in order not having to model any loop ab-initio or introduce any change in the existing linear backbone.

- Both proteins are solved by Xray, with a resolution better than 2.5 Å, and an R-factor better than 0.320 (which indicates that at least some refinement has taken place).

- Both proteins contain less than 10% suspicious residues according to the BIOTECH protein structure verification server [61].

PDB	r	R	RMSd	SID %	Class	RMSe	Description
1poh	2.00	0.14				1.978	Phosphotransferase (E. coli)
1ptf	1.60	0.16	1.244	35.29	mixed	1.977	Phosphotransferase (S. faecalis)
1nhk	1.90	0.17				2.410	Nucleoside Diphosphate Kinase (M. xanthus)
1ndc	2.00	0.18	1.554	43.75	mixed	2.082	Nucleoside Diphosphate Kinase (D. discoideum)
1bpt	2.00	0.17				2.003	Pancreatic Trypsin Inhibitor (BPTI) (B. taurus)
1aap	1.50	0.18	0.973	44.64	mixed	1.984	PInh. Domain Of Alzheimer's Protein (H. sapiens)
5pal	1.54	0.17				1.626	Parvalbumin (T. semifasciata)
1omd	1.85	0.17	0.776	44.86	alpha	1.375	Oncomodulin (R. norvegicus)
1pza	1.80	0.18				1.752	Pseudoazurin (A. faecalis)
1pmy	1.50	0.20	0.995	45.00	beta	1.807	Pseudoazurin (M. extorquens)
1thbB	1.50	0.20				1.972	Hemoglobin (H. sapiens)
1pbxB	2.50	0.18	1.240	45.21	alpha	1.983	Hemoglobin (P. bernacchii)
5hvpB	2.00	0.18				1.716	HIV-1 Protease (HIV Type 1)
1ivpA	2.50	0.20	0.892	48.48	beta	1.531	HIV-2 Protease (HIV Type 2)
2sam	2.40	0.19				1.496	SIV-1 Protease (SIV Type 1)
4phvB	2.10	0.18	1.030	51.52	beta	1.863	HIV-1 Protease (HIV Type 1)
2cro	2.35	0.20				1.872	434 Cro Protein (Phage 434)
2or1L	2.50	0.18	0.825	52.38	alpha	1.882	434 Repressor (Phage 434)
1crb	2.10	0.19				1.423	Cellular Retinol Binding Protein (R. rattus)
1opbC	1.90	0.17	0.718	56.39	beta	1.436	Cellular Retinol Binding Protein II (R. rattus)
1fkf	1.70	0.17				1.287	FK-506 Binding Protein (H. sapiens)
1yat	2.50	0.18	0.818	57.01	beta	1.189	Fk-506 Binding Protein (S. cerevisiae)
1pvaA	1.65	0.20				1.244	Parvalbumin (E. lucius)
1cdp	1.60	0.16	0.702	62.04	alpha	1.130	Parvalbumin (C. carpio)
2ycc	1.90	0.20				1.390	Cytochrome C (S. cerevisiae)
5cytR	1.50	0.16	0.574	62.14	alpha	1.386	Cytochrome C (T. alalunga)
1azrA	2.40	0.17				1.469	Azurin (Pseudomonas aeruginosa)
1aizA	1.80	0.17	0.982	63.28	mixed	1.443	Azurin (Alcaligenes denitrificans)
4azuA	1.90	0.18				1.387	Azurin (Pseudomonas aeruginosa)
1azcA	1.80	0.16	0.960	63.78	mixed	1.332	Azurin (A. denitrificans)
1mrj	1.60	0.17				1.291	Alpha-trichosanthin (T. kirilowii maxim)
1mom	2.16	0.19	0.626	65.04	mixed	1.350	Momordin (M. charantia)
1cad	1.80	0.19				0.999	Rubredoxin (P. furiosus)
8rxnA	1.00	0.15	0.604	66.67	mixed	1.001	Rubredoxin (D. vulgaris)
1tadB	1.70	0.21				1.636	Transducin-alpha (B. taurus)
1gia	2.00	0.17	1.139	69.35	alpha	1.576	Gi Alpha 1 (R. rattus)
1hsaA	2.10	0.20				1.736	Human Class I HSA (H. sapiens)
1vaaA	2.30	0.17	1.176	72.63	mixed	1.829	MHC Class I (M. musculus)
1gbt	2.00	0.16				0.798	Beta-trypsin (B. taurus)
1brcE	2.50	0.17	0.424	73.09	beta	0.865	Trypsin Variant (R. rattus)
1babB	1.50	0.16				0.968	Hemoglobin Thionville (H. sapiens)
1fdhG	2.50	0.32	0.513	73.29	alpha	0.933	Hemoglobin (H. sapiens)
1dhfA	2.30	0.18				1.397	Dihydrofolate Reductase (H. sapiens)
1dr7	2.40	0.16	0.775	75.27	mixed	1.242	Dihydrofolate Reductase (G. gallus)
8dfr	1.70	0.19				1.335	Dihydrofolate Reductase (G. gallus)
2dhfA	2.30	0.19	0.738	75.27	mixed	1.456	Dihydrofolate Reductase (H. sapiens)
1hna	1.85	0.23				1.611	Glutathione S-transferase (H. sapiens)
3gstB	1.90	0.16	1.025	75.58	alpha	1.431	Glutathione S-transferase (R. rattus)
1ala	2.25	0.20				1.042	Annexin V (G. gallus)
1avr	2.30	0.18	0.445	77.85	alpha	0.882	Annexin V (H. sapiens)
1bra	2.20	0.16				0.999	Trypsin (R. rattus)
1mct	1.60	0.17	0.421	79.82	beta	1.044	Trypsin (S. scrofa)
4p2p	2.40	0.21				2.099	Phospholipase A2 (S. scrofa)
2bpp	1.80	0.19	1.152	84.17	alpha	1.922	Phospholipase A2 (B. taurus)
135l	1.30	0.19				1.213	Lysozyme (M. gallopavo)
1hhl	1.90	0.17	0.732	86.82	alpha	1.184	Lysozyme (N. meleagris)
2gbp	1.90	0.15				0.891	Galactose binding protein (E. coli)
3gbp	2.40	0.16	0.518	94.43	mixed	0.918	Galactose binding protein (S. typhimurium)
1emy	1.78	0.15				1.330	Myoglobin (E. maximus)
1ymc	2.00	0.13	0.691	87.58	alpha	1.324	Sulfmyoglobin (E. caballus)
1ovb	2.30	0.20				1.593	Ovotransferrin (Duck)
1nnt	2.30	0.16	1.091	90.57	mixed	1.572	Ovotransferrin (G. gallus)
2lalA	1.80	0.19				0.970	Lentil Lectin (L. culinaris)
2ltnA	1.70	0.18	0.322	92.27	beta	0.977	Pea Lectin (P. sativum)
2chf	1.80	0.18				1.955	Chey (S. typhimurium)
1chn	1.76	0.19	1.376	97.62	mixed	1.963	Chey (E. coli)
1etb1	1.70	0.16				0.678	Transthyretin (H. sapiens)
1ttcA	1.70	0.18	0.255	98.31	beta	0.534	Transthyretin mutant (H. sapiens)

Table 2. Structures used to study model quality[135]. RMSd: Root mean square displacement between equivalenced atoms in the two molecules. RMSe: Root mean square atomic misplacement between the model and the real structure. SID: percentage sequence identity between a pair of sequences. R: crystallographic R-factor. r: resolution.

Additionally the dataset should be "representative" for the universe of globular water soluble protein structures that are amenable to modelling by homology.

Roughly equally many all-alpha, all-beta and mixed alpha-beta proteins were chosen, and they were distributed equally over the 35-98% pairwise sequence identity range in all these three classes. Table 2 lists the pairs of proteins used, as well as some vital statistics.

Most modelling procedures use the backbone of the template as the backbone of the model, and add the sidechains onto this backbone. The RMSe of the backbone will therefore be the same as the RMSd between the model and template backbone. We call this the starting error. Obviously, under normal conditions the final all atom RMSe will always be bigger than this starting error. Energy based calculations are not yet refined enough to improve the results significantly (see next paragraph). Statistical methods can indicate "where" backbone modifications are likely to be needed, but except for some simple cases, we can not yet predict "how" to modify the backbone.

Loops normally have roughly a similar conformation in similar structures. A weak correlation is found between differences in loop conformations and mutations involving proline or glycine. However, if loops are not predicted well, this is most often the result of differences in symmetry contacts between these loops in the model and the template structure.There is a basic error of around 1.0 Å in the backbone of every model, just as a result of differences between experimental structures. Surface located residues and structural changes caused by symmetry contacts add on average another 0.5 Å to the RMS. In the core the error is normally much less than 1.0 Å. At the surface itb is often more than 2.0 Å. Of course some models will have lower RMS errors, but the problem is that in practical cases one cannot know how good the models are, one can only gamble [61].

Energy minimisation

All 68 models were energy minimised using GROMOS [156] (other programs give the same or similar results) and after a fixed number of energy minimisation steps the half minimised structures were evaluated. The RMSe was measured, and all 68 RMSe values measured after 100, 200, etc., energy minimisation steps were bluntly averaged. The results are summarised in table 3. Two things are clearly seen. 1) The improvements than can be achieved are minimal, and 2) The energy minimisation run should be short, after a while the models get worse again. Table 3 is only an average, but inspection of all individual numbers shows that the optimum is in all but three cases between 50 and 300 energy minimisation steps. Inspection of some individual energy minimisation processes indicates that during the first steps the largest errors (such as two atoms being a bit to close to each other, or a hydrogen bond that does not have optimal geometry, or a backbone angle that was already not perfect in the template, etc.) are removed. At every step, however many, many very small errors are introduced. In the beginning removal of the big errors outweighs the introduction of the many small errors. If after a while all larger problems are solved, the only thing that still happens is the introduction of many small errors.

Steps	Ave. RSMe
0	1.4622
100	1.4542
200	1.4529
300	1.4335
500	1.4552
2000	1.4553
8000	1.4553

Table 3. Average RMSe after a fixed number of energy minimisation steps. The average RMSe was calculated averaging the RMSe of the 68 individual structures in each of the energy minimisation runs.

Modeling without homology

Most of the above deals with modeling in three dimensions. That is, it is assumed that a good model can be built. Other techniques such as secondary structure prediction can help in this case. It is often not clear why predicted secondary structures are at all published, but in the hands of a biocomputing expert some information can be extracted from the prediction. The best secondary structure prediction program that is available (this is written on august 17 1996) is without doubt PHD. This program can be used via the WWW (see below).

Future developements in protein modelling are the use of other information than homology to build models. Such information can essentially be anything. Predicted secondary structure, accessibility or contacts can equally well be used as observed cysteine bridges, proteolytic cleavage sites or accessibilities.

Concluding remarks

Model building by homology is a young field. Many improvements can still be made and much work still needs to be done to make these improvements. Our modeller can still learn a lot from the professional gambler, but we expect that improvements in energy calculation based software will within 10 years lead to a breakthrough. We would not be surprised if untill this happens improving the odds of present day methods by inclusion of information from multiple templates, the design of new algorithms and heuristics, better and larger databases, the rapid growth of the PDB, and a few more factors that we cannot yet predict, will step by step create the progress in homology modelling that is needed to close the structure gap.

WWW addresses

Secondary structure prediction:

http://swift.embl-heidelberg.de/predictprotein/

Protein structure quality:

http://swift.embl-heidelberg.de/pdbreport/

http://biotech.embl-heidelberg.de:8400/

Protein structure comparison:

http://www.embl-heidelberg.de/dali/

Acknowledgements

We thank Chris Sander, Rob Hooft, Glay Chinea, Enzo de Filippis, Hans Doeberling and his team, Brigitte Altenberg, Karina Krmoian for stimulating discussions and practical help. We appologise to the people working on other good modelling programs (especially Ruben Abagyan and Andrej Sali) for not having enough space to explain their methods and programs in detail. We appologise to the numerous crystallographers who made all this work possible by depositing structures in the PDB for not referring to each of the 4000 very important articles describing these structures.

References

1. Chothia, C., Lesk, A.M., (1986) The relation between the divergence of sequence and structure in proteins *EMBO J.*, 5 823-836.

112

2) Sander, C., Schneider, R (1991) Database of homology-derived protein structures and the structural meaning of sequence alignment.., *PROTEINS*, **9** 56-68.

3) Swindells, M.B., Thornton, J.M., (1991) Modelling by homology. *Curr.Op.Struct.Biol.*, **1** 219-223.

4) Hilbert, M., Böhm, G., Jaenicke, R (1993), Structural relationships of homologous proteins as a fundamental principle in homology modeling.., *PROTEINS*, **17**, 138-151.

5) Lesk, A.M., Chothia, C., (1980) How different amino acid sequences determine similar protein structures: the structure and evolutionary dynamics of the globins. *J.Mol.Biol.*, **136**, 225-270.

6) Kabsch, W., Sander, C., (1984) On the use of sequence homologies to predict protein structure: identical pentapeptides can have completely different conformations. *PNAS*, **81**, 1075-1078.

7) Chothia, C., Lesk, A.M., (1982) Evolution of proteins formed by b-sheets. I. Plastocyanin and Azurin. *J.Mol.Biol.*, **160**, 309-323.

8) Bajorath, J., Stenkamp, R., Aruffo, A., (1993) Knowledge-based model building of proteins: concepts and examples. *Prot.Sci.*, **2**, 1798-1810.

9) Lesk, A.M., Boswell, D.R., (1992) Homology modelling: inferences from tables of aligned sequences. *Cuur.Op.Struc.Biol.*. **2**, 242-247.

10) Havel, T.F., Snow, M.E., (1991 A new method for building protein conformations from sequence alignments with homologues of known structure. *J.Mol.Biol.*, **217**, 1-7.

11) Reid, L.S., Thornton, J.M., (1989) Rebuilding flavodoxin from Ca coordinates: a test study. *PROTEINS*, **5**, 170-182.

12) Greer, J., (1991) Comparative modeling of homologous proteins. *Meth.Enzym.*, **202**, 239-252.

13) Sudarsanam, S., March, C.J., Srinivasan, S., (1994) Homology modeling of divergent proteins *J.Mol.Biol.*, **241**, 143-149.

14) Lee, R.H (1992) Protein model building using structural homology.., *Nature*, **356**, 543-544.

15) Sali, A., Blundell, T.L., (1993) Comparative modelling by satisfaction of spatial restraints. **234**, 779-815.

16) Summers, N.L., Karplus, M., (1990) Modelling of globular proteins. A distance based search procedure for the construction of insertion regions and pro <--> non-pro mutations. *J.Mol.Biol.*, **216**,991-1016.

17) Schiffer, C.A., Caldwell, J.W., Kollmann, P.A., Stroud, R.M., (1990) Prediction of homologous protein structures based on conformational searches and energetics. *PROTEINS*, **8**, 30-43.

18) Swindells, M.B., Thornton, J.M., (1991 Modelling by homology. *Curr.Op.Struc.Biol.*, **1**, 219-223.

19) Moult, J., Pedersen, J.T., Judson, R., Fidelis, K., (1995) A large scale experiment to assess protein structure prediction methods. *PROTEINS*, **23**, 2-4.

20) Mosimann, S., Meleshko, R., James, N.G., (1995) A critical assessment of comparative molecular modeling of tertiary structures of proteins. *PROTEINS*, **23**, 301-317.

21) Harrison, R.W., Chatterjee, D., Weber, I.T., (1995) Analysis of six protein structures predicted by comparative modelling techniques. *Proteins*, **23**, 463-471.

22) Cardozo, T., Totrov, M., Abagyan, R., (1995) Homology modelling by the ICM method. *PROTEINS*, **23**, 403-414.

23) Church, W.B., Palmer A.,, Wathey, J.C., Kitson, D.H., (1995) Homology modelling of histidine-containing phosphocarrier protein and eosinophil-derived neurotoxin: construction of models and comparison with experiment. *PROTEINS*, **23**, 422-430.

24) Samudrala, R., Pedersen, J.T., Zhou, H.-B., Luo, R., Fidelis, K., Moult, J., (1995) Confronting the problem of interconnected structural changes in the comparative modeling of proteins. *PROTEINS*, **23**, 327-336.

25) Sali, A., Potterton, L., Yuan, F., Vlijmen, H. van, Karplus, M., (1995) Evaluation of comparative protein modeling by MODELLER. *PROTEINS*, **23**, 318-326.

26) Sali, A., (1995) Modelling mutations and homologous proteins. *Curr.Op.Struc.Biol.*, **6**, 437-451.

27) Vriend, G., Sander, C., (1991) Detection of common three dimensional substructures in proteins. *PROTEINS* **11**, 52-58.

28) Russell, R.B., Barton, G.J., (1992) Multiple protein structure alignment from tertiary structure comparison: assignment of global and residue confidence levels. *PROTEINS* **14**, 309-323.

29) Bowie, J.U., Clarke, N.D., Pabo, C.O., Sauer, R.T., (1990) Identification of protein folds: Matching hydrophobicity patterns of sequence sets with solvent accessibility patterns of known structures. *PROTEINS* **7**, 257-264.

30) Grindley, H.M., Artymiuk, P.J., Rice, D.W., Willett, P., (1993) Identification of tertiary structure resemblance in proteins using a maximal common subgraph isomorphism algorithm. *J.Mol.Biol.*, **229**, 707-721.

31) Zuker, M., Somorjai, R.L., (1989) The alignment of protein structures in three dimensions. *Bull. Math.Biol.* **51**, 55-78.

31) A rapid method for protein structure alignment. *J.Theor.Biol.*, (1990) **147**, 517-551.

32) Orengo, C.A., Taylor, W.R., Overington, J.P., (1992) Comparison of three-dimensional structures of homologous proteins. *Curr.Op.Struc.Biol.*, **2**, 394-401.

33) Z.-Y., Sali, A., Blundell, T.L., (1992) A variable gap penalty function and feature weights for protein 3-D structure comparisons. *Prot.Engin.*, **5**, 43-51.

34) Orengo, C.A., Brown, N.P., Taylor, W.R., (1992) Fast structure alignment for database searching. *PROTEINS* **14**, 139-167.

35) Zhu, Maiorov, V.N., Crippen, G.M., (1995) Size independent comparison of protein three dimensional structures. *PROTEINS*, **22**, 273-283.

36) Alexandrov, N.N., Takahashi, K., Go, N., (1992) Common spatial arrangements of backbone fragments in homologous and non-homologous proteins. *J.Mol.Biol.*, **225**, 5-9.

37) Fisher, D., Bachar, O., Nussinov, R., Wolfson, H., (1992) An efficient automated computer vision based technique for detection of three dimensional structural motifs in proteins. *J.Biolol.Struct.&Dyn.*, **9**, 769-789.

38) Pepperrell, C., Willett, P., (1991) Techniques for the calculation of three dimensional structural similarity using inter-atomic distances. *J.Comp.-Aid.Mol.Des.*, **5**, 455-474.

39) Maiorov, V.N., Crippen, G.M., (1994) Significance of root-mean-square deviation in comparing three-dimensional structures of globular proteins. *J.Mol.Biol.*, **235**, 625-634.

40) Brown, N.P., Orengo, C.A., Taylor, W.R., (1996) A protein structure comparison methodology. *Comp.Chem* **20**, 359-380.

41) Taylor, W.R., Orengo, C.A., (1988) Protein structure alignment. *J.Mol.Biol.*, **208**, 1-22.

42) Sali, A., Blundell, T.L., (1990) Definition of general topological equivalence in protein structures. *J.Mol.Biol.*, **212**, 403-428.

43) Flores, T.P., Orengo, C.A., Moss, D.S., Thornton, J.M., (1993) Comparison of conformational characteristics in structurally similar protein pairs. *Prot.Sci.*, **2**, 1811-1826.

44) Holm, L., Sander, C., (1993) Protein structure comparison by alignment of distance matrices. *J.Mol.Biol.*, **233**, 123-138.

45) (1994) Biological meaning, statistical significance, and classification of local spatial similarities in nonhomologous proteins. *Prot.Sci.*, **3**, 866-875.

46) Brändén, C.-I., (1990) Founding fathers and families. *Nature*, **346**, 607-608.

47) Holm, L., Ouzounis, C., Sander, C., Tuparev, G., Vriend, G., (1992) A database of protein structure families with common folding motifs. *Prot.Sci.*, **1**, 1691-1698.

48) Holm, L., Sander, C., (1994) Searching protein structure databases has come of age. *PROTEINS*, **19**, 165-173.

49) Murzin, A.G., Brenner, S.E., Hubbard, T., Chothia, C., (1995) SCOP: A structural classification of proteins database for investigation of sequence and structures. *J.Mol.Biol.*, **247**, 536-540.

50) Murzin, A.G (1993) OB (oligonucleotide/oligosaccharide binding)-fold: common structural and functional solution for non-homologous sequences.., *EMBO*, **12**, 861-867.

51) Russell, R.B., Barton, G.J., (1994) Structural features can be unconserved in proteins with similar folds. *J.Mol.Biol.*, **244**, 332-350.

52) Laurents, D.V., Subbiah, S., Levitt, M., (1994) Different protein sequences can give rise to highly similar folds through different stabilizing interactions. *Prot.Sci.*, **3**, 1938-1944.

53) Kamphuis I.G., Drenth, J., Baker, E.N., (1985) Thiol proteases. Comparative studies on the high resolution structures of papain and actinidin, and on amino acid sequence information for cathepsins B and H, and stem bromelian., **182**, 317-329.

54) Pearl, L (1993) Similarity of active-site structures.., *Nature*, **362**, 24.

55) Fisher, D., Wolfson, H., Lin, S.L., Nussinov, R., (1994) Three-dimensional, sequence order-independent structural comparison of a serine protease against the crystallographic database reveals active site similarities: potential implications to evolution and to protein folding. *Prot.Sci.*, **3**, 769-778.

56) Perry, K.M., Fauman, E.B., Finer-Moore, J.S., Montfort, W.R., Maley, G.F., Maley, F., Stroud, R.M., (1990) Plastic adaptation toward mutation in proteins: structural comparison of thymidilate synthases. *PROTEINS*, **8**, 315-333.

57) Park, J.E., Rice, D.W., Willett, P., (1992) Three dimensional structural resemblance between leucine aminopeptidase and carboxypeptidase A revealed by graph-theoretical techniques. *FEBS Lt.*, **303**, 48-52.

58) Swindells, M.B., Orengo, C.A., Jones, D.T. Pearl, L.H., Thornton, J.M., (1993) Recurrence of a binding motif? *Nature*, **362**, 299.

59) PROCHECK: (1993) a program to check the stereochemical quality of protein structures. R.A., MacArthur, M.W., Moss, D.S., Thornton, J.M., *J.Appl.Cryst.*, **26**, 283-291.

60) Laskowski, Morris, A.L., MacArthur, M.W., Hutchinson, E.G., Thornton, J.M., (1992) Stereochemical quality of protein-structure coordinates. *PROTEINS*, **12**, 345-364.

61) Hooft, R.W.W., Vriend, G., Sander, C., Abola, E.E (1996) Errors in protein structures.., *Nature*, **381**, 272.

62) Sippl, M.J., (1993) Recognition of errors in three dimensional structures of proteins. *PROTEINS*, **17**, 355-362.

63) Lüthy, R., Bowie, J.U., Eisenberg, D., (1992) Assessment of protein models with three dimensional profiles. *Nature*, **356**, 83-85.

64) Novotny, J., Rashin, A.A., Brucoleri, R.E., (1988) Criteria that discriminate between native proteins and incorrectly folded models. *PROTEINS*, **4**, 19-30.

65) Blundell, T.L., Sibanda, B.L., Sternberg, M.J.E., Thornton, J.M., (1987) Knowledge-based prediction of protein structures and the design of novel molecules. *Nature*, **326**, 347-352.

66) Naor, D., Fisher, D., Jernigan, R.L., Wolfson, H.J., Nussinov, R., (1996) Amino acid pair interchanges at spatially conserved locations. *J.Mol.Biol.*, **256**, 924-938.

67) Overington, J., Donnelly, D., Johnson, M.S., Sali, A., Blundell, T.L (1992) Environment-specific amino acid substitution tables: tertiary templates and prediction of protein folds., *Prot.Sci.*, **1**, 216-226.

68) Abagyan, R., Frishman, D., Argos, O., (1994) Recognition of distantly related proteins through energy calculations. *PROTEINS*, **19**, 132-140.

69) Bryant, S.H., Lawrence, C.E., (1993) An empirical energy function for threading protein sequence through the folding motif *PROTEINS*, **16**, 92-112.

70) Ouzounis, C., Sander, C., Scharf, M., Schneider, R., (1993) Prediction of protein structure by evaluation os sequence structure fitness. *J.Mol.Biol.*, **232**, 805-825.

71) Madej, T., Gibrat, J.-F., Bryant, S.H., (1995) Threading a database of protein cores. *PROTEINS*, **23**, 356-369.

72) Lemer, C.M.-R., Rooman, M.J., Wodak, S.J., , (1995) Protein structure prediction by threading methods: evaluation of current techniques. *PROTEINS* **23**, 337-355.

73) Hubbard, T.J., Park, J., Fold (1995) recognition and ab initio structure predictions using hiddem markov models and b-strand pair potentials. *PROTEINS*, **23**, 398-402.

74) Lathrop, R.H., Smith, T.F., Sciences (1994) A branch-and-bound algorithm for optimal protein threading with pairwise (contactpotential) amino acid interactions. Proc. 27-th Hawaii Intl. Conf. on System IEEE *Comp. Soc. Press*. 365-374.

75) Johnson, M.S., Overington, J.P., (1993) A structural basis for sequence comparisons. *J.Mol.Biol.*, **233**, 716-738.

76) Stultz, C.M., White, J.V., Smith, T.F., (1993) Structural analysis based on state-space modeling. *Prot.Sci.*, **2**, 305-314.

77) Bowie, J.U., Lüthy, R., Eisenberg, D., (1991) A Method to identify protein sequences that fold into a known three dimensional structure. *Science*, **253**, 164-170.

78) Pearson, W.R (1990) Rapid and sensitive comparison with FASTA and FASTP.., *Meth.Enzym.*, **183**, 63-98.

79) Altschul, S.F., Gish, W., Miller, W., Myers, E.W., Lipman, D.J., (1990) Basic local alignment search tool. *J.Mol.Biol.*, **215**, 403-410.

80) Delarue, M., Koehl, P., (1995) Atomic environment energies in proteins defined from statistics of accessible and contact surface areas. *J.Mol.Biol.*, **249**, 675-690.

81) Holm, L., Sander, C., ., (1992) Evaluation of protein models by atomic solvation preference. *J.Mol.Bio* **l225**, 93-105.

82) Taylor, W.R., (1986) Identification of protein sequence homology by consensus template alignment. *J.Mol.Biol.*, **188**, 233-258.

83) Vingron, M., Argos, O., (1989) A fast and sensitive multiple sequence alignment algoritm. *CABIOS*, **5**, 115-121.

84) Subbiah, S., Harrison, S.C., (1989) A method for multiple sequence alignment with gaps. *J.Mol.Biol.*, **209**, 539-548.

85) Lüthy, R., Xenarios, I., Bucher, P., (1994) Improving the sensitivity of the sequence profile *method*. *Prot.Sci.*, **3**, 139-146.

86) Smith, R.F., Smith, T.F., (1992) Pattern-induced multi sequence alignment (PIMA) algorithm employing secondary structure-dependent gap penalties for use in comparative protein modelling. *Prot.Engin.*, **5**, 35-41.

87) Higgins, D.G., (1992) Sequence ordinations: a multivariate analysis approach to analysing large sequence data sets. *CABIOS*, **8**, 15-22.

88) Barton, G.J., Sternberg, M.J.E., (1987) A strategy for the rapid multiple alignment of protein sequences. Confidence levels from tertiary structure comparisons. *J.Mol.Biol.*, **198**, 327-337.

89) Yi, T.-M., Lander, E.S., (1994) Recognition of related proteins by iterative template refinement. *Prot.Sci.*, **3**, 1315-1328.

90) Zhang, K.Y.J., Eisenberg, D., (1994) The three dimensional profile method using residue preference as a continuous function of residue environment. *Prot.Sci.*, **3**, 687-695.

91) Brown, W.J., North, A.C.T., Phillips, D.C., Brew, K., Vanaman, T.C., Hill, R.C., (1969) A possible three-dimensional structure of bovine a-lactalbumin based on that of hen's egg-white lysozyme. *J.Mol.Biol.*, **42**, 65-86.

92) Warme, P.K., Momany, F.A., Rumball, S.V., Scheraga, H.A., (1974) Computation of structure of homologous proteins: a-lactalbumin from lysozyme. *Biochemistry* **13**, 768-782.

93) Laughton, C.A., Prediction of protein side-chain conformations from local three dimensional homology reletionships. (1994) *J.Mol.Biol.*, **235**, 1088-1097.

94) McGregor, M.J., Islam, S.A., Sternberg, M.J.E., (1987) Analysis of the relationship between side-chain conformation and secondary structure in globular proteins. *J.Mol.Biol.*, **198**, 295-310.

95) Ponder, J.W., Richards, F.M., (1987) Tertiary templates for proteins. *J.Mol.Biol.*, **193**, 775-791.

96) Schrauber, H., Eisenhaber, F., Argos, O., (1993) Rotamers, to be or not to be? *J.Mol.Biol.*, **230**, 592-612.

97) Holm, L., Sander, C., (1992) Fast and simple Monte Carlo algorithm for side chain optimization in proteins: application to model building by homology. *PROTEINS*, **14**, 213-223.

98) Summers, N.L., Karplus, (1991) M., Modelling of side chains, loops and insertions in proteins. *Meth.Enzym.*, **202**, 156-205

99) Summers, N.L., Karplus, M., (1989) Construction of side-chains in homology modelling. Application to the C-terminal lobe of rhizopuspepsin. *J.Mol.Biol.*, **210**, 785-811.

100) Eisenmenger, F., Argos, O., Abagyan, R., (1993) A method to configure protein side-chains from the main-chain trace in homology modelling. *J.Mol.Biol.*, **231**, 849-860.

101) Desmet, J., Maeyer, M. De., Hazes, B., Lasters, I (1992) The dead-end elimination theorem and its use in protein side-chain positioning.., *Nature*, **356**, 539-542.

102) Taylor, W , (1992) New paths from death ends.., *Nature*, **356**, 478-480.

103) Filippis, V.de, Sander, C., Vriend, G., (1994) Predicting local structural changes that result from point mutations *Prot.Engin.*, **7**, 1203-1208.

104) Dunbrack, R.L.Jr., Karplus, M., (1993) Backbone-dependent rotamer library for proteins. Application to side-chain prediction. *J.Mol.Biol.*, **230**, 543-574.

105) Stites, W.E., Meeker, A.K., Shortle, D (1994) Evidence for strained interactions between side-chains and the polypeptide backbone.., *J.Mol.Biol.*, **235**, 27-32.

106) Dunbrack, R.L.Jr., Karplus, (1994) Conformational analysis of the backbone dependent rotamer preferences of protein side chains. *Nature Struc.Biol.*, **5**, 334-340.

107) Chinea, G., Padron, G., Hooft, R.W.W., Sander, C., Vriend, G., (1995) The use of position specific rotamers in model building by homology. *PROTEINS*, **23**, 415-421.

108) Totrov, M.M., Abagyan, R.A., (1994) Detailed ab initio prediction of lysozyme-antibody complex with 1.6 A accuracy. *Nature Struct. Biol.*, **1**, 259-265.

109) Lee, C., Levitt, M., (1991) Accurate prediction of stability and activity effects of site directed mutagenesis on a protei₁ core. *Nature* **352**, 448-451.

110) Gunsteren, W.F. van, Mark, A.E., (1992) Prediction of the stability and activity effects of site directed mutagenesis. *J.Mol.Biol.*, **227**, 389-395.

111) Simonson, T., Brunger, A.T., (1992) Thermodynamics of protein peptide interactions in the ribonuclease S system studied by molecular dynamics and free energy calculations. *Biochemistry* **31**, 8661-8674.

112) Vriend, G., Eijsink, V.G.H., (1993) Prediction and analysis of structure, stability and unfolding of thermolysin like proteases. *J.Comp.-Aid Mol.Des.* **7**, 367-396.

113) Vriend, G., Sander, C., Stouten, P.W.F (1994) A novel search method for protein sequence-structure relations using property profiles.., *Prot.Engin.* **7**, 23-29.

114) Jones, T.A., Thirup, S., (1986) using known substructures in protein model building and crystallography. *EMBO, J.*, **5**, 819-823.

115) Hobohm, U., Scharf, M., Schneider, R., Sander, C., (1992) Selection of representative protein data sets. *Prot.Sci.*, **1**, 409-417.

116) Hooft, R.W.W., Sander, C., Vriend, G., Verification of protein structures: side-chain planarity. *Cabios*, accepted.

117) Parsaye K., Chignell, M., Khoshafian, S., Wong, H., ., (1989). Intelligent databases. *John Wiley and sons, Inc*

118) Bryant, S.H., (1989) PKB: A program system and data base for analysis of protein structure. *PROTEINS* **5**, 233-247.

119) Vriend, G., (1990) Parameter relation rows: a query system for protein structure function relationships. *Prot.Engin.*, **4**, 221-223.

120) (1989) A relational data base of protein structures designed for flexible enquiries about conformation. *Prot.Engin.*, **2**, 431-442.

121 Gray, P.M.D., Paton, N.W., Kemp, G.J.L., Fothergill, J.E., (1990) An object oriented database for protein structure analysis. *Prot.Engin.*, **3**, 235-243.

122) Huysmans, M., Richelle, J., Wodak, S.J., (1991) SESAM: A relational database for structure and sequence of macromolecules. *PROTEINS*, **11**, 59-76.

123) Bernstein, F. C., Koetzle, T. F., Williams, G. B., Meyer, E. F. Jr.,Brice, M. D., Rodgers, J. R., Kennard, O., Shimanouchi, T. ; Tatsumi, M. (1977) The protein data bank: A computer based archival file for macromolecular structures. *J.Mol.Biol.* **112**, 535-542.

124) Schultze-Kremer, S., King, R.D (1992) IPSA-Inductive protein structure analysis.., *Prot.Engin.*, **5**, 377-390.

125) Read, R.L., Davison, D., Chappelear, J.E., Garavelli, J.S., (1992) GBPARSE: a parser for the GenBank flat-file format with new feature table format. *CABIOS*, **8**, 407-408.

126) Lesk, A.M., Boswell, D,R., Lesk, V.I, Lesk, V.E., Bairoch, A., (1989) A cross reference table between the protein data bank of macromolecular structures and the national biomedical research foundation protein identification resource amino acid sequence data bank. *Prot.Seq.Data.Anal.*, **2**, 295-308.

127) Stoehr, P.J., Cameron, G.N (1991) The EMBL data library., *NAR*, **19**, 2227-2230.

128) Thorton, J.M., Gardner, S.P (1989) Protein motifs and database searching.., *TIBS*, **14**, 300-304.

129) Kamel, N.N., (1992) A profile for molecular biology databases and information resources. *CABIOS*, **8**, 311-321.

130) Jia, Z., Quail, J. W., Waygood, E. B., Delbaerre L. T. J. (1993) To be published., Deposited in the *PDB*.

131) Rodriguez, R., Vriend, G., Limits to modelbuilding by homology. *to be submitted*.

132) Jia, Z., Vandonselaar, M., Hengstenberg W., ,Quail, J. W., Delbaerre L. T. J. (1993), To be published. Deposited in the *PDB*.

133) Fitzgerald, P. M. D., Mc Keever, B. M., Van Middlesworth, J. F., Springer, J. P., Heimbach, J. C., Leu, C. T., Herber, W. K., Dixon, R. A. F., Darke, P. L. (1990) Crystallographic analysis of a complex between human immunodeficiency virus type 1 protease and acetyl pepstatin at 2.0 Angstrom resolution. *J. Biol. Chem.* **265**, 14209-.

134) Huang, Q., Liu, S., Tang, Y. (1993) Refined 1.6 A resolution crystal structure of the complex formed between porcine β-trypsin and MCTI-A, a trypsin inhibitor of the squash family. J. *Mol. Biol.* **229**, 1022-.

135) G. Vriend (1990) WHAT IF: A molecular modelling and drug design program., *J.Mol.Graph.* **8**, 52-56.

136) Hubbard, R.E., , (1986) In: Computer Graphics and molecular modelling. *Edt. Fletterick, R.J., Zoller, M.*, Cold Spring Harbor 9-12.

137) Jones, T.A., (1978) A graphics modelbuilding and refinement system for macromolecules. *J.Appl.Cryst.* 268-272.

138) Dayringer, H.E., Tramontano, A., Fletterick, R.J., (1986) Interactive program for visualization and modelling of proteins, nucleic acids and small molecules. *J.Mol.Graph.* **4**, 82-87.

139) Jones, T.A., Zou, J.Y., Cowan, S.W., Kjelgaard, M., (1991) Improved methods for buildin protein models in electron density maps and the location of errors in these models. *Acta Cryst A* **47**, 110-119.

140) Schomburg, D., Reichelt, J., (1988) BRAGI: A comprehensive protein modelling program system. *J.Mol.Graph.* **6**, 161-165.

141) Moult, J., James, M.N.G., (1986) An algorithm for determining the conformation of polypeptide segments in proteins by systematic search. *PROTEINS* **1**, 146-163.

142) Bruccoleri, R.E., Karplus, M., (1987) Prediction of the folding of short polypeptide segments by uniform conformational sampling. *Biopolymers* **26**, 137-168.

143) Fine, R.M., Wang, H., Shenkin, P.S., Yarmush, D.L., Levinthal, C (1986) Predicting antibody hypervariable loop conformations. II: minimization and molecular dynamics studies of MCPC603 from many randomly generated loop conformations.., *PROTEINS* **1**, 342-362.

144) Havel, T.F., Snow, M.E., (1990) A new method for building protein conformations from sequence alignments with homologues with know structure. *J.Mol.Biol.* **217**, 1-7.

145) Sippl, M.J., Hendlich, M., Lackner, P., (1992) Assembly of polypeptide and backbone conformations from low energy ensambles of short fragments. *Prot.Sci.* **1**, 625-640.

146) Simon, I., Glasser, L., Scheraga, H.A., (1991) Calculation of protein conformation as an assembly of stable overlapping segments: application to BPTI. *PNAS* **88**, 3661-3665.

147) Ripoll, D.R., Scheraga, H.A., (1990) On the multiple minima problem in the conformational analysis of polypeptides. *Biopolymers* **30**, 165-176.

148) Moult, J., Judson, R., Fidelis, K., Pedersen, J.T., (1995) A large scale experiment to assess protein structure prediction methods. *PROTEINS* **23**, ii-iv.

149) Baumann, G., Froemmel, C., Sander, C., (1989) Polarity as a criterion in protein design. *Prot.Engin.* **2**, 329-334.

150) Bryant, S.H., Amzel., L.M., (1987) Correctly folded proteins make twice as many hydrophobic contacts. *Int.J.Pept.Prot.Res.* **29**, 46-52.

151) Hendlich, M., Lackner, P., Weitcus, S., Floeckner, H., Froschauer, R., Gottsbacher, K., Cassari, G., Sippl, M.J., (1990) Identification of native protein folds amongst a large number of incorrect models. *J.Mol.Biol.* **216**, 167-180.

152) Morris, A.L., MacArthur, M.W., Hutchinson, E.G., Thorton, J.M., (1992) Stereochemical quality of protein structure coordinates. *PROTEINS 12*, 3456-364.

153) Eisenberg, D., McLachlan, A.D., (1986) Solvation energy in protein folding and binding. *Nature,* **319**, 199-203.

154) Gregoret, L.M., Cohen, F.E., (1990) Novel method for the rapid evaluation of packing in protein structures. *J.Mol.Biol.* **211**, 959-974.

155) Vriend, G., Sander, C., (1993) Quality control of protein models: directional atomic contact analysis. *J.Appl.Cryst.* **26**, 47-60.

156) Van Gunsteren, W.F., Berendsen, H.J., (1987) GROMOS. BIOMOS, Biomolecular software, Lab. Phys. Chem., Uni., Groningen, The Netherlands.

MOLECULAR MODELING OF GLOBULAR PROTEINS : STRATEGY 1D ⇒ 3D :

Secondary Structures and Epitopes

Alain J.P. ALIX
Laboratoire de Spectroscopies et Structures Biomoléculaires (LSSBM); Université de Reims Champagne Ardenne (URCA); Institut Fédératif de Recherches "Biomolécules" IFR 53; INSERM Unité 314 "Conformations Cellulaires et Moléculaires" ; Faculté des Sciences, B.P. 1039, 51 687 Reims Cedex 02, France (alain.alix@univ-reims.fr).

1. Introduction

In recent years a very considerable body of strategies [1-6] have been developed which makes it possible to built relatively good molecular models for large proteins. The purpose of that paper is to present coherently some essential elements of one of these strategy in the " twilight zone " where standard sequence and/or structure homology methods are not available but is not meant to be an exhaustive survey of the literature which has now grown to a considerable size. In order to hold this paper to a reasonable size, only predictions of secondary structures [7-56] and epitopes [57-71] are treated in detail. However, essential and/or very recent references are reported here (classified by topics) which will help the reader whose interest embraces either one or both of the described fields, say structure [7-35] and spectroscopy [36-56].

1.1. STRATEGY 1D ⇒ 3D AND / OR 1D ⇒ 4D

In the study of biomolecules, for many years, a type of research work specially involving protein structure determination, used the general strategy: from the sequence, of the gene and/or of the protein [1,6] (1D-level), to the three-dimensional macromolecular structures [3-5] (static conformation 3D-level and/or dynamic conformations 4D-level) in order to be able to explain structure-function and/or structure activity relationships.
In the field of the Quantitative Structure Activity Relationships (QSAR) and/or Quantitative Structure Properties Relationships (QSPR) the 4-D level corresponds to choose as 4-th parameter, any quantitative property which can be plot (through the conversion : numerical values ⇒ coloured spectrum) in the 3D usual space. This is typically the case of the Three-Dimensional Molecular Hydrophobicity Potential (3D-MHP) which value (colour) can be plot in any point in the 3D-space and this especially on the surface of the molecule [72-75].

G. Vergoten and T. Theophanides (eds.), Biomolecular Structure and Dynamics, 121–150.
© *1997 Kluwer Academic Publishers.*

The main goals leading from the strategy 1D $\Rightarrow$ 3D are achieved on the one hand by :

(i) treatments of the informations contained in the sequence (primary structure [1, 6-7]) giving a lot of 1D-profiles of the type " biophysical properties versus the number of the amino acid in the sequence " : e.g., surface probability [62, 67-68], hydrophobicity [57-59, 63-65, 69-71], flexibility [60-61]...; secondary structures predictions (local assignments of the structural state of each amino acid residue side chain [17-33]), predictions of the epitopes as potential antigenic determinants [4, 57-71]. This level also includes comparisons of sequences (alignments, homologies...[1, 6, 12]),

(ii) treatments of the 2D-information : secondary structures and sequence/structure homologies from the Hydrophobic Cluster Analysis [34, 35] applied to a series of molecules extracted from a 1D-sequence database (SWISS-PROT, PIR [1,6-7]...) and/or from a 3D-structure database (Protein Data Bank [1, 6, 8-12]),

(iii)3-D molecular modeling either without (*de novo*) or under constraints (1D + 2D informations and/or experimental constraints) [1-6].

On the other hand, optical spectroscopies, vibrational and/or electronic ones (Raman scattering [36-42], Infrared absorption [43-45] and Circular Dichroisms [46-54]) which use allows us to derive from spectroscopic data a lot of local and/or general structural informations. These are for instance percentages of secondary structure, micro-environments of aromatic (tyrosines, trytophans) residues, conformations of the disulfide bridges, folding type (secondary structure class)... Moreover, these technics give us simple experimental data which may be of use for improving the local pure secondary structure predictions (2D) [55-56] and building a 3D macromolecular model under constraints.

This paper deals with:

(i) experimental secondary and/or tertiary structure estimations: i.e., characterizations of peptides, native proteins or recombinant proteins by optical spectroscopies in order to study their structure / function and/or the structure / activity relationships and to provide constraints for molecular modeling,

(ii) theoretical secondary structure and epitope predictions: i.e., characterizations of proteins by the treatment of the informations contained in the primary structure for either only one or all the sequences,

(iii) macromolecular modeling of the 3D structures of the proteins by using molecular mechanics for the static stable structures and molecular dynamics for the trajectories; 3D-molecular hydrophobicity potential approach...

1.2. SECONDARY STRUCTURES AND LINEAR EPITOPES

1.2.1. *Secondary Structures*

The usual theoretical standard methods for predicting the secondary structure assignment of each amino acid residue (hereafter defining the **local secondary structures**) are mostly based on the usual Chou-Fasman procedures [17-23] and/or the more sophisticated GOR methods [24-31].

The standard experimental methods for estimating the quantitative secondary structure contents of a globular protein (hereafter defining the **global secondary structures**) are based on the processing of optical spectroscopic data [36-54]. The obtained results also permit us to define the folding types (class of proteins) in terms of secondary structural content.

Our method LINK gives the missing link closing the gap between the standard and/or derived predictions methods on the one hand and optical spectroscopic data on the other hand. It will permit to adjust the set of "pure" local secondary structure predictions by taking into account the experimental spectroscopic estimation of the global secondary structures [55, 56].

1.2.2. *Epitopes*

A single small segment (sequence recognition) or domain (conformation recognition) of a protein could act as an antigen (antigenic determinant) versus an antibody. Epitopes of the first kind being continuous segment along the sequence (linear), generally bent with a typical non-ordered structure (turns and/or loop), can be predicted from the knowledge of the primary structure [57-58, 60-66, 69, 76].

Our program P.E.O.P.L.E. (Predictive Estimation Of Protein Linear Epitopes) uses combined prediction methods taking into account the basic fundamental properties corresponding to what should be an ideal epitope: bent (secondary structure mainly β-turns), surface accessible, hydrophilic and mobile and/or flexible [76].

2. Secondary Structures :

2.1. WHAT IS THE PROBLEM ? : CONNECTING TWO WORLDS

Optical spectroscopists (using UV-, visible-, infrared-Raman inelastic scattering and/or electronic-, infrared-absorptions) involved in the quantitative determination of the global secondary structure of a protein from experimental data are generally not familiar with the world of structuralists (biophysicists, biochemists, molecular biologists...) and reciprocally. This is also clearly seen by screening the titles of the usual Journals in which are publishing the two communities and which are reported in different sections of the Current Contents (Physical, Chemical and Earth Sciences on the one hand, and Life Sciences on the other hand).

2.1.1. *The World of Spectroscopists : Biomolecular Optical Spectroscopy*

The first fundamental constraint for the experiments is: **Get the protein** (some mg) **!!**
The spectroscopic data follows from the other constraint: **Get the spectrum !!**

UV-resonance Raman- (UVRR-), visible normal Raman- (R-) [36-42], Near Infrared Fourier Transform Raman- (NIR-FT-), Fourier Transform InfraRed- (FT-IR-) [43-45], Vibrational (VCD-) and Electronic Circular Dichroism- (ECD-) [46-54] spectroscopies are now very useful as specific and complementary sophisticated techniques (see the other related papers in that book).

The processing of the data of either the whole spectrum or of the structural marker bands (Amide I, II, III bands related to the CONH peptide bond vibrations) by using mathematical sophisticated treatments (band decomposition, Fourier self-decomposition, second derivative, maximum entropy, reference profiles, reference spectra, principal components method, factor analysis...) leads to a estimation of the global secondary structure percentage contents related to specific choices of conformational states (see below for their definitions, [13-16]). Generally the number of structural states are reduced to 3 (α- helices, β-strands and/or β-sheets and random coil) or 4 (α-helices, β-strands and/or β-sheets, turns and random coil).

It must be pointed out that specific optical spectroscopies can be used to characterize on the one hand, the global secondary structure of a native protein in any state (solid powder, crystal, lyophilized, in aqueous buffer solution or in any solvent in presence or in absence of a cosolvent...) and this in any different experimental conditions (temperature, pH, ionic force...) permitting one to study the global modifications of conformation. On the other hand the interactions of the studied protein (or enzyme) with other molecule or biomolecule (activator, inhibitor, substrate) could be done in order to precise the structure-function and the structure-activity relationships.

Raman spectroscopy. Here we present in Figure 1, the Near Infrared Fourier Transform Raman spectrum of a very fluorescent (even in the visible range) insoluble protein elastin. The use of the excitation line in the infrared range (1.06 μm = 1006 nm) permits us to get a spectrum free of fluorescence background.

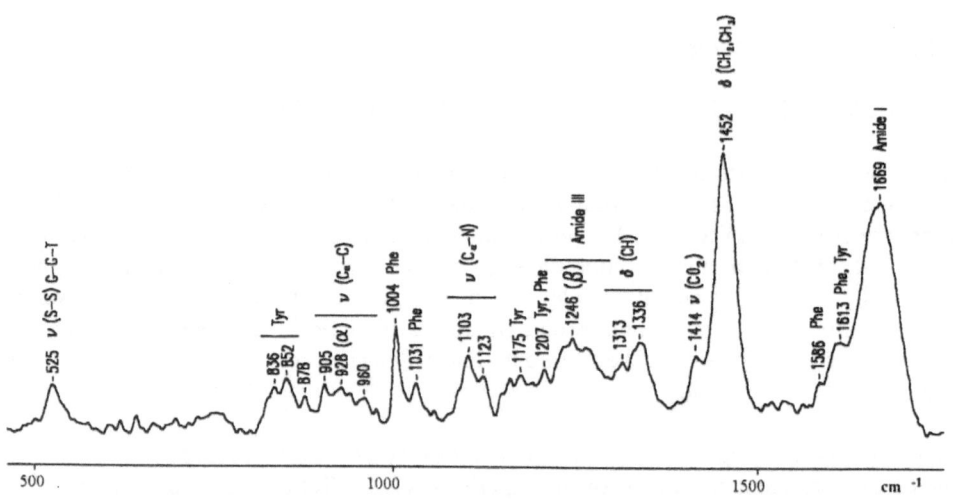

Figure 1. Near InfraRed Fourier Transform - Raman spectrum of human aortic elastin (powder, excitation at 1.06 μm, power 300 mW, 400 scans.) [79-80].

Infrared spectroscopy. Figure 2 shows the Fourier Transform Infrared spectrum in the Amide I and II regions of one soluble precursor of elastin, the bovine tropoelastin.

Circular dichroism. The CD spectrum of the bovine tropoelastin in aqueous solution is shown in Figure 3

For details on the material and methods, see below sections 4.1.1. and 4.2. and [76-80].

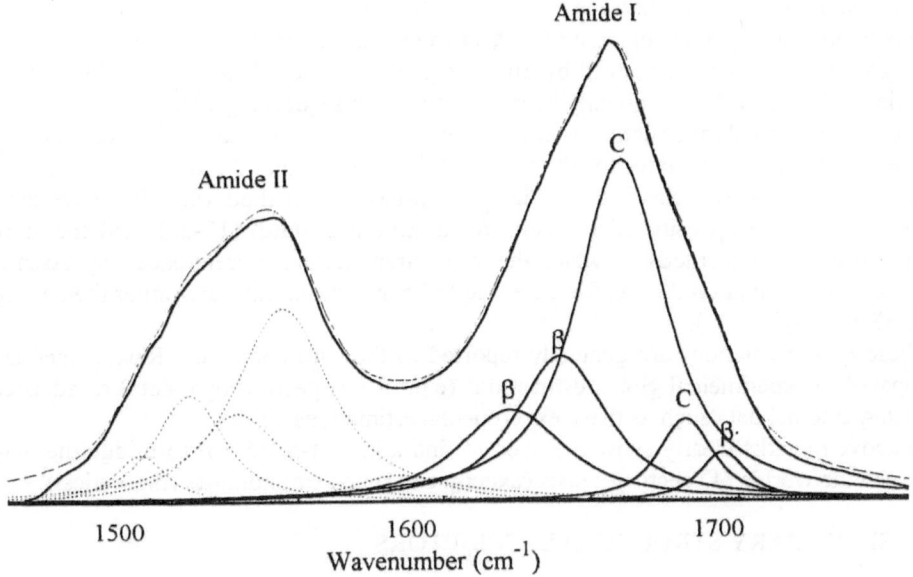

Figure 2. Amide I and II regions of the Fourier Transform InfraRed spectrum of bovine tropoelastin (powder on ZnSe window, 1000 interferograms) [79- 80].

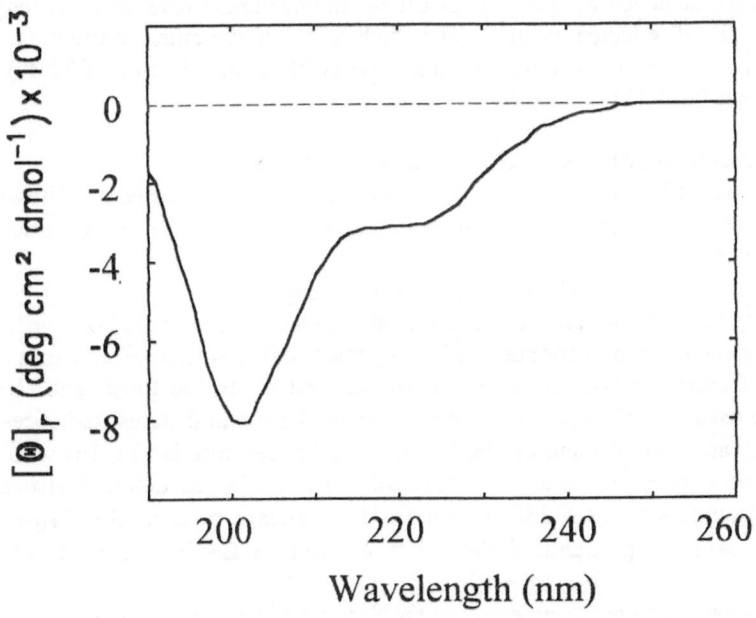

Figure 3. Circular dichroism spectrum of bovine tropoelastin (10^{-6} M) [77- 78, 80].

2.1.2. *The World of Structuralists : Structural Biology*

The only one fundamental constraint is : **Get the sequence of the protein !!**
This goal is now usually achieved by knowing the sequence of the gene coding for the protein (and/or now, less frequently by standard protein sequencing.) [1, 6-7].

Then using secondary structure prediction methods, one gets the local assignments of the secondary structural states for the amino acid residues along the sequence. The most used methods are the classical standard Chou-Fasman method (in which the main parameter is the propensity of a residue for a structural state) [17-23], and the more sophisticated GOR methods (in which the main parameter is the preference, expressed in terms of the information theory, for a residue to be in a structural state rather then not to be in that state) [24-31].

These pure predictions are generally reported as they are computed. Rarely, they are compared to experimental global estimations (e.g., from spectroscopic data) or adjusted by using external data such as these experimental estimations.
The above remarks clearly show the need to find a link [4-5, 55-56] to bridge the gap between the worlds of spectroscopists (experimental) and structuralists (theoretical).

2.2. SECONDARY STRUCTURE DESCRIPTORS

2.2.1. *Definition of a Non-Redundant Reference Protein Data Set*

A statistical analysis of the relations between the sequence (1D-level [7]) and the structure (3D-level [8]) requires to define a non-redundant reference protein data set. This is easily achieved by fixing a cut-off for the maximum level of homology accepted for any pair of selected proteins and such as each structural protein family is just represented. The up to date representative set (with a cut-off value of 30 %) has nearly 275 proteins [6, 9-12].

2.2.2. *Definitions of the Secondary Structural States S*

The principal definitions of the secondary structural states are based on different criteria:
- ϕ, ψ dihedral angles defining the orientation in space of the residue side chain with respect to the backbone,
- C_α-C_α distances and C_α-C_α-C_α-C_α torsion angles,
- Hydrogen bonds (in terms of geometrical distance or in terms of bond-energy).

Two different algorithms for automatic assignment of the structural state of a residue in a protein of known 3D-structure are respectively based on the Levitt and Greer criteria [13] who used only the C_α coordinates to define 4 structural states (α-helices, β-sheets, turns and random coil) and on the Kabsch and Sander criteria [14, 16] who use all the atomic coordinates as found in a standard PDB file [8] to define 8 structural states (standard outputs of the DSSP program : H = standard α-helix, G = 3_{10}-helix, I = π-helix; E = extended β-strand / β-sheet; T = reverse turn; B = bridge, S = bend, C = coil.).

2.2.3. *Statistics of the Occurrences of the Secondary Structural States S*

Having defined a reference protein data set on the one hand and the criteria for defining the different secondary structural states on the other hand, one may count the numbers of each type of residue (i = 1 to 20) in each state S (and $ for not-S).

TABLE 1. Statistics of occurrences for a structural secondary state S ($S = \alpha, \beta, ...$)

AA(i)	State S	State $	Σ
1	n(1,S)	n(1,$)	n(1) = n(1,S) + n(1,$)
.	.	.	.
i	n(i,S)	n(i,$)	n(i) = n(i,S) + n(i,$)
.	.	.	.
20	n(20,S)	n(20,$)	n(20) = n(20,S) + n(20,$)
Σ_i	N(S)	N($)	N

In Table 1, N(S) is the total (i.e., for the whole data set containing N residues) number of residues in the state S; $N(\$) = 1 - N(S)$, is the complementary number of residues in the complementary state $, meaning state not- S (for instance for $S = \alpha$, $\$ = \beta \cup t \cup c$).

2.2.4. *Theory of Probabilities and Theory of Informations*
The fundamental principle used for calculating explicitly the secondary structure descriptors is based on the use of the observed frequencies of some events (i.e., statistics of occurrences) for defining their corresponding probabilities: frequency $\Rightarrow$ **probability**.

Probability. Using the following direct probabilities

$$\textbf{p(i and S)} = f(i \text{ and } S) = n(i,S)/N, \quad \textbf{p(S)} = f(S) = N(S)/N, \quad \textbf{p(i)} = f(i) = n(i)/N,$$

one defines the conditional probability **p(S/i)** of to be in the state S knowing the type i of the residue (in other words, it answer the question : knowing the type of a residue, what is its probability to find it in the state S ?) as follows

$$\textbf{p(S/i)} = \textbf{p(i and S)} / \textbf{p(i)} = f(S/i) = n(i,S)/N(i). \tag{1}$$

Propensity. The propensity $\textbf{P}_i{}^S$ for a residue of the type i to adopt a conformational state S, is defined as the ratio of the probability, knowing the type i of a residue, to find its state S (**p(S/i)**), on the probability **p(S)** to find that state S (i.e., independently of the type of the residue):

$$\textbf{P}_i{}^S = \textbf{p(S/i)} / \textbf{p(S)} = \textbf{p(i and S)} / (\textbf{p(i) p(S)}). \tag{2}$$

The second part of Equation (2) shows clearly that the propensity expresses the degree of coupling between the events **i** (to be a residue of type i) and S (to be in the state S). If the events are independent, then of course the propensity is unity. A residues with a propensity value larger, equal to, or less then one is respectively named former, indifferent, or breaker of a structural state S. The propensity $\textbf{P}_i{}^\$$ for a residue i **not** to adopt the state S (i.e., to adopt $ = any one of the other states; for instance: not-α means β + turn + coil) can easily be derived in the same manner.

We would like to point out that, according to Equation (2), the propensity has fundamentally to be considered as "multiplicative" and not "additive" as commonly used in Chou-Fasman like methods [17-18].

Information. The quantity of information (which can be expressed in **nats**, i.e. natural units, or in centinats **cnats**) that a residue of type **i** carries on its own conformational state S is defined by :

$$I_i^S = \ln (p(i \text{ and } S) / [p(i) \, p(S)]) = \ln (p(S/i) / p(S)). \tag{3}$$

Comparing Equations (2) and (3) permits us to derive the fundamental following result

$$I_i^S = \ln P_i^S. \tag{4}$$

The information I_i^S is then directly connected to the propensity P_i^S. The latter is the exponential of the quantity of information I_i^S which confirms its status of multiplicative parameter.

In terms of informations, a residue is said to be former, indifferent or breaker of a conformation for I_i^S values respectively positive, null or negative (this criterion is seen to be the same as the one defined in terms of propensities as shown by Equation (4).).

Preference. In a similar way, it is obvious to derive the quantity of information that residue **i** carries on the preference for its own conformation S rather than for \$,

$$I(S:\$, i) = I_i^{\Delta S} = I_i^S - I_i^\$. \tag{5}$$

Using Equations (4), (3) and (2), one respectively gets

$$I_i^{\Delta S} = \ln P_i^S - \ln P_i^\$ = \ln [P_i^S / P_i^\$], \tag{6}$$

$$I_i^{\Delta S} = \ln [(p(S/i) / p(S)) / (p(\$/i) / p(\$))], \tag{7}$$

$$I_i^{\Delta S} = \ln [(n(i,S) / N(S)) / (n(i,\$) / N(\$))]. \tag{8}$$

The expression in the bracket of Equation (8) represents the reciprocal of the ratio of the molar fractions of residues **i** in the states \$ and S, which in fact is exactly the partition coefficient, noted f_i^S, between the states S and \$.
Thus, it follows

$$I_i^{\Delta S} = - \ln f_i^S, \text{ where } f_i^S = P_i^\$ / P_i^S. \tag{9}$$

Free Energy of Transfer. One may define for each residue of type **i**, its free energy of transfer from the state \$ to S as

$$\Delta G_i^S = - RT \ln [(n(i,S) / N(S)) / (n(i,\$) / N(\$))] = RT \ln f_i^S = - RT \, I_i^{\Delta S}. \tag{10}$$

This theoretical result is of fundamental importance because some research works (experimental and or theoretical) are now reporting scales of values for ΔG_i^S which can be compared with values derived from Equation (10).

2.2.5. *Reduced Expressions for the Descriptors of Secondary Structural States*

Finally, by analogy to the partition coefficient for a residue of type i, f_i^S, it is possible to define

$$F^S = p(\$) / p(S) = (N(\$) / N) / (N(S) / N) = N(\$) / N(S), \qquad (11)$$

as the partition coefficient between the states S and $\$$ for the whole reference protein data set. The partition coefficients, f_i^S and F^S, are both shown to be very sensitive to the size and composition of the used protein database as well as to the criterion used for defining secondary structures (ϕ, ψ torsion angles, Levitt-Greer algorithm [13], DSSP from Kabsch-Sander [14], Garnier *et al.* criterion [24-28]...).

Thus, using as parameters the partition coefficients defined for each residue of type i, f_i^S, and for the whole reference protein data set, F^S, one gets

$$p(S/i) = 1 / (1 + f_i^S F^S), \qquad (12)$$

$$P_i^S = (1 + F^S) / (1 + f_i^S F^S), \qquad (13)$$

$$I_i^S = \ln P_i^S, \qquad (14)$$

$$I_i^{\Delta S} = - \ln f_i^S, \qquad (15)$$

$$\Delta G_i^S = RT \ln f_i^S. \qquad (16)$$

2.3. STANDARD PREDICTION METHODS

A prediction of the local secondary structure consists in, firstly having chosen a secondary structure descriptor, secondly in calculating for each residue along the sequence, its predicted conformational state which is the one having the largest descriptor value (over all the possible states).

At this stage it must be pointed out that the prediction of the assignment of the secondary structural state of a residue will strongly depend on which descriptor is used: Sup $p(S/i)$; Sup P_i^S which is equivalent to Sup I_i^S; Sup $I_i^{\Delta S}$ which is equivalent to Inf ΔG_i^S (over all the states S).

2.3.1. *Chou-Fasman-Like Methods*

The Chou-Fasman-like methods [17-23] uses as secondary structure descriptor the single residue information : propensity P_i^S.

The first step consists in calculating, from a chosen reference protein data set, for each residue of type i, the propensities P_i^S for each defined secondary structural state S. For example, in Table 3 (see below) are given the amino acid residue propensities P_i^S calculated according to the following choices:

(i) a reference data filtered set of 67 proteins which, taken by pairs, have less than 50 % of homology [25-26, 28] and whom the Brookhaven PDB codes [8] are given in Table 2. That reference data set will also here be used for other examples of illustration of different methods.

TABLE 2. Reference data set of 67 proteins (homology < 50 %) : Brookhaven PDB codes [8]

4APE	2APP	2ACT	3WGA	4ADH	2ALP	4ATC	1AZA	2ABX	1CPV
3CIB	2CAB	5CPA	8CAT	5CHA	2CTS	1CRN	1GCR	3CYT	1CCR
2CCY	2CYP	3C2C	2C2V	351C	3FDR	2EST	2EBX	1ECD	1FDX
3FXC	3FXN	2FD1	1GP1	1MMQ	2HHB	2LBH	1HIP	1MCP	1FB4
1REI	2RHE	2PKA	4LDH	1LH1	2LZM	1LZ1	1MLT	1MBN	1SN3
1OVO	1PPD	1BP2	1PCY	2PAB	2SGA	3RP2	1RN3	5RNX	2SNS
1SBT	2SOD	3TLN	1TPO	4PTI	2STV	4SBV			

(ii) choice of 3 simple secondary structure conformational states (S = α, β, **coil**) obtained by gathering the eight states, output of the program DSSP [14], according to the criteria of Gibrat *et al.* [25]:

- α-helices = H + G (except for a series of 3G which defined a 3_{10} specific turn),
- β-strands and/or β-sheets = E,
- random coil = I + T + B + S + C.

TABLE 3. Amino acid residues propensities P_i^S

CODES		Helices	Strands	Coils
G	Gly	0.4244	0.7352	1.4494
A	Ala	1.5125	0.7432	0.8101
V	Val	1.0212	1.7691	0.6595
L	Leu	1.3906	1.1899	0.6907
I	Ile	1.1318	1.6400	0.6500
S	Ser	0.6703	0.8738	1.2465
T	Thr	0.8042	1.2753	0.9970
D	Asp	0.9769	0.3985	1.2701
E	Glu	1.4019	0.6900	0.8975
N	Asn	0.7244	0.5817	1.3395
Q	Gln	1.1744	0.9338	0.9263
K	Lys	1.2094	0.7625	0.9789
H	His	1.0026	0.8648	1.0562
R	Arg	0.9923	1.0849	0.9683
F	Phe	1.2372	1.3356	0.7182
Y	Tyr	0.8870	1.4288	0.8831
W	Trp	1.2204	1.3977	0.7015
C	Cys	0.6971	1.3944	1.0090
M	Met	1.2765	1.2925	0.7136
P	Pro	0.5826	0.4176	1.4925
ref. set %S		29.1 %	21.2 %	49.7 %
AA nbN(S)		3226	2355	5520

Thus, the application of a Chou-Fasman like method is straightforward [17-18]:

- assign to each amino acid residue AA_j of type i (considered as the first residue of a tetrapeptide) and located at position j in the sequence of the studied protein, the average of the propensity $<P_j^S>$ computed over the 4 residues of the defined tetrapetide (moving average method), and this for each state S,

$$< P_j^S > = (\sum P_i^S \text{ for } AA_j \text{ to } AA_{j+3}) / 4, \quad S = \alpha, \beta, ..., \tag{17}$$

- the predicted local secondary structural state of the residue AA_j is given by the largest value (over S) of $< P_j^S >$.

2.3.2. GOR Methods

The GOR method [24-28] uses as secondary structure descriptor the directional informations of the type $I_i^{\Delta S}$ defined not only for the AA_j, the state of which is to be predicted, but also for some of its neighbours. Thus, the state of one residue is shown to implies 17 residues (influence of the local sequence).

The preference for a conformational state S of a residue AA_j (of the type i) located at position j in the protein chain and having 16 neighbours (defined by m-values ranging from - 8 to - 1 and from + 1 to + 8) is defined and calculated as follows,

$$I_j^{\Delta S} = I(S : S ; i) = {}_m\Sigma I(\text{Table } S : \text{Row name of } AA_{j+m}, \text{ Column } j + m), \tag{18}$$

where $m : -8$ to $+8$ ($m = 0$ corresponds to the single residue information of AA_j).
Having calculated for the jth residue in the sequence ($j = 1$ to n, $n = $ total number of AA in the calculated protein) its preferences for each state S (which now are shown to depend also on the local sequence), the largest value of the preference (over the states) gives the predicted conformational state.

The original GOR-method, (GOR I, set up in 1978 [24]) used a small not-filtered set of 25 reference proteins whose the four states (helix, extended, turns and coil) of secondary structures were defined in terms of ϕ , ψ angles and permit to construct four Tables of directional informations (17 columns, 20 rows). Moreover, in the original method, decision constants, DCs, were introduced for taking into account experimental results (essentially CD data at that time). The preference informations were modified according to :

$$I_i^{\Delta S}(DC) = I(S : S ; i; DC^S) = I_j^{\Delta S} - DC^S = I(S : S ; i) - DC^S. \tag{19}$$

Typical values of the decision constants were defined for different classes of proteins. GOR II [26] used a larger non redundant set of 67 reference proteins (see above, Table 2) whose secondary structures by gathering the 8 states issued from the DSSP program into four states (helices = H + G + I, strands/sheets = E, turn = T, coil = S + B + C); GOR III [25] used only three states (see above section 2.3.1.) but took into account informations for pairs of residues. All these methods used the preference information as secondary structure descriptor and criteria for the prediction.

Recently, Garnier [30] proposed to assign a state not by using Sup $I_i^{\Delta S}$ over all the states S, but by using Sup $p(S/i)$ which is not equivalent and consequently changes some of the assignments of secondary structures.

So using the formalism stated above and especially by combining Equations (12) and (15) permits one to get

$$p(S/i) = 1 / (1 + e^{-I(S \, : \, \$, \, i)} F^S), \text{ with the constraint } _S\Sigma \, p(S/i) = 1. \quad (20)$$

Finally, we would like to point out the close connection between the Chou-Fasman method (single residue information expressed by its propensity: see Equations (2, 13)) and the GOR method (directional information of residues expressed by their preferences: see Equations (5, 15)) when the latter is used without taking into account the 16 neighbours (which is totally equivalent to only have m = 0 in Equation (18).).
So, combining Equations (13) and (15) leads to

$$P(S/i) = (1 + F^S) / (1 + e^{-I(S \, : \, \$, \, i)} F^S). \quad (21)$$

In other words, it means that, applying on the one hand the Chou-Fasman method or applying on the other hand any GOR-method using a local sequence reduced to only one single residue (both methods being defined with the same choices' of reference data set and secondary structure criteria) is completely equivalent. Some standard GOR programs permitting one to choose the width (2 m + 1) of the window describing the local sequence, it is sufficient to put in m = 0. Equation (21) permits one also to derive any Table of propensity from the knowledge of the GOR Tables.

2.4. THE MISSING LINK CLOSING THE GAP BETWEEN PREDICTIONS AND OPTICAL SPECTROSCOPIC DATA

We would like to point out the major problem concerning both internal and external consistency of the used experimental (spectroscopy) and/or theoretical (prediction) methods. One has to remind that all the protocols involved in the calculations and in the comparisons of the results must taken into account the same basic definitions of secondary structures [13-16].

As we were concerned, we had to face such problem of consistencies when firstly, using the Broohaven PDB [8] and some dictionaries of protein structures (Levitt and Greer [13], Kabsch and Sander [14]), then secondly, in using standards and/or variants Chou-Fasman [17-23] and GOR methods [24-31] and, finally in computing results from standard and/or modified versions of programs.
For instance we modified CONTIN, the program of Provencher [46-47], originally set up for the determination of the global secondary structures of soluble globular proteins from circular dichroism data and which previously used Levitt and Greer secondary structure assignments [13].
We introduced reference protein secondary structures now derived from the 8 states, output data of DSSP [14], reduced to either a simple three- or a four- states description (e.g.; α–helix; β–strand; β-turn; coil; or turn + coil) according to the Gibrat et al. criterion [25] (see above section 2.3.1. (ii)).
So CONTIN-LG (LG3, LG4) and CONTIN-KS (KS3, KS4) are respectively used with comparison of results with corresponding LG (Levitt-Greer) or KS (Kabsch-Sander) assignments (either actual or predicted structures).

2.4.1. *Principle of the Method LINK*

The first serious attempt to use experimental global results to improve local predictions was done in 1978 by Garnier *et al.* [24] with introduction in their original method (GOR I) of some Decision Constants (DCs, fitted by using 25 protein structures in terms of the ϕ and ψ torsion angles) related to specific ranges of % contents of secondary structures. We note that this approach was totally given up in GOR II and GOR III [25-26, 27-28, 55]).

We introduce the new very simple method LINK [55-56] for closing the gap between predictions and experimental methods. LINK can be implemented in any variant of Chou-Fasman [42] and/or GOR methods. For a given protein, the adjustments of the initial pure predictive local secondary structure results are simply done by calculating some correction factors, CFs, (and/or decision constants, DCs) which are closely related to the experimental global secondary structure results (percentage contents.).

First, one of the fundamental application of LINK is obviously to adjust the **pure** (more or less accurate, especially in terms of global percentages of secondary structure) predictions of the local secondary structures, in such a manner to improve (LINK-1) or to ensure (LINK-2) the **identity** between "observed" and "predicted" percentages of secondary structures and consequently to improve the local secondary structure predictions. That way, a better description of the fine local secondary structure is obtained. Second, this is also a way to detect the regions involved in conformational transitions when the percentages of the associated changes are obtained after spectroscopic experiments.

The accuracy of LINK, in improving the local conformational assignments of the residues (AA), was checked by using, as experimental data, the actual structures (taken from the Brookhaven PDB [8]) for a very large filtered set of nearly two hundred proteins.). Different criterion for secondary structure definitions were also used (Levitt and Greer [13], Kabsch and Sander [14], Gibrat *et al.* [25]) for prediction methods as well as for experimental quantitative estimation of the percentage contents of secondary structure [42, 55-56]. For each protein on the one hand and for the whole reference protein set on the other hand, the improvement of the " pure predictions " was evaluated by comparing the actual distribution of the amino acid structures along the sequence and the adjusted ones. Quantitative criteria of accuracy defined in terms of quality indices [32, 33]), were derived from the parameters reported in Table 4 :

TABLE 4 : Statistical parameters of occurrences used for checking the accuracy of **LINK**

State S	Observed (S)	Not observed ($)	Σ
predicted (S)	a	b	a + b
not predicted ($)	c	d	c + d
Σ	a + c	b + d	n = a + b + c + d

a, b, c, d, are the numbers of counted residues; n is the total number of residues for a given protein. Statistics of occurrences (predicted and/or observed) may also be gathered in one Table for all the studied proteins (whole data set for which N is the sum of all n values). As we are concerned, we essentially used as quality indices Q_S (range [-1, +1], the Pearson coefficient of correlation between positive predicted and observed as well as negative predicted and nonobserved, this for each state S, and Q_Σ (range [0, 1]) respectively defined by:

$$Q_S = (ad-bc)/[(a+b)(a+c)(d+b)(d+c)]^{1/2}, \quad Q_\Sigma = (\Sigma_S a)/n.$$

2.4.2. LINK-1

LINK-1 Used to Improve Chou-Fasman-Type Pure Predictions. The fundamental principle used in LINK-1, which connects the experimental results, expressed as global percentage contents of secondary structures ($\%\alpha = f(\alpha)$, $\%\beta = f(\beta)$, $\%t = f(t)$,...) to the Chou-Fasman pure local secondary structure predictions, consists in replacing the partition coefficients (see above Equation (11))

$$F^S = p(\$) / p(S) = f(\$) / f(S) = N(\$) / N(S), \quad (\text{for } S = \alpha, \beta, t, \text{coil}; \$ = \text{not-S}),$$

calculated for the whole reference protein data set by the corresponding experimentally **observed** values obtained for the studied protein :

$$^{prot}F^S = {}^{prot}f(\$) / {}^{prot}f(S) = \%^{prot}\$ / \%^{prot}S = (100 - \%^{prot}S) / \%^{prot}S, \quad (22)$$

this, in the equations giving the probability $p(S/i)$ and/or the propensity P_i^S (see above Equations (12-13).). Thus one gets

$$^{prot}p(S/i) = 1 / (1 + f_i^S {}^{prot}F^S) = {}^{prot}f(S) / ({}^{prot}f(S) + f_i^S {}^{prot}f(\$)). \quad (23)$$

We must notice that these adjusted values $^{prot}p(S/i)$ have first to be renormalized to give the sum over the states S **equal to unity** before to be used (for adjusted propensity calculation for example).

$$^{prot}p(S/i) = (1 + {}^{prot}F^S) / (1 + f_i^S {}^{prot}F^S) = 1 / ({}^{prot}f(S) + f_i^S {}^{prot}f(\$)). \quad (24)$$

It can easily be shown that the above process is also equivalent to only modify the original values of the partition coefficients f_i^S : i.e., rewriting Equation (23) as follows,

$$^{prot}p(S/i) = 1 / [1 + f_i^S ({}^{prot}F^S / F^S) F^S] = 1 / (1 + {}^{prot}f_i^S F^S), \quad (25)$$

which also reads

$$^{prot}p(S/i) = 1 / [1 + f_i^S CF^S F^S]. \quad (26)$$

Equation (26) clearly shows that the correction factor CF^S can act either on f_i^S or on F^S giving respectively either $^{prot}f_i^S$, see Equation (25), or $^{prot}F^S$, see Equation (23), i.e.,

$$^{prot}f_i^S = f_i^S CF^S, \quad {}^{prot}F^S = F^S CF^S, \quad \text{with } CF^S = {}^{prot}F^S / F^S. \quad (27)$$

Finally, in order to define the CF^Ss corresponding to specific classes of protein, we defined the limits of $^{prot}F^S$ for a non-S protein with $f(S) \cong 1$ % ($^{prot}F^S = 99/1 = 99$) and for a all-S protein with $f(S) = 100$ % ($^{prot}F^S = 0/100 = 0$) [42].

LINK-1 used to improve GOR-type pure predictions ($DC^S{s} = 0$). The improvement of GOR pure local predictions is straightforward as it is easily shown that the relation between the DCs and the CFs is quite simple. Using Equation (19) leads to :

$$\exp [- I(S : S ; i; DC^S)] = {}^{prot}f_i S = \exp [- I(S : S ; i) + DC^S],$$

$$= \exp [-I(S : S ; i)] \cdot \exp DC^S. \tag{28}$$

Now, remembering Equation (15) which reads $f_i S = \exp -I(S : S ; i)$, one finally obtains

$$\exp (DC^S) = CF^S \Leftrightarrow DC^S = \ln CF^S = \ln ({}^{prot}F^S / F^S). \tag{29}$$

The application of the method LINK-1 is very simple. The Decision Constants DC^Ss are calculated from spectroscopic (${}^{prot}F^S$) and structural (F^S) data.

2.4.3. *LINK-2*

LINK-2 [56] is an automatic procedure (based on a direct search method) which permits ones by using as input the **observed global** secondary structures to ensure identical **predicted** and **observed** percentages of secondary structures and consequently to improve the adjusted predictions.

We like to point out that, in LINK-2, we use as a fundamental constraint

$$[\text{predicted} \%] = (a + b) / n \equiv [\text{observed} \%] = (a + c) / n, \tag{30}$$

which is formally equivalent to $(a + b) \equiv (a + c)$. If follows $b = c$, which means that (i) the number b of overpredictions (predicted but not observed) is equal to the number c of underpredictions (observed but not predicted); (ii) as in the general cases b and c are not null, even if one strictly ensures the fulfilling of Equation (30), whatever the method used, one can not have the guarantee of a perfect prediction (for which one should have $b = c = 0$). That remark of fundamental importance was completely overlooked by the majority of experimentalists.

In practice, the numbers (integers) of residues are used instead of the percentages. The principle of the direct search is first based on a choice of the values of all the DCs for which the adjusted predictions are computed. For each state S , their ranges are [- 750, + 750] cnats; and the used steps are 10, 5, 1.., depending on how accurate Equation (30) has to be fulfilled. The deviation between predicted and observed global contents is simply calculated from:

$$\text{Deviation}^2 = D^2 = \Sigma_S [(a + b) - (a + c)]^2 = \Sigma_S (b - c)^2 \quad S = \alpha, \beta, ... \tag{31}$$

Its minimum value (not always null, depending on the chosen step) defines the optimal choice of the DCs. One may use, as a variant, not to scan the DCs like for a grid but to start (either from zero or from the point corresponding to the DCs calculated from LINK-1) and to turn around that point by increasing the step (spiral moving step).

We present in Table 5, the application of LINK-2 to the 760 AA sequence of the human tropoelastin for which the observed (Raman data) global secondary structure is 59 α, 280 β and 421 coil residues (the best adjustment gives 60 α, 274 β and 421 coil residues for a 5 cnats step; thus $D^2 = 1^2 + 6^2 + 5^2 = 62 \Rightarrow D < 8$ for 760 residues).

TABLE 5 : Application of the method LINK-2 to the sequence of the human tropoelastin (HTE)

GOR III (pure predictions), Link-2 (adjusted predictions);

0 = α-helices, **/** = β-strands/sheets, **-** = random coil

```
HTE   1   GGVPGAIPGG VPGVFYPGA GLGALGGGAL GPGGKPLKPV PGGLAGAGLG AGLGAFPAVT FPGALVPGGV ADAAAAYKAA KAGAGLGGVP
GOR III   -----/---- ---///--- -////----- --------- ---/-/0-/0 0---/--00/ ----/----0 0000000000 000----/--
Link-2    -----/---- ---///--- -/////---- --------- ----/--/// -------/// ----/----/ /----///// /----/

HTE   91  GVGGLGVSAG AVVPQPGAGV KPGKVPGVGL PGVYPGGVLP GARFPGVGVL PGVPTGAGVK PKA=GVGGAF AGIPGVGPFG GPQPGVPLGY
GOR III   //////// ///--/--- --------- -----//// -----//// -----//// -----/--/ -/---/---- -----/----
Link-2    //////// ///----- --------- ----//// ----//// ----//// ---/--/ //--/--- //-/----

HTE  181  PIKAPKLPGG YGLPYTTGKL PYGYGPGGVA GAAGKAGYPT GTVGVGPQAA AAAKAAAKF GAGAAGVLPG VGGAGVPGVP GAIPGIGGIA
GOR III   ---------- ---//////- -/---//// ////////- -/--00000 0000000000 000//////// ------/--- ------////
Link-2    ---------- --//////// -/-----// ////////// -/--00000 0000000-0 000/////// -----/---/ -----/////

HTE  271  GVGTPAAAAA AAAAKAAAKY GAAAGLVPGG PGFGPGVVGV GTVGVGPGGV PGAGIPVVPG AGIPGAAVPG VVSPEAAAKA AAKAAKYGAR
GOR III   //--000000 0000000000 000//////- -------/// -------/// -------/// ---/---/// /-0000000- 0000000---
Link-2    //----000 0000000-/ --/////-- ------//// ------//// ---/-//// ---/--/// -/-0000000 0000000---

HTE  361  PGVGVGGIPT YGVGAGGFPG FGVGVGGIPG VAGVPSVGGV PGVGGVPGVG ISPDAQAAAA AKAAKYGAAG AGVLGGLVPG PQAAVPGVPG
GOR III   -/////--// /////----/ /////----/ //////---/ /////--/// -00000000 0000000000 0//////// -//------
Link-2    -//////--// /////--// /////--// //////--/ ////--/// -00000000 000//---/ ////////// -//------

HTE  451  TGGVPGVGTP AAAAKAAAK AAQFGLVPGV GVAPGVGVAP GVGVAPGVGL APGVGVAPGV GVAPGVGVAP GIGPGGVAAA AKSAAKVAAK
GOR III   ---/-----0 0000000000 00000/---- //--//// /-///--// //////--// //////--// ------0000 0000000000
Link-2    ---/----- -0000000- 00000/--- //--//// /--///--/ //////--// //////--// -----0000 0000000000

HTE  541  AQLRAAAGLG AGIPGLGVGV GVPGLGVGAG VPGLGVGAGV PGFGAGADEG VRRSLSPELR EGDPSSSQHL PSTPSSPRVP GALAAAKAAK
GOR III   000000000- --------// //-/////- -////----/ //--------/ ////---/- //////---- ---------0 0000000000
Link-2    0/////// -------// //-/////- -//////--/ //-------/ ////---/-/ //////---- ---------0 0000000000

HTE  631  YGAAVPGVLG GLGALGGVGI PGGVVGAGPA AAAAAKAAA KRAQFGLVGA AGLGGLGVGG LGVPGVGGLG GIPPAAAAKA AKYGAAGLGG
GOR III   000////// /////--// /-////--00 0000000000 0000000000 /-/////// /////////// ----000000 0000000///
Link-2    -----////- -////--// --///// -000000000 0-0000000 /--///// //-///// ----000000 0000000///

HTE  721  VLGGAGQFPL GGVAARPGFG LSPIFPGGAC LGKACGRKRK
GOR III   ///--/--// --/--/// ----------
Link-2    ///----//- --////// ----------
```

3. Epitopes

3.1. PREDICTION METHODS FOR CONTINUOUS LINEAR BEND EPITOPES

Several algorithms (1D-level) attempt to locate antigenic motifs in a protein [57-66, 76]. The most popular one was that of Hopp and Woods [57-58] who used only their own scale of hydrophilicity. Later, Kyte and Doolitle proposed a similar method based on their defined scale of hydropathy [59], and Karplus and Schultz introduced the concept of flexibility [60]. Then, combined methods using in a weighted manner different biophysical characteristics (hydropathy, surface accessibility, flexibility and secondary structure) were proposed [62-64, 76]. The only differences in these methods are in the choice of different scales [69-71].

3.2. P.E.O.P.L.E. (Predictive Estimation of Protein Linear Epitopes)

This method [76], which is an extension and combination of the methods of Parker *et al.* [63] and of Jameson and Wolf [64], yields a composite profile used as antigenic index.

3.2.1. *Parameters (scales) used in the antigenic index*
Four classes of basic biophysical parameters are considered for the determination of the antigenic index AG.

Secondary structure. The first class to be considered is the type of secondary substructure of residues, predicted using the above described methods. Only the β–turns and coil conformations are considered to be favourable for antigenicity and thus the residues predicted in these conformations are the most likely to be antigenically predicted.

Hydrophilicity. The second class of parameters considered is the hydropathy of residues. Four scales are used: Hopp-Woods [57], Kyte-Doolitle [59], Parker *et al.* [63] and Efremov-Alix [72-73]. The latter one (see Table 5), MHPS derived from a statistical analysis of Three Dimensional Molecular Hydrophobicity Potential (3D-MHP) data obtained for 23 proteins with known 3D structure, reflects the influence of water surrounding (including accessible surface area and configuration of water-filled space near the residue) and correlates well with some commonly used scales [69-71] obtained in complementary approaches.

TABLE 5. **MHPS** : Molecular Hydrophobicity Potential(water) derived Scale (%)

ILE	- 0.0	TRP	- 12.9	GLY	- 52.7	GLN	- 71.0
PHE	- 1.1	VAL	- 12.9	THR	- 59.1	SER	- 72.0
LEU	- 5.4	TYR	- 22.6	ASN	- 65.6	PRO	- 77.4
CYS	- 6.5	HIS	- 24.7	ARG	- 66.7	GLU	- 78.5
MET	- 7.5	ALA	- 39.8	ASP	- 67.7	LYS	- 100.0

Surface accessibility. The third class is the surface accessibility of residues. Two scales are used: the fractional probabilities of Emini *et al.* [62] recalculated and corrected by Alix *et al.* [unpublished results] and the interior to surface transfer energy of Janin [68-69] updated by us with a larger database (by using our program SURF).

The whole formalism presented above for secondary structure predictions (see sections 2.2.3. to 2.2.5.) is valid for the prediction of the surface accessibility of residues.

One has just to define the state S for " to be on the surface " (accessible, or exposed, or exterior...) and S = not-S for " not to be on the surface " (non accessible, or buried, or interior...). Different criteria are used for defining an exposed (or a buried) residue [67-70]. Thus the surface fractional probabilities of Emini *et al.* [62] which were derived from the data of Janin *et al.* [67] are nothing else then the usual propensities P_i^S (see above Equations(2)). As example, we reported in Tables 6-7 two of the corresponding scales recalculated by us when using the 67 reference proteins given Table 2 and as criterion for a residue to be on the surface : to have more then 20 Å^2 of exposed surface (equivalent to have the possibility to be in contact with 2 water molecules).

TABLE 6. Surface fractional probability Scale = Scale of Propensity to be exposed (P_i^S)

CYS	0.394	GLY	0.714	SER	1.115	ASN	1.296
ILE	0.603	MET	0.714	HIS	1.180	GLN	1.348
LEU	0.603	TRP	0.808	THR	1.184	GLU	1.445
VAL	0.606	ALA	0.815	PRO	1.236	ARG	1.475
PHE	0.695	TYR	1.089	ASP	1.283	LYS	1.545

TABLE 7. Interior to surface transfer energy Scale (ΔG_i^S)

CYS	0.97	GLY	0.43	SER	- 0.19	ASN	- 0.55
ILE	0.60	MET	0.43	HIS	- 0.31	GLN	- 0.69
LEU	0.60	TRP	0.29	THR	- 0.32	GLU	- 1.01
VAL	0.60	ALA	0.28	PRO	- 0.42	ARG	- 1.14
PHE	0.46	TYR	- 0.15	ASP	- 0.52	LYS	- 1.62

Each prediction of surface accessibility is first treated independently, and the residues predicted as surface accessible are considered more likely to be antigenic.

Flexibility. The last class concerns the flexibility of residues [60-61], with the flexible residues more likely to be antigenic.

3.2.2. Computation of the AG index

In each class of parameters, each individual prediction is filtered and/or smoothed to give a specific 1D-profile; in the same class, these specific profiles are then averaged to give a class profile.

The antigenic index profile is finally defined as a linear combination of the four class profiles (the weights of these classes are respectively 0.40, 0.30, 0.15 and 0.15 [64]). The final ordered predictions of the antigenic segments are revealed by the ordered heights of the peaks of the antigenic index [76].

4. Examples of Applications

4.1. MATERIAL AND METHODS

4.1.1. *Optical Biomolecular Spectroscopy*

The Near InfraRed Fourier Transform-Raman spectra of powder samples were recorded at Lille, on a Brücker FRA 106 system coupled to a Brücker IFS 88 spectrometer (the infrared laser excitation line was 1.06 µm) (see Figure 1, [77-80, 82]).

The Fourier Transform InfraRed Spectra (on ZnSe window or in KBr pellets) were respectively recorded at Reims on a BOMEM MB100 and at Lille on a Brucker IFS 48 spectrometer (see Figure 2, [77, 78, 80, 82]).

The Circular Dichroism spectra in the 240-190 nm range were recorded at Reims on a MARK III (Jobin & Yvon) dichrograph (see Figure 3, [77, 78, 80, 82]).

The standard visible normal macro- and/or micro- Raman spectra were recorded at Reims, on a PHO (Coderg) double monochromator mono-channel spectrometer and on a MOLE S3000 (Jobin Yvon, ISA) triple monochromator mono- and multi-channel spectrometer (excitation wavelengths 488 and 514 nm) respectively [82].

4.1.2. *Secondary Structures and Epitopes*

The secondary structure quantitations from experimental data were performed by using either CONTIN-LS or CONTIN-KS programs (CD data), our RIP-method (normal Raman [36-39]), and by decomposition methods (FT-IR and NIR-FTR) using Fourier self-deconvolution, second derivative and maximum entropy methods (SPOV program developed in Shemyakin Institute in Moscow; CURVEFIT module of the Labcalc package from Galactics Industries).

For 1D $\Rightarrow$ 3D analyses, calculations were done on a PC computer with home-made and commercial available programs:

1D-level: CFGOR (for Chou-Fasman and GOR predictions); LINK1 and LINK2 (for adjusted predictions); PEOPLE (linear epitope prediction),
2D-level: HCA (Hydrophobic Cluster Analysis), Doriane Cie, Paris, France,
3D-level: SURF (surface accessibility predictions).

4.1.3. *Macromolecular modeling*

For the 3D structure determinations (molecular mechanics and/or molecular dynamics), all computations were performed on a Silicon graphics workstation (R4400 - Elan) using BIOSYM products: HOMOLOGY and DISCOVER softwares. All the different structures were minimized using DISCOVER 3.1 (BIOSYM package) and the AMBER force field. The initial models were optimized by steepest descent and then by conjugate gradient energy minimisation methods until the maximum derivatives was less than 0.001 kcal / (mol Å). 3D-Molecular Hydrophobicity Potential was calculated by using HIPPO. Solvent accessibility associated with specific molecular mechanics (*in vacuum*; in non-explicit solvent ($\varepsilon = 80$ and ε-distance dependent 4r) and in explicit solvent) were also used to characterize the models.

The schematic plots of proteins structures were obtained by using the program MOLSCRIPT [87].

4.2. A PSEUDO - 3D MOLECULAR MODEL FOR TROPOELASTIN

Elastin is an insoluble, highly hydrophobic and fluorescent, cross linked macropolymer of tropoelastins which forms fibres endowing tissues with elasticity. Corresponding soluble elastin peptides are prepared by solubilisation of the insoluble polymers by KOH yielding κ-elastins. The 760 amino acid residues sequence (~ 70 kDa) of the human precursor (tropoelastin) is fully atypical and give it the status of a unique type of molecule. The same stands for the bovine tropoelastin molecule (734 AA). Thus no homology methods are usable (" twilight zone ").

For both bovine and human tropoelastins, secondary structure predictions (CFGOR) yielded 11 major regions in which the pleated conformation was predominant, separated by 10 strong helical segments of various lengths located within alanyl rich regions of the chains. The overall structure is estimated to contain 18 % ± 6 % α-helices, 63 % ± 17 % β-sheets, 13% ± 13 % β--turns and 6 % ± 6 % coil giving the status of an all-β globular protein, consistent with a liquid drop architecture for the tridimensional elastin network.

Antigenicity predictions indicated the presence of seven decapeptide epitopes [76, 80]. For human tropoelastin, two of them are species specific (Exon 26 : 600-609, 610-619) and one other (Exon 36 : 751-760) characterises the highly charged C-terminal part which contains the unique disulfide bridge.

The Raman spectra of both bovine and human (see Figure 1) elastin, reveal the presence of a band centred around 525 cm^{-1} which demonstrates the presence of S-S bridges (mainly in **ggt** local conformation) and thus strongly suggests the existence of similar intrachain bondings in tropoelastins [77-80]. Further analysis of the Raman and/or infrared spectroscopic features also permitted us to estimate the global secondary structures of all our samples (bovine, human, insoluble, soluble) by decomposition of their respective amide I bands (see Figure 2). The CD spectrum of the soluble tropoelastin (see Figure 3) is consistent with the presence of very short strands alternating with turns and random coil motives and/or strands packed in β-barrels.

From all our optical spectroscopic results, it seems that the global secondary structures of tropoelastins and/or elastins are nearly the same : ~ 10 % **α**-helices, 40 % **β**-sheets and 50 % of reverse **β**-turns and undefined conformations. So we conclude that elastin is also a fibrous protein constituted of all-β globular tropoelastin molecules (liquid drop model of elastin architecture).

Application of LINK-2 to the sequences of the tropoelastins (see Table 5 for human tropoelastin) permits us to describe the elastic regions (outside the helical regions) as rather short and/or irregular strands alternating with reverse turns and/or coil regions which strongly correlated with spectroscopic observations.

Comparisons for all our samples of all our predicted and observed results, allows us to make the assumption that the polymerization of the tropoelastins into elastins should induce minimal structural changes. A pseudo 3D molecular model (see Figure 4) has been proposed for the tropoelastin molecules in which mainly pleated "elastic" regions (sheets folded in β-barrels + turns + coil) alternate with helical crosslinking "rigid" domains. Human exon 26 is the main species specific epitope, and exon 36 corresponds to the highly hydrophobic charged pocket stabilized by the **ggt** disulfide bridge [77, 79].

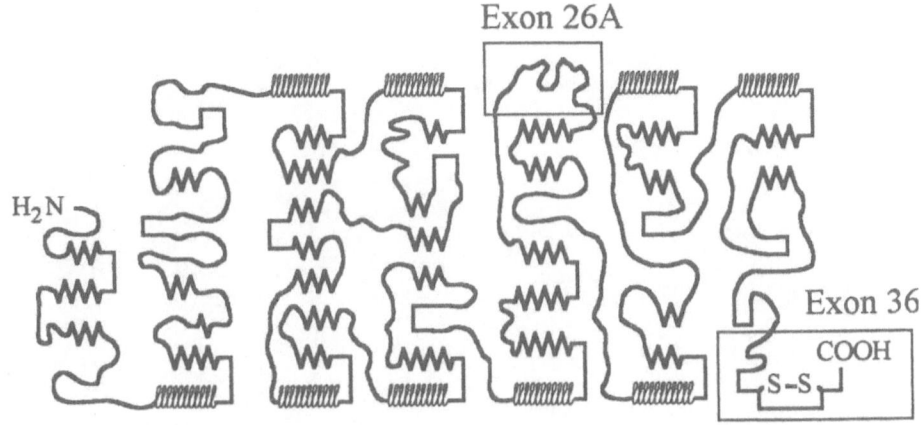

Figure 4. Pseudo 3D-molecular model of the tropoelastin molecule
$\alpha \sim 10\%$, $\beta \sim 40\%$, $t \sim 20\%$, $u \sim 30\%$, ($U = t + u \sim 50\%$.).

4.3. A MOLECULAR MODEL FOR THE RECOMBINANT ELAFIN

Elafin is a strong elastase-specific inhibitor which has been recently identified in human airway secretions. This inhibitor is a single peptide chain of 57 amino acid residues (~ 6.0 kDa) belonging to the class of 4-disulfide-bridge core proteins and containing neither tyrosine nor tryptophan residues. Elafin is thought to play an important protective role against destructive degradation, by excessive elastase, of the structural integrity of elastin-containing tissues.

The three dimensional molecular model of the recombinant elafin (r-elafin) is shown in Figure 5 [81]. This model is determined from:
- 1D data: secondary structure predictions methods using the primary sequence, alignment and comparisons of known proteins belonging to the 4-disulfide bridge family and the two domains of SLPI : Specific Leukocyte Protease Inhibitor),
- 2D structure: Hydrophobic Cluster Analysis and experimental global and/or local secondary structures obtained respectively by Circular Dichroism, visible normal Raman, infrared spectroscopies (r-elafin is an all-β protein) and applications of the methods LINK)
- 3D structure: local conformations of the disulfide bridges (obtained from the treatment of the specific marker band of the S-S bond in the Raman spectrum of r-elafin) and overall 3D-structure (*de novo* and/or homology with SLPI) were used to reach a full molecular model of r-elafin consistent with the experimental optical spectroscopic data.

As fundamental structure-function relationship results, a specific turn is shown to be essential and PRO28 has an important role in the conformation of the binding loop (segment 22-27) with elastase, which was very recently confirmed by us using NMR spectroscopy [82] and by Tsunemi *et al.* using X-ray crystallography [83].

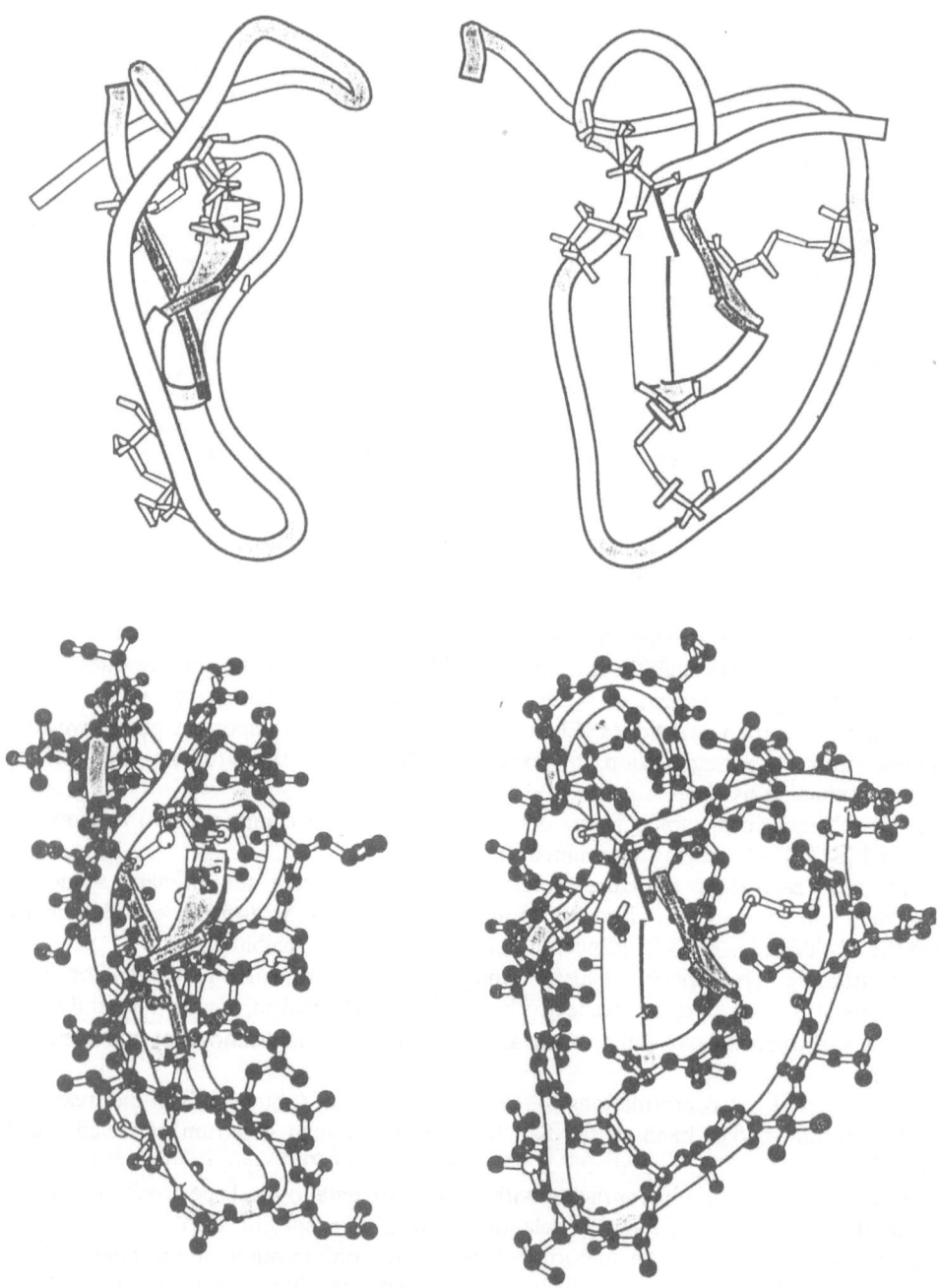

Figure 5. Molecular model of the recombinant elafin

4.4. A MOLECULAR MODEL FOR A IRON SUPER OXIDE DISMUTASE

The3D molecular model of one monomer (197 AA) of the dimeric iron-superoxide dismutase (Fe-SOD) from *Plasmodium Falciparum* is shown in Figures 6a, and 6b [86]. This model is determined from;
-1D data: secondary structure predictions; alignments and comparisons of 65 known Fe/Mn-SOD sequences),
-2D structure: Hydrophobic Cluster Analysis,
-3D homology-based computer assisted modeling with known X-ray 3D_structures extracted from the PDB (Fe-SOD of *Pseudomonas ovalis* and of *Escherichia coli*; and Mn-SOD of *Thermus thermophilus*).

The initial 3D-model was obtained by transferring the well known Structurally Conserved Regions (SCR) and then building variable regions. This model was then optimized using standard molecular mechanics procedures (*in vacuum* and in non-explicit solvent with either ε = 80 and 4r)). During the minimization process, a specific attention was drawn on the behaviour of the SCDs , on the residues involved in the chelation of the metal, and on the interface with the second monomer.

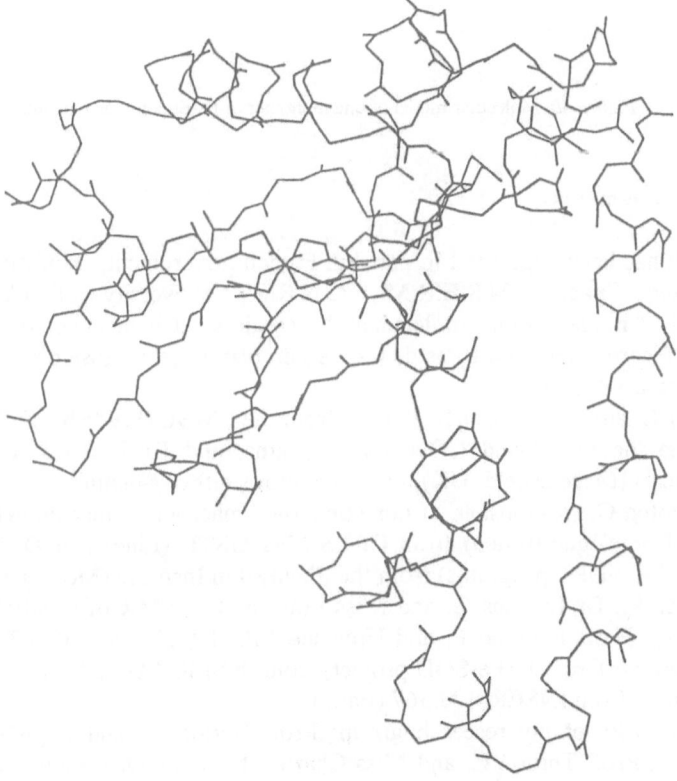

Figure 6a. Molecular model of one monomer of Fe-Superoxide Dismutase

144

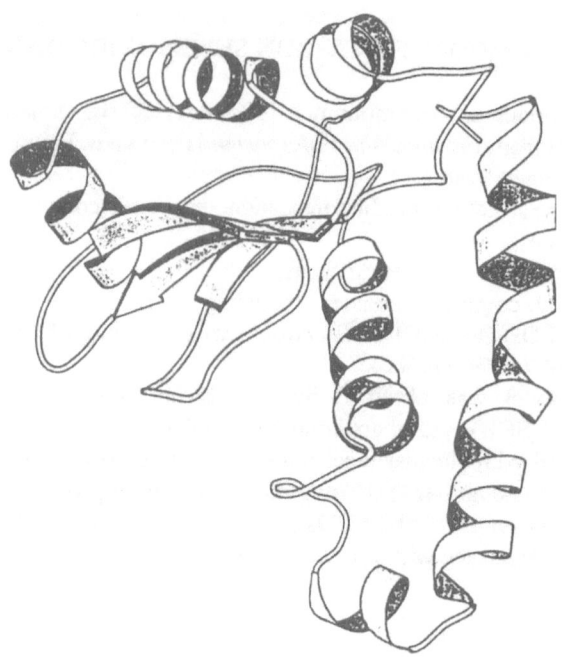

Figure 6b. Molecular model of one monomer of Fe-Superoxide Dismutase

5. Acknowledgements

Our research has been supported in part by: French government, Ministry of Education Research and Space, I.N.S.E.R.M., C.N.R.S., University of Reims, « Contrat interrégional " modélisation moléculaire " (régions Champagne-Ardenne, Haute-Normandie, Picardie, and Nord-Pas de Calais) du plan Bassin Parisien » .
All my thanks are due to:

Drs Sander C. and Provencher S.W. both from the EMBL (Heidelberg) for respectively providing me the DSSP and CONTIN Programs, and Dr Fitton J. from ZENECA pharmaceuticals (Macclesfield, UK) for his generous gift of r-elafin;

Prof. Vergoten G. (responsible of our common " macromolecular modeling network " Lille-Reims-Compiègne-Rouen) from CRESIMM USTL (Lille) and Dr Efremov R.G. (HIPPO, SPOV, SURF programs) from the Shemyakin Institute (Moscou);

Prof. Tartar A., Dr Lippens G. and Miss Francart C. (NMR of r-elafin) from Institut Pasteur (Lille); Profs Legrand P. and Huvenne J.P. (FT-IR and NIR-FT Raman) from LASIR (Lille); Dr Dive D. (Fe-SOD project) from INSERM U.42 (Lille); Dr Jacob M.P. (Elastin project) from INSERM U.367 (Paris),

the collaborators of my research group: Prof. Berjot M. and Dr Marx J. (Raman spectroscopy), Prof. Thirion C. and Miss Charton F. (CD), Dr Dauchez M. (Molecular modeling, r-Elafin and Fe-SOD projects) and our post-doctoral and doctoral students Dr Debelle L. (Elastin project) and Mr Goulyaev D (HIPPO, LINK, SURF programs).

6. References

INTRODUCTION GOALS

1. Methods in Enzymology (1990) R.F. Doolitle (ed), *Molecular evolution: Computer analysis of protein and nucleic sequences*, Academic press, San Diego, **183**, pp. 1-690

2. Monge, A., Friesner, R.A. and Hönig, B. (1994) An algorithm to generate low-resolution protein tertiary structures from knowledge of secondary structure, *Proc. Natl. Acad. Sci. U.S.A.* **91**, 5027-5029.

3. Rost, B., Sander, C. and Schneider, R. (1994) Redefining the goals of protein secondary structure prediction, *J. Mol. Biol.* **235**, 13-26.

4. Eisenhaber, F., Persson, B. and Argos, P. (1995) Protein structure prediction: recognition of primary, secondary, and tertiary structural features from amino acid sequence, *Crit. Rev. Biochem. Mol. Biol.*, **30**, 1-94.

5. Rost, B. and Sander, C. (1996) Bridging the protein sequence-structure gap by structure predictions, *Ann. Rev. Biophys. Biomol. Struct.* **25**, 113-136.

6. Methods in Enzymology (1996) R.F. Doolitle (ed), *Computer methods for macromolecular sequence analysis*, Academic press, San Diego, **266**, pp 1-680.

DATA BASES

7. Bairoch, A. and Boeckmann, B. (1994) The SWISS-PROT protein sequence data bank: current status, *Nucleic Acids Res.* **22**, 3578-3580.

8. Bernstein, F.C., Koetzle, T.F., Williams, G.J.B, Meyer, E.F., Brice, M.D., Rodgers, J.R., Kennard, O., Schimanouchi, T. and Tasumi, M. (1977) The protein data bank: a computer based archival file for macromolecular structures, *J. Mol. Biol.* **112**, 535-542.

9. Hobohm, U., Scharf, M., Schneider, R. and Sander, C. (1992) Selection of representative protein data sets, *Protein Science* **1**, 409-417.

10. Pascarella, S. and Argos, P. (1992) A data bank merging related protein structures and sequences, *Prot. Eng.* **5**, 121-137.

11. Holm, L. and Sander, C. (1994) Searching protein structure databases has come of age, *Proteins* **19**, 165-173.

12. Sander, C. and Schneider, R. (1994) The HSSP database of protein structure-sequence alignment, *Nucleic Acids Res.* **22**, 3597-3599.

SECONDARY STRUCTURE DEFINITIONS

13. Levitt, M. and Greer, J. (1977) Automatic identification of secondary structure in globular proteins, *J. Mol. Biol.* **114**, 181-239.

14. Kabsch, W. and Sander, C. (1983) Dictionary of protein secondary structure: pattern recognition of hydrogen bonded and geometrical features, *Biopolymers* **22**, 2577-2637.

15. Pancowska, P., Blazek, M. and Keiderling, T.A. (1992) Relationships between secondary structure fractions for globular proteins. Neural network analysis of crystallographic data sets, *Biochemistry* **31**, 10250-10257.

16. Colloc'h, N., Etchebest, C., Thoreau, E., Henrissat, B. and Mornon J.P. (1993) Comparison of three algorithms for the assignment of secondary structure in proteins: the advantages of a consensus assignment, *Preen Engage* **6**, 377-382.

CHOU-FASMAN METHODS

17. Chou, P.Y. and Fasman, G. (1974) Conformational parameters for amino acids in helical, beta-sheet and random coil regions calculated from proteins, *Biochemistry* **13**, 211-221.

18. Chou, P.Y. and Fasman, G. (1974) Prediction of protein conformation, *Biochemistry* **13**, 222-245.

19. Chou, P.Y. and Fasman, G. (1978) Predictions of the secondary structure of proteins from their amino acid sequence, *Adv. Enzymol.* **47**, 45-148.

20. Argos, P., Hanei, M. and Garavito, R.M. (1978) The Chou-Fasman secondary structure prediction method with an extended data base, *FEBS Lett.* **93**, 19-24.

21. Fasman, G.D. (1989) The developments of the prediction of protein structure, in G.D. Fasman (ed), *Prediction of protein structure and the principles of protein conformation*, Plenum Press, New York and London, pp. 193-316.

22. Prevelige, P. and Fasman, G.D. (1989) Chou-Fasman prediction of the secondary structure of proteins: the Chou-Fasman-Prevelige algorithm, in G.D. Fasman (ed), *Prediction of protein structure and the principles of protein conformation*, Plenum Press, New York and London, pp. 391-416.

23. Chou, P.Y. (1989) Predictions of protein structural classes from amino acid compositions, in G.D. Fasman (ed), *Prediction of protein structure and the principles of protein conformation*, Plenum Press, New York and London, pp. 549-586.

GOR (I, II, III) METHODS

24. Garnier, J., Osguthorpe, D.J. and Robson, B. (1978) Analysis of the accuracy and implications of simple methods for predicting the secondary structure of globular proteins, *J. Mol. Biol.* **120**, 97-120.

25. Gibrat, J-F, Garnier, J. and Robson B. (1987) Further developments of protein secondary structure prediction using information theory. New parameters and consideration of residues pairs, *J. Mol. Biol.* **198**, 425-443.

26. Biou, V., Gibrat, J.F., Levin, J.M., Robson, B. and Garnier, J. (1988) Secondary structure prediction: combination of three different methods, *Prot. Eng.* **2**, 185-191.

27. Robson, B. and Garnier J. (1988) *Introduction to proteins and proteins engineering*, Elsevier, Amsterdam.

28 Garnier, J. and Robson, B. (1989) The GOR-method for predicting secondary structures in proteins, in G.D. Fasman (ed), *Prediction of protein structure and the principles of protein conformation*, Plenum Press, New York and London, pp. 417-465.

29. Garnier, J. (1990) Protein structure prediction, *Biochimie* **72**, 513-524.

30. Garnier, J. and Levin, J.M. (1991) The protein structure code: what is the present status?, *Comput. Appl. Biosci.* **7**, 133-142.

31. Levin, J.M., Pascarella, S., Argos, P. and Garnier, J. (1993) Quantification of secondary prediction improvement using multiple alignments, *Prot. Eng.* **6**, 849-854.

SECONDARY STRUCTURE PREDICTION ACCURACY

32. Kabsch, W. and Sander, C. (1983) How good are predictions of protein secondary structure, *FEBS Lett.* **155**, 179-182.

33. Schultz, G.E. (1988) A critical evaluation of methods for prediction of protein secondary structures, *Ann. Rev. Biophys. Chem.* **17**, 1-21.

HYDROPHOBIC CLUSTER ANALYSIS (HCA)

34. Gaboriaud, C., Bissery, V., Benchetrit, T. and Mornon. J.P. (1987) Hydrophobic cluster analysis: an efficient way to compare and analyse amino acid sequences, *FEBS Lett.* **224**, 149-155.

35. Lemesle-Varloot, L., Henrissat, B., Bissery, V., Morgat, A. and Mornon, J.P. (1990) Hydrophobic cluster analysis: procedures to derive structural and functional information from 2-D representation of protein sequences, *Biochimie* **72**, 555-574.

SPECTROSCOPIC ESTIMATION OF SECONDARY STRUCTURE CONTENTS

VISIBLE NORMAL RAMAN SPECTROSCOPY

36. Alix, A.J.P., Berjot, M. and Marx, J. (1985) Determination of the secondary structure of proteins, in A.J.P. Alix, L. Bernard and M. Manfait (eds), *Spectroscopy of biological molecules*, Wiley, Chichester, pp. 149-154.

37. Berjot, M., Marx, J. and Alix, A.J.P. (1987) Determination of the secondary structure of proteins from the Raman Amide I band : the Reference Intensity Profiles Method, *J. Raman Spectrosc.* **18**, 289-300.

38. Alix, A.J.P., Pedanou, G. and Berjot, M. (1988) Fast determination of the quantitative secondary structure of proteins from Raman Amide I band, *J. Mol. Struct.* **174**, 159-164.

39. Alix, A.J.P., Berjot, M., Marx, M. and Pedanou, G. (1988) Determination of the secondary structure of proteins : from X - ray crystallography to Raman spectroscopy. in E.D. Schmid, F.W. Schneider and F. Siebert (eds), *Spectroscopy of Biological Molecules - New Advances*, Wiley, Chichester, pp 51-56.

40. Bussian, B.N. and Sander, C. (1989) How to determine protein secondary structure in solution by Raman spectroscopy; Practical guide and test case DNA I, *Biochemistry* **28**, 4271-4277.

41. Alix, A.J.P. and Pedanou, G. (1994) Fast quantitative determination of protein secondary structures from Raman Amide I band, in N.T. Yu and X.Y. Li (eds), *XIVth International Conference on Raman Spectroscopy*, John Wiley & Sons, New York, Singapore, pp 122-123.

42. Pedanou, G. K. (1994) Etudes quantitatives de la structure secondaire des protéines: méthodes prédictives et spectroscopiques, Thèse de Doctorat Nouveau Régime de l'Université de Reims Champagne Ardenne, Reims, France, pp. 1-270.

FOURIER TRANSFORM INFRARED SPECTROSCOPY

43. Arrondo, J.L.R., Muga, A., Castredana J. and Goni, F.M. (1993) Quantitative Studies of the structure of proteins in solution by Fourier Transform Infrared Spectroscopy, *Prog. Biophys. Molec. Biol.* **59**, 23-56.

44. Surewicz, W.K., Mantsch, H.H. and Chapman, D. (1993) Determination of protein secondary structure by Fourier Transform Infrared Spectroscopy, *Biochemistry* **32**, 389-394.

45. Jackson, M. and Mantsch, H.H. (1995) The use and misuse of FTIR spectroscopy in the determination of protein structure, *Crit. Rev. Biochem. Mol. Biol.*, **30**, 95-120.

ELECTRONIC AND VIBRATIONAL CIRCULAR DICHROISMS

46. Provencher, S.W. and Glöckner, J. (1981) Estimation of globular protein secondary structure from circular dichroism, *Biochemistry* **20**, 33-37.

47. Provencher, S.W. (1982) CONTIN version users manual, *Technical Report EMBL* **DAO5**, Heidelberg.

48. Manavalan, P. and Johnson, W.C.Jr. (1983) Sensitivity of circular dichroism to protein tertiary structure class, *Nature* **305**, 831-832.

148

49. Johnson, W.C.Jr. (1990) Protein secondary structure and circular dichroism: a practical guide, *Proteins* **7**, 205-214.

50. Sreerama, N. and Woody, R.W. (1993) A self-consistent method for the analysis of protein secondary structure from circular dichroism, *Anal. Biochem.* **209**, 32-44.

51. Sreerama, N. and Woody, R.W. (1994) Protein secondary structure from circular dichroism spectroscopy, *J. Mol. Biol.* **242**, 497-507.

52. Geourgon, C. and Deleage, G. (1994) SOPM: a self-optimized method for protein secondary structure prediction, *Prot. Eng.* **7**, 157-164.

53. Greenfield; N.J. (1996) Review: methods to estimate the conformation of proteins and polypeptides from circular dichroism data, *Anal. Biochem.* **235**, 1-10.

54. Baumruk, V., Pancoska, P. and Keiderling T.A. (1996) Predictions of secondary structure using statistical analyses of electronic and vibrational circular dichroism and Fourier transform infrared spectra of proteins in H_2O, *J. Mol. Biol.* **259**, 774-791.

LINK METHODS

55. Alix, A.J.P., (1993) Molecular modeling of protein secondary structure: the missing link closing the gap between prediction methods and optical spectroscopic data, in J. Anastassopoulou , N. Fotopoulos and T. Theophanides (eds) *Fifth international conference on the spectroscopy of biological molecules*, Kluwer **Academic Publishers, Dordrecht, pp 13-16, 1993**

56. Alix, A.J.P., Goulyaev, D.I., Efremov, R.G. (1995) Global and local secondary structures of globular proteins: linking spectroscopies and predictions, in J.C. Merlin. S. Turrell and J.P. Huvenne (eds), *Spectroscopy of biological molecules*, Kluwer Academic Publishers, Dordrecht, pp 89-90.

EPITOPES

57. Hopp, T.P. and Woods, K.R. (1981) Prediction of protein antigenic determinants from amino acid sequences, *Proc. Natl. Acad. Sci. U.S.A.* **78**, 3824-3828.

58. Hopp, T.P. (1993) Retrospective: 12 years of antigenic determinant predictions and more, *Peptide Res.* **6**, 183-190.

59. Kyte, J. and Doolitle, R.F. (1982) A simple method for displaying the hydrophobic character of a protein, *J. Mol. Biol.* **157**, 105-132.

60. Karplus, P.A. and Schultz, G. (1985) Prediction of chain flexibility in proteins, *Naturwissenschaften* **72**, 212-213.

61. Vihinen, M., Torkkila, E. and Riikonen, P. (1994) Accuracy of protein flexibility predictions, *Prot.: Struct. Funct. and Genet.* **19**, 141-149.

62. Emini, E., Hughes, J.V., Perlow, D.S. and Boger, J. (1985) Induction of hepatis A virus-neutralizing antibody by a virus-specific synthetic peptide, *J. Virol.* **55**, 836-839.

63. Parker, J.M.R., Guo, D. and Hodges, R.S. (1986) New hydrophilic scale derived from high-performance liquid chromatography peptide retention data: correlation of predicted surface residues with antigenicity and X-ray-derived accessible sites, *Biochemistry* **25**, 5426-5432.

64. Jameson, B.A.. and Wolf, H. (1988) The antigenic index: a novel algorithm for predicting antigenic determinants, *Comput. Appl. Biosci.* **4**, 181-186.

65. Van Regenmortel, M.H.V. and Daney de Marcillac, G. (1988) An assessment of prediction methods for locating continuous epitopes in proteins, *Immunol. Lett.* **17**, 95-108.

66. Thornton, J.M., Barlow, D.J. and Edwards, M.S. (1989) Antigenic recognition, in R.W. Graham (ed), *Computer-aided molecular design*, IBC Technical Services Ltd, London, pp. 187-196.

67. Janin, J., Wodak, S.J., Levitt, M. and Maigret, M. (1978) Conformation of amino acid side-chains in proteins. *J. Mol. Biol.* **125**, 357-386.

68. Janin, J. (1979) Surface and inside volumes in globular proteins, *Nature* **277**, 491-492.

69. Cornette, J.L. Cease, K.B., Margalit, H., Spouge, J.L., Berzofsky, J.A. and DeLisi C. (1987) Hydrophobic scales and computational techniques for detecting amphiphatic structures in proteins, *J. Mol. Biol.* **195**, 659-685.

70. Rose, G.D. and Wolfenden, R. (1993) Hydrogen bonding, hydrophobicity, packing and protein folding, *Annu. Rev. Biophys. Biomol. Struct.* **22**, 381-415.

71. Po₁ ¡uswamy, P.K. (1993) Hydrophobic characteristics of folded proteins, *Prog. Biophys. Molec. Biol.* **59**, 57-103.

MOLECULAR HYDROPHOBICITY POTENTIAL

72. Efremov, R.G. and Alix, A.J.P. (1993) Environmental characteristics of residues in protein : Three-dimensional Molecular Hydrophobicity Potential Approach, *J. Biomol. Struct. Dynamics* **11**, 483-507.

73. Efremov, R.G., Gulyaev, D.I., Feofanov, A.V., Vergoten, G. and Alix, A.J.P. (1993) Environmental characteristic of residues in proteins: UV resonance Raman spectroscopy and 3D molecular hydrophobicity potential approach, in J. Anastassopoulou , N. Fotopoulos and T. Theophanides (eds) *Fifth international conference on the spectroscopy of biological molecules*, Kluwer Academic Publishers, Dordrecht, pp 13-16.

74. Efremov, R.G., Golovanov, A.P., Vergoten, G., Alix, A.J.P., Tsetlin, V.I. and Arseniev, A.S. (1995) Detailed assessment of spatial hydrophobic and electrostatic properties of 2D NMR-derived models of neurotoxin II, *J. Biomol. Struct. Dyn.* **12**, 971-991.

75. Efremov, R., Golovanov, A., Alix, A.J.P., Tsetlin, V., Arseniev, A. and Vergoten, G. (1995) Detailed assessment of spatial hydrophobic and electrostatic properties of 2D NMR - derived models of neurotoxin II, in J.C. Merlin, S. Turrell and J.P. Huvenne (eds), *Spectroscopy of biological molecules*, Kluwer Academic Publishers Dordrecht, pp 99-100.

TROPOELASTIN, ELASTIN : EPITOPES AND PSEUDO 3-D MOLECULAR MODEL

76. Debelle, L., Wei, S.M., Jacob, M.P., Hornebeck, W. and Alix A.J.P. (1992) Predictions of the secondary structure and antigenicity of human and bovine tropoelastins, *Eur. Biophys. J.* **21**, 321-329.

77. Debelle, L. and Alix, A.J.P. (1995) Optical spectroscopic determination of bovine tropoelastin molecular model, *J. Mol. Struct.* **348**, 321-324.

78. Debelle, L., Alix, A.J.P., Berjot, M., Jacob, M.P., Huvenne, J.P., Sombret, B. and Legrand, P. (1995) Bovine elastin and κ-elastin secondary structure determination by optical spectroscopies, *J. Biol. Chem.* **270**, 26099-26103.

79. Alix, A.J.P., Dauchez, M. and Debelle, L. (1995) A critical disulfide bridge in elastins : modeling from spectroscopic data to 3D structures, in J.C. Merlin, S. Turrell and J.P. Huvenne (eds), *Spectroscopy of biological molecules*, Kluwer Academic Publishers Dordrecht, pp 97-98.

80. Debelle, L. (1995) Etudes biophysiques des structures et modifications conformationnelles des tropoélastines et élastines bovines et humaines, Thèse de Doctorat Nouveau Régime de l'Université de Reims Champagne Ardenne, Reims, France, pp. 1-275.

r-ELAFIN MOLECULAR MODEL

81. Alix, A.J.P., Dauchez, M., Berjot, M., Charton, F., Thirion, C. and Jacquot, J. (1995) Optical spectroscopies and molecular modeling of r-elafin: an elastase-specific inhibitor, in J.C. Merlin. S.

Turrell and J.P. Huvenne (eds), *Spectroscopy of biological molecules*, Kluwer Academic Publishers, Dordrecht, pp 93-94.

82. Francart, C.. Dauchez. M., Alix. A.J.P. and Lippens, G. (1997) Solution structure of r-elafin, a specific inhibitor of elastase, *J. Mol. Biol.* **accepted.**

83. Tsunemi, M., Matsuura, Y., Sakakibara, S. and Katsube, Y. (1996) Crystal structure of an elastase-specific inhibitor elafin complexed with porcine pancretic elastase at 1.9 Å resolution, *Biochemistry* **35,** 11570-11576

IRON SUPER OXIDE DISMUTASE

84. Hassan, H.M. (1989) Microbial superoxide dismutases, *Advances in genetics* **26,** 65-67.

85. Lah, M.S., Dixon, M.M., Partridge. K.A., Stallings, W.C., Fee, J.A. and Ludwig, M.L. (1995) Structure-function in *Escherichia coli* iron superoxide dismutases: comparisons with the manganese enzyme from *thermus thermophilus, Biochemistry* **34,** 1646-1660.

86. Dauchez, M., Alix, A.J.P. and Dive, D. (1995) Three dimensional molecular modeling of iron-SOD of *Plasmodium falciparum, Second European Congress of prositology and Eight European Conference on ciliate biology,* Clermont-Ferrand, France, 25-5-1995.

MISCELLANEOUS

87. Kraulis, P. (1991) MOLSCRIPT: a program to produce detailed and schematic plots of protein structures, *J. Appl. Crystallogr.* **24,** 946-950.

88. Laskowski, R.A., MacArthur, M.W., Moss, D.S. and Thornton, J.M. (1993) PROCHECK, a program to check the stereochemical quality of protein structure, *J. Appl. Crystallogr.* **26,** 283-291.

PHYSICOCHEMICAL PROPERTIES IN VACUO AND IN SOLUTION OF SOME MOLECULES WITH BIOLOGICAL SIGNIFICANCE FROM DENSITY FUNCTIONAL COMPUTATIONS

T. MARINO, T. MINEVA, N. RUSSO AND M. TOSCANO

Dipartimento di Chimica, Università della Calabria, I-87030 Arcavacata di Rende (CS), Italy

ABSTRACT. A number of molecules with biological significance have been selected to show the potentialities and the limits of gradient corrected density functional theory in predicting the physicochemical properties in vacuo and in solution for these class of compounds.
The studied systems are:
— alanine
— 1,4-dihydronicotinamide
— 4-thiouracil
Results for the considered properties are in agreement with available experimental data and competitive with those obtained by the most sophisticated post-Hartree-Fock methods. The speed of the computations and the reliability of the results allow us to suggest the density functional methods as tools for the study of large and complex systems such as the biological molecules.

1. Introduction

The rapid development of computer power and the availability of very efficient algorithms in the sophisticated post-Hartree-Fock approaches (i.e. multireference configuration interaction, CI, and coupled cluster, CC) coupled with large basis sets are able to

151

G. Vergoten and T. Theophanides (eds.), Biomolecular Structure and Dynamics, 151–178.

provide results of good chemical accuracy for many properties (i.e. conformational equilibria, spectroscopic constants, binding energies and thermochemical parameters) regarding a wide range of chemical systems [1,2]. For these methods, the computational time increase with the number of electron (N) and with basis orbitals and is very high (in current practice it grows up about as N^7 or N^8), so the applications in the field of molecules with biological and pharmaceutical significance is prohibitive (the computations for systems with $N \geq 10$ are difficult to perform). In' addition, the interpretation of the results in terms of chemical concepts is not simple and requires complicate and accurate analysis.

In the last five years methods rooted in the so-called density functional (DF) theory, and in particular those based on the Kohn-Sham approach, have grown considerably [3-14]. The development of generalized gradient approximations (GGA) [3,4] has significantly improved the results obtained by the local density (LDA) and its spin polarized version (LSD) [15]. The applications in many field of modern chemistry show the reliability of GGA both for closed and open shell systems along the whole periodic table [6-14].

The main advantage of DF theory consist in the use of density $\rho(r)$ as the basic variational object instead of the wave function. This is an enormous simplification of the many-body problem. In fact, DF methods are able to include a large amount of correlation using the same, often less, computational resources required by the Hartree-Fock ones. Furthermore the results coming from the single-determinant DF computations are easy to interpret.

Nowadays, with the available computer codes (i.e. Gaussian 94 [16], CADPAC [17], DGAUSS [18], DMOL [19], DeMon [20], ADF [21]) in which modern density functional approaches are implemented, it is possible to obtain:

— geometrical structures with an accuracy of about 0.002 Å for the interatomic distances, 1 degree for the valence angles and few degrees for the dihedral angles;

— vibrational frequencies reasonable (the error is less than 10%);

— binding energies with an error (typically $\geq 5\%$) acceptable for chemical applications;

— EPR, NMR, UPS and XPS spectral properties with an accuracy that often falls in the experimental error range

Of course not all chemical problems have been solved until now and the applications of DF methods in systems containing hydrogen bonds or in Van der Waals complexes do not give results comparable with the more sophisticated high level post-Hartree-Fock approaches. Recent development in the so-called hybrid functionals is very promising and the next few years can become significative to increase the reliability of DF computations in biology, chemistry and physics. Notwithstanding the massive use of DF methods in many field of modern chemistry and physics, few applications concern the biomolecular systems [22-28].

In this contribution, we will show that the linear combination of gaussian-type orbitals-density functional method using the gradient corrected functionals for the exchange-correlation contribution is able to predict correctly, for a series of systems (alanine, uracil, 5-fluorouracil, 4-thiouracil, 1,4-dihydronicotinamide), several physicochemical aspects and parameters (conformational behaviour and tautomerization processes, proton affinities, ionization potentials and electron affinities, hydration free energies, dipole and quadrupole moments, diamagnetic susceptibility, proton transfer path) that are of fundamental importance for understanding better the functions of the biomolecules.

2. Brief resume of density functional theory and computational details

Modern density functional theory is based on the fundamental works of Hohenberg and Kohn [29] and Kohn and Sham [30] by showing that the electron density could be used as fundamental quantity for developing a rigorous many-body theory applicable to the ground state of any kind of atomic or molecular system. Comparison with the more familiar Hartree-Fock (HF) method is useful to understand better the difference between the two approaches and to appreciate the DF concepts. In DF theory the total energy of the system depends on the position of the atoms (R) and on the total electron density, while in HF the total energy is expressed as a function of R and of the total wave function:

HF	**DF**
$E = E[\Psi, R]$	$E = E[\rho, R]$

The expression of total energy in the two methods is consequently different. In fact, in HF the exact ground state total energy is given as an expectation value of an exact (non-relativistic) Hamiltonian while in DF the Hamiltonian is decomposed in three terms (Eq. 1) (kinetic ($T[\rho]$), electrostatic or Coulomb ($U[\rho]$) and many-body ($E_{XC}[\rho]$)) with the latter not exactly known. The kinetic term correspond to the kinetic energy of a system of non-interacting particles that yield the same density as the original electron system. The many-body term includes the exchange and correlation contributions and the correlation part of the kinetic energy:

$$\begin{matrix} \mathbf{HF} & \mathbf{DF} \end{matrix}$$

$$E = \int \Psi^* \left[\sum_i h_i + \sum_{i>j} \frac{1}{r_{ij}} \right] \Psi \, d\tau \qquad E = T[\rho] + U[\rho] + E_{XC}[\rho] \qquad (1)$$

HF method uses the Slater determinant as an approximation for the total wave function. In DF theory, the total density is decomposed into single-particle densities originated from one-particle wave function:

$$\begin{matrix} \mathbf{HF} & \mathbf{DF} \end{matrix}$$

$$\Psi = \left| \psi_1(1), \psi_2(2) \ldots \ldots \psi_n(n) \right| \qquad \rho(\mathbf{r}) = \sum_{occ} \left| \psi_i(\mathbf{r}) \right|^2 \qquad (2)$$

HF require that upon the variation of the Ψ the total energy assumes a minimum while in DF the minimum is found upon the variation of total electron density [3]:

$$\begin{matrix} \mathbf{HF} & \mathbf{DF} \end{matrix}$$

$$\frac{\partial E}{\partial \Psi} = 0 \qquad\qquad \frac{\partial E}{\partial \rho} = 0 \qquad (3)$$

Applying the variational principles on the Slater determinant the Hartree-Fock one-particle eigenvalue equations (eq. 4) are obtained:

HF

$$\left[-\tfrac{1}{2}\nabla^2 + V_C(\mathbf{r}) + \mu_X^i(\mathbf{r})\right]\psi_i = \varepsilon_i\psi_i \qquad (4)$$

In DF, the variation of the total electron density with the associated total energy minimum leads to conditions for the Kohn-Sham effective one-particle Schrödinger equations:

DF

$$\left[-\tfrac{1}{2}\nabla^2 + V_C(\mathbf{r}) + \mu_{XC}(\mathbf{r})\right]\psi_i = \varepsilon_i\psi_i \qquad (5)$$

The equations 4 and 5 are formally similar. The main difference concerns the μ term. In HF method μ represents the exchange effects and depends on the ψ_i, while in DF theory this term contains all the many-body effects and is the unknown part of the energy functional. The kinetic energy of the noninteracting particles and the coulomb potential operator are the same in HF and DF theories.

The DF hamiltonian (eq. 5) is an effective one-electron operator that contains the kinetic, the coulomb V_C (including the electron-electron, electron-nuclei and nuclei-nuclei electrostatic interactions) and the exchange-correlation potential terms. Practically, the HF method solves the exact hamiltonian with an approximate many-body wavefunction, while Kohn-Sham equations solve an approximate many-body Hamiltonian with exact wavefunctions. For this reason DF method provides, in a unic set of calculation, the exchange and correlation contributions. On the contrary, to take into account these effects with the HF method, systematic improvements in the form of the many-body wavefunction (e.g. by configuration interaction expansions) are necessary.

In DF theory the convergence to the exact solution of the Schrödinger equation is obtained, , by improving the description of the exchange-correlation potential that is defined as:

$$\mu_{XC}(\mathbf{r}) = \frac{\partial E_{XC}[\rho]}{\partial\rho(\mathbf{r})} + C \qquad (6)$$

The functional derivative in eq. 6 is determined to within a constant C, since the constrain for fixed N is:

$$\int C\partial\rho(\mathbf{r})d\mathbf{r} = 0 \quad \text{for any } C$$

Today, the most common approximation to the exchange-correlation energy is the LDA or its spin-polarized generalization.

$$E_{XC}[\rho] = \int \rho(\mathbf{r})\, \varepsilon_{XC}[\rho(\mathbf{r})]d\mathbf{r} \tag{7}$$

where the $\varepsilon_{XC}[\rho]$ is the exchange-correlation energy per electron in an interacting electrons system of constant density ρ.

Different LDA or LSD functional have been proposed and applied in finite electron systems [15]. Notwithstanding the local nature of these functionals, good results have been obtained in the reproduction of many spectroscopic parameters in a wide range of systems with chemical and physical significance. Generally the LDA or the LSD overestimate the exchange-correlation contributions and conseguently the energetic parameters. The introduction of the so-called nonlocal corrections improves strongly the results.

Many nonlocal correction schemes have been implemented in the Kohn-Sham method. One of the most used is the gradient expansion approximation (GEA) [2]. The GEA consists in the truncating of the asymptotic Taylor series of Exc about ρ (r) at the lowest finite order beyond the LSD:

$$E_{XC}[\rho] = E_{XC}^{LDA} + \int d\mathbf{r}\, C_{XC}(\rho(\mathbf{r}))[\nabla(\mathbf{r})]^2 \tag{8}$$

where the coefficient C_{XC} is initially assumed independent on the density gradient itself.

In all our computations the exchange and correlation potentials proposed by Perdew and Wang (PW) [31] and Perdew (P) [32] respectively have been used. Their expressions are:

$$E_X^{PW} = -\frac{3}{4}\left(\frac{3}{\pi}\right)^{1/3} \int d\mathbf{r}\, \rho^{4/3}(\mathbf{r}) F_X^{PW}(s(\mathbf{r}))$$

$$\tag{9}$$

$$F_X^{PW}(s) = \left(1 + 1.296 s^2 + 14 s^4 + 0.2 s^6\right)^{1/15}$$

and

$$E_C^P = E_C^{LSD} + \int d^{-1}(\xi)\exp- \Phi C_C(\rho)\frac{|\nabla\rho|^2}{\rho^{4/3}}\, dr$$

(10)

$$\Phi = 1.745f\left[C_C(\infty)\big/C_C(\rho)\right]\frac{|\nabla\rho|}{\rho^{7/6}}$$

Where $d(\xi)$ is a spin-interpolating function and $C_C(\rho)$ is the beyond-RPA gradient expansion coefficient for the correlation energy.

The local functional used in this study is that proposed by Vosko, Wilk and Nusair (VWN) [33].

For taking into account the solvent effect, the self-consistent reaction field (SCRF) approach has been used [34].

In this method, the interaction free energy is given by:

$$\Delta G = -\frac{1}{2}\sum_{l=0}^{\infty}\sum_{m=-1}^{1} g_l^m \langle M_l^m\rangle\langle M_l^m\rangle$$

(11)

The coefficients g_l^m, called reaction field factors, have analytical expressions depending only on the cavity shape (sphere [35], spheroid [36,37], ellipsoid [38,39]). If the shape is assumed to be spherical the g_l^m is

$$g_l = \frac{(l+1)(\varepsilon-1)}{1+\varepsilon(l+1)}\frac{1}{a_0^{2l+1}},$$

(12)

$$M_l^m = \left(\frac{4\pi}{2l+1}\right)^{\frac{1}{2}}\int \rho r^l Y_l^m(\delta\varphi)$$

(13)

where $Y_l^m(\delta\varphi)$ are the spherical harmonics, a_0 is the cavity radius and $\langle M_l^m\rangle$ are the multipolar values.

In order to account for the polarization of the electronic distribution of the solute due to the electrostatic potential of the liquid the KS equations must include the perturbation term $\Phi(r)$. In the scheme one minimizes the energy of the solute plus the free energy of solvation:

$$E^s = E^o - \frac{1}{2} \sum_{lm} g_l \langle M_l^m \rangle \langle M_l^m \rangle , \tag{14}$$

In this context the Fock matrix elements of the system become

$$F_{\mu\nu} = F_{\mu\nu}^o + \sum_{lm} \langle R_l^m \rangle \langle \varphi_\mu | M_l^m | \varphi_\nu \rangle , \tag{15}$$

where

$$\langle R_l^m \rangle = \sum_{l=o}^{\infty} \sum_{m=-l}^{l} g_l \langle M_l^m \rangle \tag{16}$$

Different strategies have been proposed for the practical solution of the continuum dielectric reaction field problem. We have used the Onsager approach [40] in which only the dipole term in the multipole expansion is used [41]. The solvation free energy is now given by

$$\Delta G^1 = -\frac{1}{2} g_1 \mu^2$$

Full energy optimizations have been performed by using the VWN+PWP gradient corrected exchange-correlation potentials (hereafter indicated as NLSD) for all the presented systems treated in vacuo. Then the obtained geometries have been used for single point computations in water solvent ($\varepsilon=80$).

The employed orbital basis sets are of TZVP quality and have the following expansion [42]:

H   (41/1*)
C, N, O, F   (7111/411/1*)
S   (311/211/1)

The corresponding auxiliary basis set for fitting the coulomb and the exchange-correlation potentials has the form [42]:

H   (5,1;5,1)
C, N, O, F   (4,4;4,4)
S   (2,3;2,3)

For sulphur atom a S(+6) model core potential that allows to treat explicitly the $3s^2 3p^4$ valence electrons has been used. The high quality of the employed orbital basis set exhibits a small basis set superposition error since the core orbitals are represented well.

The adiabatic ionization potential and electron affinity have been computed by the ΔSCF procedure.

3.   Selected   applications

The applications that we have chosen concern molecules with great biological significance and give indications on the state-of-the-art of the theoretical methods.

3.1. ALANINE PROPERTIES

The importance of the amino acid systems in biochemistry is documented well in literature. For computational chemistry these compounds are an attractive target because contain a variety of intramolecular interactions that can be treated well only if the correlation effects are considered. In addition, theoretical studies are useful to predict unknown parameters that are difficult to measure because many amino acids suffer decomposition under particular experimental conditions. Different ab-initio studies on the α-amino acids exist in the recent literature, but the major part have been performed without the introduction of correlation contributions. The most studied system is the glycine for which accurate HF [43], CI [44] and DF [24] studies have been published recently . As a part of a systematic DF investigation on the physicochemical properties of the most common 20 α-amino acids, we report here the results for alanine. In a range of few kcal/mol ten conformations have been characterized at HF and MP2 levels of theory [45].

In this study we have considered the first two low energy conformations found by Gronert and O'Hair [45] (structures **a** and **b** of Figure 1).

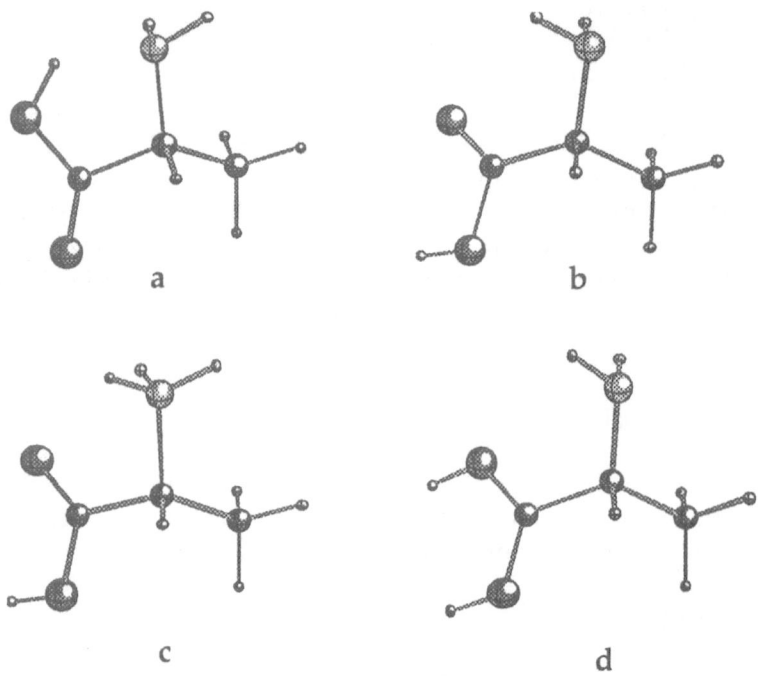

a   b

c   d

Figure 1. NLSD optimized structures of alanine conformations **a** and **b** and protonated alanine on nitrogen (**c**) and oxygen (**d**).

The calculations indicate that the more stable structure is **a** with **b** that lies at 1.7 kcal/mol (Table 1). At HF and MP2 levels structure **b** is the minimum and **a** is higher in energy by 2.9 and 0.9 kcal/mol respectively. A recent DF investigation employing the Becke functional [46] for the exchange and the Perdew [47] for the correlation gives the **a** conformer as the most stable and the **b** one at about 0.7 kcal/mole [47].

Because of the very small energy difference between the two conformations the question can be considered still open. Further investigations at both HF (using larger configuration interactions) and DF (using different exchange-correlation functionals) levels are necessary to solve the question. Comparison between the DF and HF geometrical parameters evidences many similarities. In table 1 are

Table 1. Theoretical and experimental physicochemical parameters for the conformations **a** and **b** (in parenthesis) of alanine.

Parameter	NLSD	HF [MP2]	Exp.
ΔE(a-b) (kcal/mol)	1.67	-2.9[a] [-0.9]	/
PA(N) (kcal/mol)	216.0	225.8[b]	215.8[c], 222.1[d] 219.2[e], 213.9[f]
PA(O) (kcal/mol)	183.8	/	/
GB(N) (kcal/mol)	208.1	/	207.4[c]
ΔH^0_{acid} (kcal/mol)	338.2	344.0[g]	341.0[g]
IP (eV)	9.01	8.46[b]	9.8[h]
AE (eV)	1.24	/	/
ΔH_{hydr} (kcal/mol)	22.5	/	23.2[i]
μ (vacuo) Debye	5.79 (1.22)	5.69 (1.50)[a]	5.1 (1.8)[l]
Q_{xx} (10^{-26} esu cm^2)	-1.95 (-5.76)	/	/
Q_{yy} (10^{-26} esu cm^2)	-3.61 (1.74)	/	/
Q_{zz} (10^{-26} esu cm^2)	5.65 (4.02)	/	/
$<r^2>$ (10^{-16} cm^2)	1627 (1669)	/	/
χ_{xx} (10^{-16} ergs/G^{2}mol)	-329.7 (-321.2)	/	/
χ_{yy} (10^{-16} ergs/G^{2}mol)	-440.1 (-615.0)	/	/
χ_{zz} (10^{-16} ergs/G^{2}mol)	-612.4 (-476.5)	/	/

a) from. ref. 45; b) from. ref. 49; c)from. ref. 55; d) from. ref. 56; e) from. ref. 57; f) from. ref. 58; g) from. ref. 62; h) from. ref. 48; i) from. ref. 63; l) from. ref. 51.

reported other significant spectroscopic parameters for the two considered conformations.

The adiabatic ionization potential (IP) is found to be, for the ground state, 9.01 eV in agreement with the experimental result of 9.8 eV [48]. HF method gives an IP value (8.46 eV) [49] that appears underestimated with respect to experimental data. From Mulliken population analysis it is shown that the ionized electron arises from the amino group.

The electron affinity (EA) value reported here for the first time, is 1.24 eV. Previous DF determinations of AE in different systems have shown the reliability of this method in reproduction of this parameter [50].

In gas phase, a dipole moment of 5.79 and 1.22 Debye for conformers **a** and **b** respectively has been found. The experimental counterparts are 5.1 and 1.8 Debye [51]. Our value agrees also with HF results of 5.7 Debye for **a** and 1.50 Debye for **b** obtained employing double-zeta quality of orbital basis sets [45]. Dipole moment of the zwitterion form is expected to be much larger than that of the neutral species. In fact, we found 12.23 Debye. A similar trend has been obtained in a previous HF study (13.45 Debye) [49].

To our knowledge, no experimental or theoretical information is available regarding the quadrupole moments (Q), second moments ($<r^2>$) and diamagnetic susceptibility ($<\chi>$) values of alanine. Our estimations for the two alanine conformers are reported in table 1. Because recent papers [52-54] have shown the reliability of DF methods in the reproduction of these constants, we can regard with confidence these results that can be useful for future experimental investigations.

The mechanism of protonation of amino acids is currently of interest due to the belief that protonated species are important intermediates for many biochemical reactions. Gas-phase proton affinity (PA) and basicity (GB) of the 20 most common α-amino acids have been measured employing several mass spectrometry techniques [55-58]. Results indicate different PA and GB scales. In this contest it is interesting to study theoretically the protonation process. The reliability of the DF methods in the evaluation of PA has been previously confirmed for a series of organic and biological systems [59,60]. Following the procedure used in our study on glycine [61], we have performed calculations also for alanine. We have considered the possibility that H^+ can attach the nitrogen or oxygen atom (see structures **c** and **d** of Figure 1). From table 1 it is evident that the protonation process on nitrogen atom is strongly favoured (32.2 kcal/mol of difference). At 298 K our calculated PA is 216.0 kcal/mol and compare well with the most recent experimental results [55-58]. The HF data (225.8 kcal/mol) [49] appears to be overestimated with respect to all the proposed experimental values. The NLSD PA value for glycine is 214.2 kcal/mol [61]. The larger value of alanine is probable due to the presence, in this system, of a methyl group that makes the nitrogen atom more attractive for the proton. Our GB value of 208.1 kcal/mol is in agreement with that proposed on the basis of mass spectrometry measurement [55].

Another significant parameter for the α-amino acids is the gas-phase acidity because its knowledge can give insight in the understanding of their reactivity. No DF studies exists on this matter. We report here the first evaluation of gas-phase basicity employing a density functional method.
Considering the following process

$$NH_2CHCH_3COOH \text{ ------>} NH_2CHCH_3COO^- + H^+$$

the computed alanine gas phase acidity at 0 K (ΔH^0_{acid}) is 338.22 kcal/mol. The experimental value proposed by O'Hair, Bowie and Gronert on the basis of the kinetic method [62] is 341.01 kcal/mol. The agreement is also good with the HF result of 344.01 kcal/mol [62].

The solution chemistry of the amino acids is rich and significant from a biological point of view. It is well known that alanine, as well as other amino acids, in water solution exists in the zwitterionic form . Our computation in gas-phase gives the neutral form more favoured, but, in solution the zwitterionic form appears to be more stable (the neutral form has positive heat of solvation).

The heat of hydratation (ΔH_{hydr}) is found to be 22.5 kcal/mol in agreement with the experimental value [63] (23.2 kcal/mol).

Proton transfer (Figure 2) is a fundamental process in the chemistry of peptides [64]. In order to have insight on this mechanism, we have calculated, for the alanine, the reaction path of the proton shift from oxygen to nitrogen . It is evident that the H-transfer occurs without energy barrier. A previous study [65] on glycine gives the same results at both DFT and MP2 levels of computation.

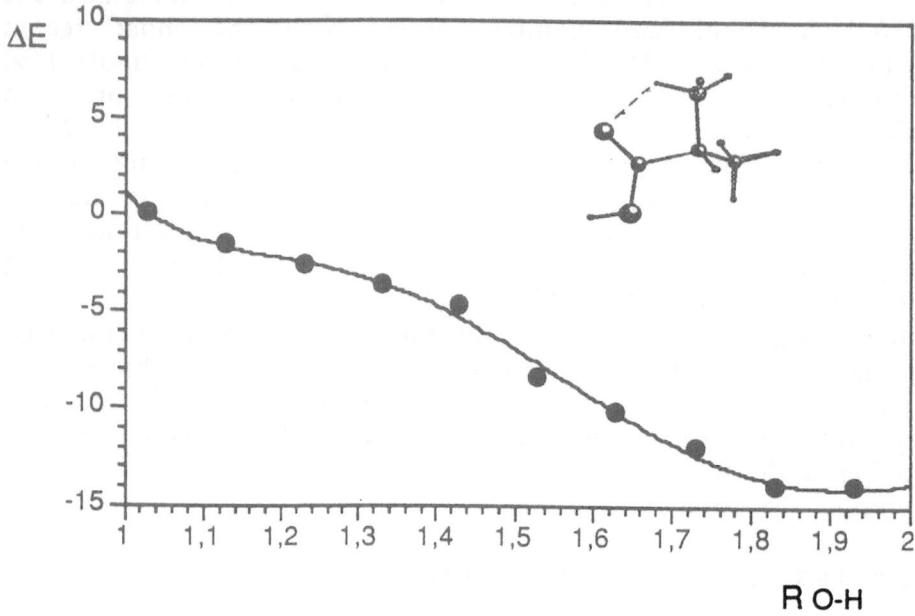

Figure 2. Relative energy changes along the proton transfer path in alanine.

3.2. 1,4-DIHYDRONICOTINAMIDE AS A MODEL FOR NADH SYSTEM

The NADH/NAD+ coenzyme dehydrogenase is involved in many biological oxidation reactions (e.g. catalytic conversion of alcohols to aldehides or ketones) [66]. In a given dehydrogenase only one of the two hydrogens of C4 atom of the 1,4-dihydronicotinamide (Figure 3) ring of NADH is transferred during the reaction and the transferring process is highly stereospecific [67]. In fact, two classes of dehydrogenase called A-specific and B-specific have been observed.   A recent review on this subject, lists 156 A- and 121 B-specific dehydrogenases [67]. Details about the elementary mechanism are not completely known, but there are many evidences that allow us to emphatize the importance of the conformational behaviour of both ring and amino moiety of the 1,4-dihydronicotinamide. It has been suggested that the amide group in the NADH-enzyme complex is twisted out of the plane of the ring in

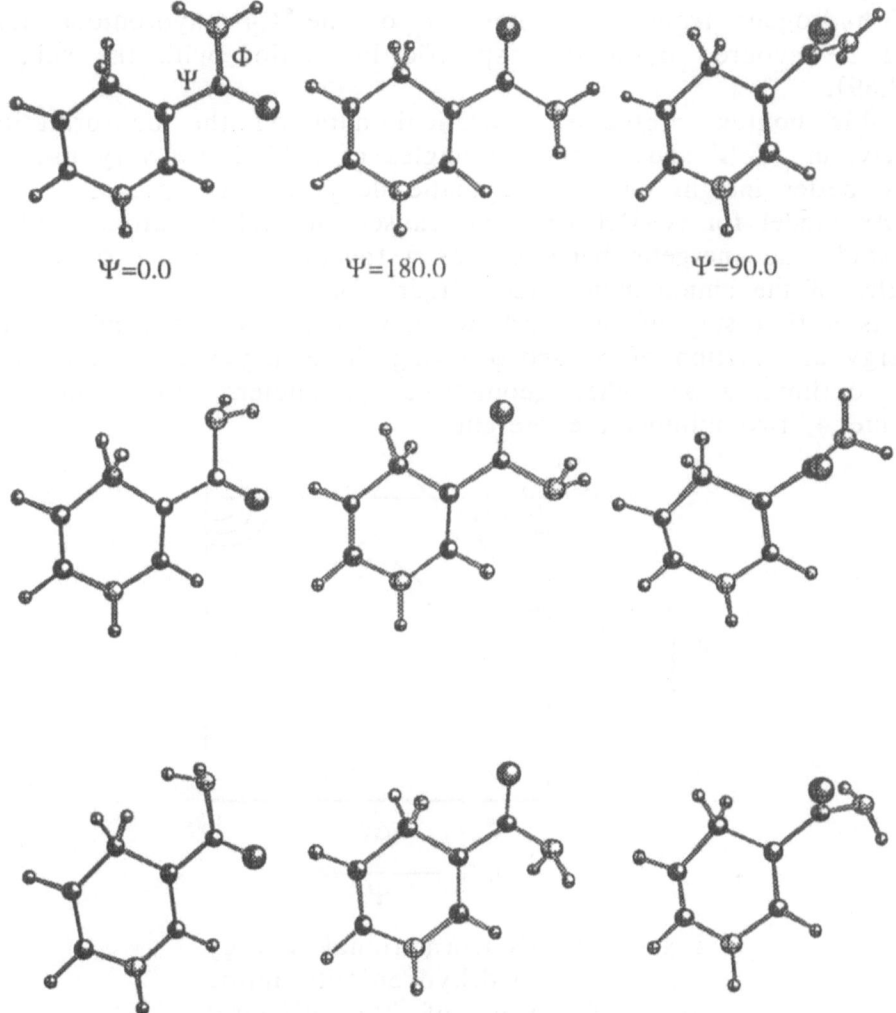

Figure 3. The optimized conformers of 1,4-hydroxynicotinamide.

the transition state of the reduction and that the migration of one of the hydrogens from the 4 position of the 1,4-dihydronicotinamide ring is favoured by a stereospecific interaction with the substrate [68,69].

In this context, reliable theoretical data on the conformational behaviour of Ψ and ϕ dihedral angles of NADH are very useful to give better insight on the enzymatic dehydrogenase process.

As model for NADH we have chosen the 1,4-dihydronicotinamide to study the energetic behaviour as a function of Ψ and ϕ torsional angles of the amide moiety (see Figure 3).

As a first step of the work we have explored the conformational energy as function of Ψ and ϕ fixing these angles at a given value and optimizing all other geometrical parameters. As is shown in Figure 4, two minima are present.

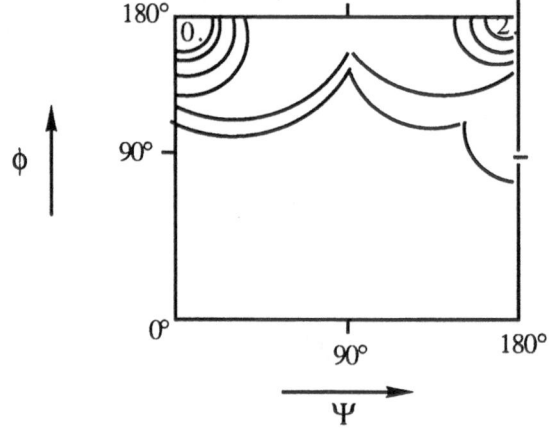

Figure 4. Conformational energy map of 1,4-dihydronicotinamide as a function of the dihedral angles Ψ and ϕ. Each line, around the zero, represent isoenergetic points in a range of 2 kcal/mol.

The absolute minimum is characterized by the $\Psi = 0°$ and $\phi = 180°$ values and the second one, that lies at about 2 kcal/mol, has $\Psi = 180°$ and $\phi = 180°$. The other regions of the conformational space are higher than 10 kcal/mol above the absolute minimum. The conformational energy map indicate that the preferred path for the

Ψ rotation is that yields to the -NH₂ group planarity. So, the rotation around Ψ with ф at 0.0 (the three first structures of Figure 3) has been examined in detail. Full optimization of cis and trans conformers together with that in which Ψ is fixed at 90° (TS) have been performed. The reaction path is depicted in Figure 5.

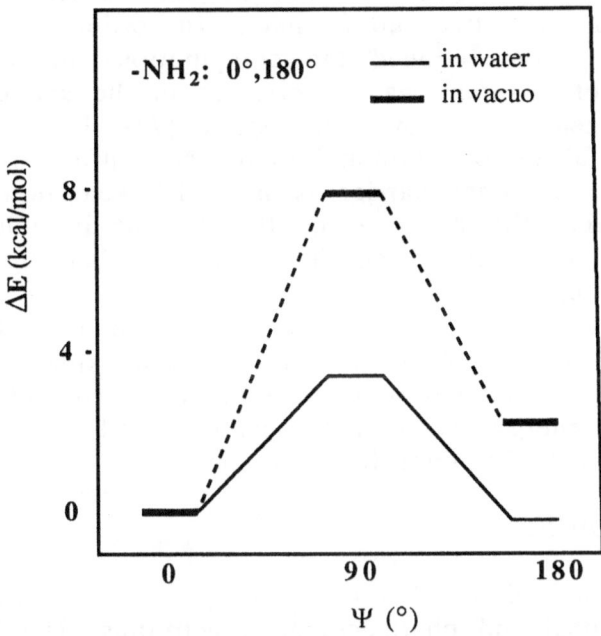

Figure 5. Relative energies in vacuo and in water for the 1,4-dihydronicotinamide. The zero corresponds to the energy of cis conformer.

In gas phase the cis conformation is preferred over the trans one by 2.11 kcal/mol. This result is consistent with those obtained at HF and MP2 levels that prefer the cis form over the trans one by 2.6 and 0.9 kcal/mol respectively [70]. The conformation at Ψ=90° is not a minimum but a transition state as revealed by the presence of one imaginary frequency in the computed vibrational spectrum.

The cis--> trans rotational barrier is of 8 kcal/mol. The corresponding HF barrier is 8.7 kcal/mol and is reduced to 3.1 kcal/mol at MP2 level [70]. Although the cis conformation is observed in the X-ray structures of a series of nicotinamide derivatives [71], the trans conformation is observed in the X-ray structure of the NAD^+ free acid [72] and in many enzyme active sites [73]. Due to the small difference between the cis and the trans forms and the relatively low rotational barrier, the presence of the solvent or the enzymatic environment can reverse the stability order between the two conformers. In order to verify this hypothesis, we have redone the computations in water solvent. The presence of different water molecules in the active site of the enzyme has been proven by X-ray study [74]. In water the trans form is the "absolute minimum" and cis conformer lies at 0.2 kcal/mol. The rotational barrier is now 3.5 kcal/mol. This result demonstrates that the presence of the solvent is able to modify both the relative stabilities and the rotational barrier with respect to the gas-phase.

The highest dipole moment of trans form (5.6 Debye) with respect to that of the cis one (3.8 Debye) can explain the inversion of the stability of the two isomers in the polar solvent. In fact, the hydration free energy (electrostatic part) is -2.15 kcal/mol for cis structure and -1.03 kcal/mol for trans ones.

3.3. 4-THIOURACIL

Thio and fluoro derivatives of nucleosides are of interest because of their biological and pharmacological activities. The 5 substituted fluoro derivative of uracil nucleoside is a potent inhibitor of thymidylate enzyme and an active antitumor and antifungal chemotherapeutic agent [75,76]. The thiated nucleobases appear naturally in various biological materials and play a significant rule in numerous metabolic processes [77]. It has been pointed out [78,79] that the rare or unusual tautomeric forms of nucleic bases determine an alteration of the normal base-base pairing, thus leading to the transition-type point mutations. The ability of uracil and their derivatives to adopt rare tautomeric forms is object of current interest. Different experimental and theoretical studies have been undertaken to characterize the tautomerization process in uracil and 5-fluorouracil [22,80-82]. Less attention have been devoted to the thio derivatives [83,84]. In previous paper we have reported the DF results for uracil (U) [25] and 5-fluorouracil (FU) [26] tautomerization process in vacuo and in water.

Herein, the extension of the study to the 4-thiouracil (SU) is reported. For pourpose of comparison, we report also the main results of uracil and 5-fluorouracil studies. The four low energy tautomers depicted in Figure 6 have been considered as in the case of uracil and 5-fluorouracil.

The optimized structures are consistent with that derived from X-ray diffraction. Some difference are observed with respect to x-ray structure for the parameters involving atoms participating in hydrogen bonding in the crystal [85]. The 2-oxo,4-thione (oxo-thione) tautomer (structure 1 of Figure 6) is the most stable one as indicated by infrared spectroscopy [83] and previous ab-initio studies [84]. The relative energies (ΔE) of the other three tautomers with respect to the oxo-thione 1 are reported in Table 2 together with those coming from previous theoretical investigations. The oxo-thione absolute minimum is followed by the 2-hydroxy,4-thione (structure 2) and by the 2-oxo,4-thiol (structure 3) forms that lies at 47.1 and 48.5 kJ/mol respectively.

The 2-hydroxy,4-mercapto (structure 4) tautomer is the less stable one and lies at 58.6 kJ/mol above 1. The NLSD relative energies does not agree with the ab initio study of Les and Adamowicz [84] that found the 2-hydroxy,4-mercapto tautomer as the most stable after the oxo-thione one. From this study it is evident that the relative stabilities depends strongly on the employed basis set quality. The greatest basis set used in the HF study are of double-zeta size (6-31G**) with polarization functions on hydrogen and heavy atoms. We have used a triple-zeta basis set. Less and Adamowicz [84] take into account the correlation contributions by using the many-body perturbation theory truncated at the second order. This procedure stabilizes strongly the tendency of the pyridine ring to adopt the aromatic structure. As the same author point out, higher order correlation effects can affect the reported tautomeric stability order. In addition, the Zero Point Energy corrections (ZPE) have been made. Our study on this way is in progress. In any case also if we use the same ZPE of ref. 84 calculated at 3-21G* level and scaled by a factor of 0.91 to reproduce better the experimental frequencies, our ΔEs are different (see Table 2) from that obtained at MBPT(2) + ZPE level. On the other hand, previous HF [80,81] and DF [22,25,26] computations on uracil and 5-fluorouracil the type 4 structure very unfavoured with respect to the 1.

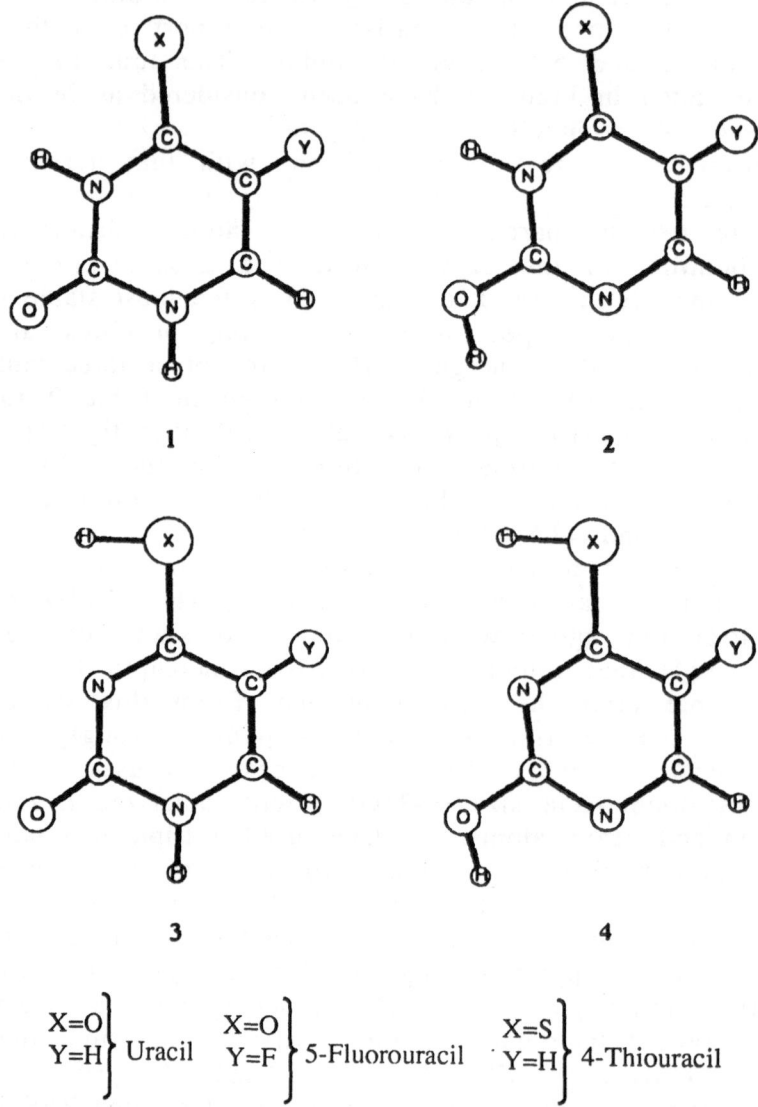

Figure 6. The four tautomers of uracil, 5-fluorouracil and 4-thiouracil.

The influence of polar environment on the tautomeric equilibria is important of because the possibility to reverse the stability order and consequently affect the metabolism of heterocycles in biological systems. An estimation of the influence of the polar solution on the tautomeric equilibrium of 4-thiouracil has been tried previously on the basis of calculated HF dipole moments [84]. However, this crude treatment is not able to give insight on this effect also because the calculated dipole moments result significantly overestimated (about 1 Debye) with respect to the experimental data [83] (see Table 2).

Table 2. Relative energies (kJ/mol) for 4-thiouracil tautomers from theory and experiment. The oxo-thione tautomer (structure **1** of Figure 6) is taken as the reference.

Method	$\Delta E(1\text{-}2)$	$\Delta E(1\text{-}3)$	$\Delta E(1\text{-}4)$	$\mu(1)$	$\mu(2)$	$\mu(3)$	$\mu(4)$
			in vacuo				
NLSD	47.1	48.5	58.6	4.91	4.05	5.07	1.58
NLSD+ZPE[a]	43.8	36.7	44.0	/	/	/	/
MBPT(2)[b]	45.2	60.1	51.4	/	/	/	/
MBPT(2)+ZPC[b]	42.2	49.4	38.1	/	/	/	/
HF/6-21G**[b]	44.3	65.6	57.9	5.54	4.89	5.60	2.28
IIF/3-21G*[b]	65.8	103.3	115.4	/	/	/	/
Exp[c]	/	/	/	4.47	/	/	/
			in water				
NLSD	55.5	49.7	76.9	5.95	4.97	5.85	1.81

a) by using the ZPE of ref. 84; b) from ref. 83; c) from ref. 83.

We have computed, for the first time, the tautomeric behaviour of 4-thiouracil in water solvent. From Table 2 it is evident that the relative energy of the tautomer 3 is not affected by the presence of the solvent, while the structure 4 is sensibly destabilized. The trend of the dipole moments (see Table 2) account for these effects.

NLSD ΔE's of uracil and substituted 5-fluorouracil and 4-thiouracil have been visualized quantitatively in Figure 7 for comparison. The following considerations can be drawn: for the all considered compounds the structure **1** is the absolute minimum; uracil and 4-thiouracil show a similar behaviour in both gas- and liquid-phase. (the tautomer **2** and **3** have almost the same energy in gas-phase and the structure **3** is more stable in water); in all cases the tautomer **4** is the highest in energy.

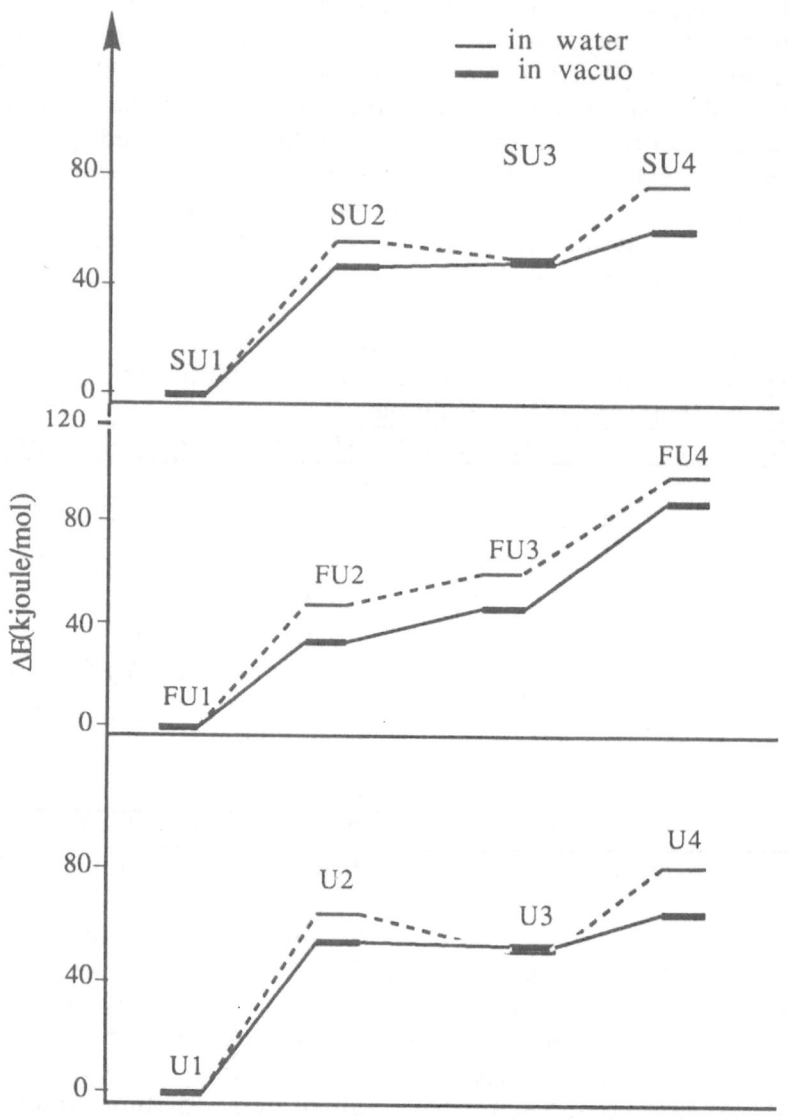

Figure 7. NLSD relative energies (ΔE) in vacuo and in water of various tautomeric forms of uracil (U), 5-fluorouracil (FU) and 4-thiouracil (SU). The numbers indicate the structures as given in Figure 6. The tautomer **1** is taken as the reference.

A contribution to the interpretation of some phenomena related to the biochemical properties of these system, can arise from the calculation of the equilibrium constant ($K = \exp[(-\Delta H + T\Delta S)/kT]$) for the tautomeric equilibrium between structures **1** and **3′**. This choice is motivated to the fact that tautomer **3** can form the characteristic N1-glycosidic bond in the nucleic acid.

The calculated K in gas-phase (water solvent) are $5.0 \ 10^{-9}$ ($2.9 \ 10^{-8}$), $4.9 \ 10^{-9}$ ($1.1 \ 10^{-9}$) and $1.0 \ 10^{-9}$ ($5.3 \ 10^{-8}$) for uracil, 5-fluorouracil and 4-thiouracil respectively (assuming an entropic contribution of 1.3 kJ/mol at 298 K as estimated experimentally [86] and k=0.008314 kJ/mol). All values fall into the region of the estimated frequency of spontaneous point mutations that is 10^{-8} - 10^{-11}. So, the values of the tautomeric relative stabilities do not exclude the possibility of the formation, in nucleic acids, of mismatches that involve the rare hydroxy (and with lower probability the mercapto) tautomeric forms in 5-fluorouracil and 4-thiouracil.

4. Conclusions

We have used the gaussian density functional method using the gradient corrected approximation to determine many physicochemical properties of three systems (alanine, 1,4-hydroxynicotinamide and 4-thiouracil) with considerable biological significance. The main general conclusions from this study are:
• significant parameters such as the dipole moments, ionization potentials, gas-phase proton affinities, gas-phase acidities, geometrical structures are reproduced with good accuracy and are close to the experimental counterparts.
• tautomeric equilibria and conformational behaviours are described well. In alanine as well as in glycine the employed nonlocal functional proposes a conformational order different from those obtained with high level correlated methods. The use of other functionals, e. g. hybrid nonlocal functionals, can improve the agreement.
• our implementation of the Onsager's self consistent reaction field coupled with gaussian density functional method seems to be a reasonable and efficient approach to investigate the solution phenomena.

• the DF methodology is highly competitive because of its low computational cost compared with the more expensive ab initio correlated methods. This characteristic is very important to study molecules with biological and pharmacological significance that involve generally a large number of atoms and require, necessarily, the computations of correlation effects to give reliable results.

Acknowledgements

This investigation was supported by CNR, MURST and CINECA.

References

1. A. Szabo and N. Ostlund, *Modern Quantum Chemistry: Introduction to Advanced Electronic Structure Theory* (MacMillan New York, 1982).
2. W.J. Hehre, L. Radom, P. v. R. Schleyer and J.A. Pople, *Ab Initio Molecular Orbital Theory* (J. Wiley & Sons, New York, 1986)
3. R. Parr, and W. Yang , *Density-Functional Theory of Atoms and Molecules* (Oxford University Press, NY, 1989).
4. N. March, *Electron Density Theory of Atoms and Molecules*; (Academic Press, New York, 1991).
5. R. O. Jones and O. Gunnarson, Rev. Mod. Physics 61, (1990) 1280.
6. T. Ziegler, Chem. Rev. 91, (1990) 651.
7. J. Labanowski and J. Andzelm (eds.) *Density Functional Methods n Chemistry* (Springer Verlag: NY, 1991).
8. D. R. Salahub and N. Russo (eds.) *Metal Ligand Interaction. From Atoms, to Clusters, to Surfaces* (Kluwer: Dordrecht, 1992).
9. J. M. Seminario and P. Politzer (eds.) *Density Functional Theory: a Tool for Chemistry* (Elsevier, New York, 1995).
10. B. G. Johnson, P. M. W. Gill and J. A. Pople, J. Chem. Phys. 98, (1993) 5612
11. A. M. Lee and N. C. Handy, J. Chem. Soc. Faraday Disc. 2, (1993) 1
12. Y. Abashkin, N. Russo and M. Toscano, Int. J. Quantum Chem. 52, (1994) 695.
13. N. Russo and D. R. Salahub (eds.) *Metal Ligand Interaction: Structure and Reactivity* (Kluwer: Dordrecht, 1995).

14. D. P. Chong (ed) *Recent Advances in Density Functional Methods.* Vol 1 (World Scientific, Singapore, 1995).
15. D. R. Salahub, Adv. Chem. Phys. 69 (1987) 447.
16. GAUSSIAN 94/DFT, Gaussian Inc., Pittsburgh, PA, 1994.
17. CADPAC5: The Cambridge Analityc Derivatives Package issue 5, Cambridge, 1992.
18. J. Andzelm and E. Wimmer, J. Chem. Phys. 96 (1992) 508.
19. B. Delley, J. Chem. Phys. 92 (1990) 508.
20. A. St. Amant, PhD Thesis, Universite' de Montreal, 1992.
21. Amsterdam Density Functional System (ADF), Department of Theoretical Chemistry, Vrije Universiteit, Amsterdam, The Netherlands; E.J. Baerends and P. Ros, Int. J. Quantum Chem. S12, (1978) 69.
22. D. A. Estrin, L. Paglieri and G. Corongiu, J. Phys. Chem. 98 (1994) 5653; ibidem Int. J. Quantum Chem. 56, (1995) 615.
23. V. Malkin, O. L. Malkina, M. E. Casida and D. R. Salahub, J. Am. Chem. Soc. 116 (1994) 5898.
23. V. Barone, C. Adamo and F. Lelj, J. Chem. Phys. 102 (1995) 364
25. R. J. Hall, N. A. Burton, I. Hillier and P. E. Young, Chem. Phys. Lett. 220, (1994) 129.
26. T. Marino, N. Russo and M. Toscano, Int. J. Quantum Chem., submitted.
27. 6] R. Santamaria, A. Quiroz-Gutierrez, C. Juarez, J. Mol. Struct. (Theochem) 357, 161 (1995) and reference therein.
28. W. G. Han and S. Suhai, J. Phys. Chem. 100, (1996) 3942.
29. P. Hohenberg and W. Kohn, Phys. Rev. 136, (1964) B864.
30. W. Kohn and L. J. Sham, Phys. Rev. A140, (1965) 1133.
31. J. P. Perdew and Y. Wang, Phys. Rev. B33, (1986) 8800.
32. J. P. Perdew, Phys. Rev. B33, (1986) 8822.
33. S. H. Vosko, L. Wilk and M. Nusair, Can. J. Phys. 58, (1980) 1200.
34. J. Tomasi and M. Persico, Chem. Rev. 94 (1994) 2027.
35. J. L. Rivail and D. Rinaldi, Chem. Phys. 18 (1976) 233.
36. S. Harrison, N. Nolte and D. Beveridge, J. Phys. Chem. 80 (1976) 2580.
37. C. Felder, J. Chem. Phys. 75 (1981) 4679.
38. J. L. Rivail and B. Terryn, J. Chem. Phys. 79 (1982) 1.
39. D. Rinaldi, M. F. Ruiz-Lopez and J. L. Rivail, J. Chem. Phys. 78 (1983) 834.

40. L. Onsager, J. Am. Chem. Soc. 58, (1936) 1486.
41. T. Mineva, N. Russo and M. Toscano, Int. J. Quantum Chem. 56, (1995) 663.
42. N. Godbout, D. R. Salahub, J. Andzelm and E. Wimmer, Can. J. Chem. 70, (1992) 560.
43. S. Vishveshwara and J. A. Pople, J. Am. Chem. Soc. 99 (1977) 2422.
44. C-H. Hu, M. Shen and H. F. Schaefer III, J. Am. Chem. Soc. 115 (1993) 2923.
45. S. Gronert and R. A. J. O'Hair, J. Am. Chem. Soc. 117, (1995) 2071.
46. A. D. Becke, Phys. Rev. A38, (1988) 3098 ; J. Chem. Phys. 88, (1988) 2547.
47. I. Topol, S. Burt, N. Russo and M. Toscano, to be published.
48. T. P. Debeis and J. W. Rabelais, J. Electron Spectrosc. 3 (1974) 315.
49. L. R. Wright and R. F. Borkman, J. Am. Chem. Soc. 102, (1980) 6207.
50. N. Russo, E. Sicilia and M. Toscano, J. Chem. Phys. 97, (1992) 5031.
51. P. D. Godfrey, S. Firth, L. D. Hatherley, R. D. Brown and A. P. Pierlot, J. Am. Chem. Soc. 115, (1993) 9687.
52. G. De Luca, N. Russo, E. Sicilia and M. Toscano, Gazzetta, 126 (1996) 441.
53. G. De Luca, N. Russo, E. Sicilia and M. Toscano. J. Chem. Phys. in press
54. P. Duffy, D. P. Chong and M. Dupuis, J. Chem. Phys., 102, (1995) 3312.
55. M. J. Locke and R. T. Mc Iver, J. Am. Chem. Soc. 105 (1983) 4226.
56. G. S. Gorman, J. P. Speir, C. A. Turner and I. J. Amster, J. Am. Chem. Soc. 114 (1992) 3986.
57. X. Li and A. G. Harrison, Org. Mass Spectrometry 28 (1993) 366.
58. M. Meot-Ner, E> P. Hunter and F. H. Field, J. Am. Chem. Soc. 101 (1979) 686.
59. G. De Luca, T. Mineva, N. Russo, E. Sicilia and M. Toscano, *in Recent Advances in Density Functional Theory*, D. P. Chong (ed), (World Scientific, Singapore, Part. II, in press).

60. A. M. Schmiedekamp, I. A. Topol, S. K. Burt, H. Razafinjanahari, H. Chermette, T. Pfaltzgraff and C. J. Michejda, J. Comput. Chem. 15 (1994) 875.
61. M. Belcastro, T. Mineva, N. Russo, E. Sicilia and M. Toscano in *Modern Density Functional Theory*, J. M. Seminario (ed), (Elsevier, New York, vol. 2, in press).
62. R. A. J. O'Hair, J. H. Bowie and S. Gronert, Int. J. Mass Spectrom. Ion Processes, 117 (1992) 23.
63. J. S. Gaffney, R. C. Pierce and L. Friedman, J. Am. Chem. Soc. 99 (1977) 4293.
64. K. Zhang, D. M. Zimmerman, A. Chung-Phillips and C. Cassady, J. Am. Chem. Soc. 115 (1993) 10812.
65. S. Burt, I. Topol and N. Russo, to be published.
66. C. Walsh, *Enzymatic Reaction Mechanisms*, (W. H. Freeman and Co., San Francisco, 1979).
67. K. You, CRC Crit. Rev. Biochem. 17 (1984) 313.
68. M. C. A. Donkersloot and H. M. Buck, J. Am. Chem. Soc. 103, (1981) 6554.
69. P. M. T. de KoK, M. C. A. Donkersloot, P. M. von Lier, G. H. W. M. Mendendijks, L. A. M. Bastiansen, H. J. G. van Hooff, J. A. Kanters and H. M. Buck, Tetrahedron 42 (1986) 941.
70. Y-D. Wu and K. N. houk, J. Org. Chem. 58 (1993) 2043.
71. A. Glasfeld, P. Zbinden, M. Dobler, S. A. Benner and J. D. Dunitz, J. Am. Chem. Soc. 110, (1988) 5152 and references cited therein.
72. R. Parthasarathy and S. M. Fridey, Science 226 (1984) 969.
73. P. C. Skarzynski, P. C. E. Moddy and A. J. Wonacott, J. Mol. Biol. 193 (1987) 171 and reference cited therein.
74. C. Abad-Zapatero, J. P. Griffith, J. L. Sussman and M. G. Rossman, J. Mol. Biol. 198 (1987) 445.
75. P. V. Danenberg, Biochim. Biophys. Acta 473 (1977) 73.
76. C. Heidelberg, P. V. Danenburg and R. G. Moran, Adv. Enzymol. 54 (1983) 57.
77. M. Altweg and E. Kubli, Nucl. Acids Res. 8 (1980) 215.
78. P. O. Lowdin, Adv. Quantum Chem. 2 (1965) 213.
79. B. Pullman and A. Pullman, Adv. Heterocycl. Chem. 13 (1971) 77.
80. M. J. Scanlan and H. Hillier, J. Am. Chem. Soc. 106, (1984) 3737.
81. A. Les and L. Adamowicz, J. Phys. Chem. 93, (1989) 7078.
82. P. Beak and J. M. White, J. Am. Chem. Soc. 104, (1982) 7073.

83. H. Rotkowska, K. Szczepaniak, M. J. Novak, J. Laszcynski, K. KuBulat and W. B. Person, J. Am. Chem. Soc. 112, (1990) 2147.
84. A. Les and L. Adamowicz, J. Am. Chem. Soc. 112, (1990) 1504.
85. B. Lesyng and W. Saenger, Z. Naturforsch. 36C (1981) 956.
86. M. Shibata, T. J. Zielinski and R. Rein, Int. J. Quantum Chem. 18, 323 (1980).

GMMX CONFORMATION SEARCHING AND PREDICTION OF NMR PROTON-PROTON COUPLING CONSTANTS

FRED L. TOBIASON[a] and GÉRARD VERGOTEN[b]
[a]Department of Chemistry, Pacific Lutheran University, Tacoma, WA
98447, U.S.A [b]CRESiMM, Department of Chemistry, University of Science
and Technology, Lille 1, 59655 Villeneuve D'ASCQ Cedex, FRANCE

Abstract: This work discusses the use of vicinal $^3J_{HH}$ coupling constant equations as applied to flavan-3-ols and flexible monosaccharides. Prediction of vicinal coupling constants for (+)-catechin and D-glucitol are discussed using the GMMX searching methodology. Other extended Karplus coupling equations are reviewed.

1. Introduction

It was known early in NMR studies that a relationship existed between coupling constants and the conformational geometry of flexible molecules [1]. Probably the greatest literature database was accumulated for cyclohexane and disubstituted ethane structures [1,2], where the $^3J_{HH}$ vicinal proton coupling constants were mostly analyzed.

    Theoretical valence bond quantum mechanics studies by Karplus [3,4] beginning in 1957 led to the set of equations (1). These equations give the relationship between a vicinal coupling constant and the dihedral angle separating the planes that contain the two interacting protons.

$$^3J_{HH} = 8.5 \cos^2 \Phi - 0.28 \quad 0^o \le \Phi \le 90^o$$
$$^3J_{HH} = 9.5 \cos^2 \Phi - 0.28 \quad 90^o \le \Phi \le 180^o \tag{1}$$

    The $^3J_{HH}$ versus Φ for these equations is plotted in Figure 1. Figure 2 shows the relationship between the dihedral angle, Φ, and planes in which the two interacting protons lie. It was understood early on that many second-order effects influenced the total magnitude of the coupling constants [1,4,5], for example, the electronegativity and orientation of substituents. These are not accounted for in the Karplus equation (1).

    Altona and Haasnoot [5,6,7,8] studied these factors for a number of years and arrived at an extended Karplus equation (2) that is now in common usage in computer programs such as MacroModel [9], PCMODEL [10] and GMMX [11]. In programs like

179

G. Vergoten and T. Theophanides (eds.), Biomolecular Structure and Dynamics, 179–186.

180

Insight II, the NMR_Refine routine uses the standard Karplus equation to set dihedrals [12]. In equation (2), A through F are empirical constants determined by regression analysis from a data base of 315 experimental vicinal coupling constants from 109 molecules.

$$^3J_{HH} = A \cos^2 \Phi + B \cos \Phi + C + \Sigma \Delta \chi_i [D + E \cos^2 (\Phi \bullet \xi_i + F | \Delta \chi_i |)] \quad (2)$$

Here Φ is the dihedral angle, $\Delta \chi_i$ is the difference in electronegativity and ξ_i is the orientation factor. The constant C was set equal to zero because there was not complete independence of the variables in the total regression fit [5].

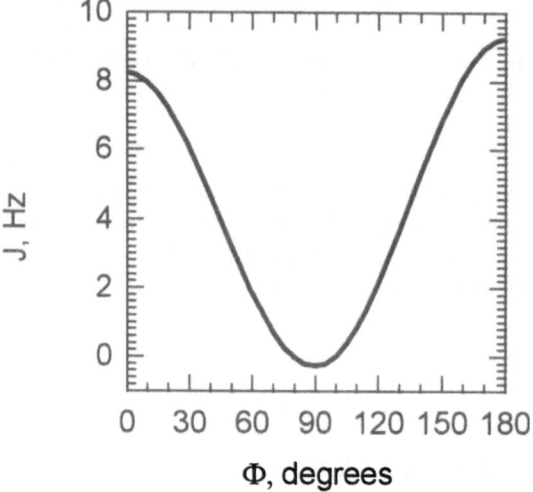

Figure 1. The variation of the vicinal proton-proton coupling constant with dihedral angle as computed from the Karplus equation.

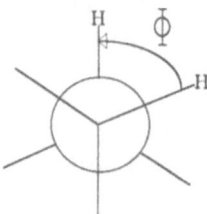

Figure 2. The definition of the dihedral angle separating the H atoms in their respective C-C-H planes.

In dealing with flexible molecules equation (2) is used to compute the vicinal coupling constant for each appropriate torsional angle in each conformer. It is then substituted into the Boltzmann summing equation (3), where P_i is the probability of finding the ith conformer in the ensemble. This is summed over the N conformers. Here, P_i is given by the Boltzmann equation (4) [13,14,15], where E_i is the energy of the ith conformer.

$$J = \sum_{i=1}^{N} P_i J_i \tag{3}$$

$$P_i = \frac{e^{-\frac{Ei}{kT}}}{\sum_i e^{-\frac{Ei}{kT}}} \tag{4}$$

Considerable interest has recently been shown in the flexible and conformational structural characteristics of proanthocyanidins (condensed tannins) [16,17] and carbohydrate structures [18,19]. The above Boltzmann techniques will be reviewed as applied to selected polyflavanoids (e.g., (+)-catechin) and carbohydrate monomers using the GMMX searching methodology coupled with the application of the Altona/Haasnoot vicinal coupling constant equation [5].

2. Methodology

The proanthocyanidin flavan-3-ol structural problem that is being examined is illustrated by the (+)-catechin molecule shown in Figure 3. Here the E-conformer and A-conformer are depicted representing the pseudoequatorial and pseudoaxial conformers, respectively. When the energy difference between the E- and A-conformers is small, e.g., less than 2.0 kcal/mol, there will be significant contribution toward the coupling constants from both conformers in the ensemble. Substitution can occur on either the α or β position and these symbols are used to describe the appropriate coupling constant as well as the orientation nature of the substituent in the nomenclature.

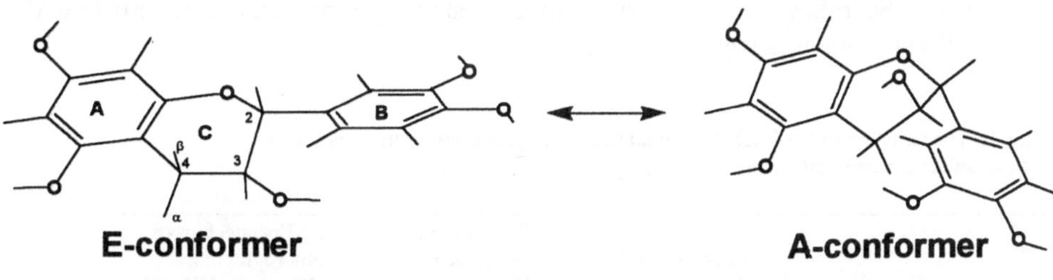

E-conformer   **A-conformer**

Figure 3. The E- and A-conformers for (+)-catechin as representative of many proanthrocyanidins. The pyran ring, the C-ring, conformation is most responsible for variation in the proton coupling constants.

The conformational searching method used in these studies is based on the GMMX protocol [11] of mixed combinations of Monte Carlo coordinate movements and bond rotations. The MMX force field was used and structures were generated with PCMODEL [10]. The electrostatic term was set with dipole/dipole interactions and the hydrogen bonding function was examined with the switch set both on and off. When examining NMR coupling constants or other solution properties, such as dipole moments, an energy cutoff of 3.0 kcal/mol is sufficient to include all of the contributing conformer populations. All degrees of rotational freedom are allowed in a search. In addition to the pyran ring angles on catechin type structures, for example, the rotation of the B-ring as well as all hydroxy and methoxy groups is allowed. For carbohydrate structures, in addition to the internal C-C or C-O bond rotation, all exocyclic hydroxy and hydroxymethyl groups are free to rotate. This technique has shown considerable success in the polyphenol proanthocyanidins, like (+)-catechin, and derivatives [14,20], and with several monosaccharides, for example, D-glucitol and D-glucopyranose [19]. See Figure 4 for the structure numbering of D-glucitol.

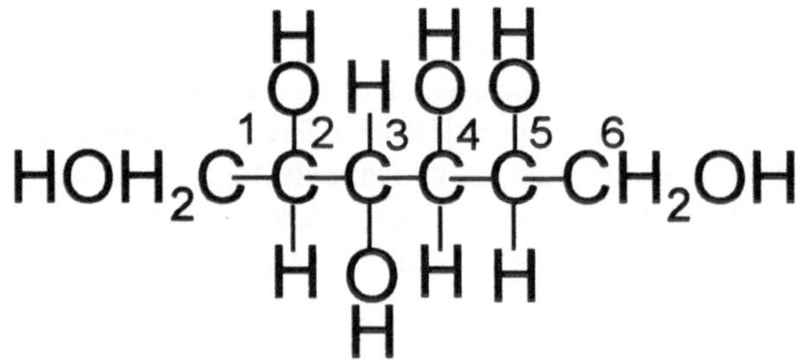

Figure 4. The numbering system for the coupling constants in D-glucitol.

3. Results and Discussion

3.1 FLAVAN-3-OLS

Some selected GMMX search results [14,20] are illustrated in Table 1. The values given here are taken for a dielectric constant of 4.0 or 5.0 using dipole/dipole interactions. Searches were run with different starting structures to insure that the same coupling parameters were obtained.

TABLE 1. The variation of the GMMX computed coupling constants with variation of proanthocyanidin structure and hydrogen bonding.

	(+)-Catechin[a]		Tetra-O-methyl-(+)-catechin		(-)-Epciatechin-4β phloroglucinol	
	HB on	HB off	HB on	HB off	HB on	HB off
Conformers searched	3192	3580	4957	7935	9149	8336
Final ensemble of conformers	451	443	347	428	1064	960
$J_{2,3}$, Hz	7.83	7.67	8.16	7.68	1.40	1.12
$J_{3,4\alpha}$, Hz	5.11	5.05	5.20	5.05	3.21	2.64
$J_{3,4\beta}$, Hz	9.46	9.26	9.88	9.33	--	--
Observed coupling constants	7.6, 5.3, & 8.0 Hz		8.1, 5.5, & 9.0 Hz		1.0, 2.0 Hz	

a) Dielectric constant was 5.0 for (+)-catechin and 4.0 for the other two compounds.

Examination of the data in Table 1 clearly shows that the ensemble of conformers found through the GMMX search gives a reasonable representation of the vicinal proton coupling constants as generated by equations (3), (4) and the Altona equation (2). In general, the coupling constant fit is better with the MMX force field having the hydrogen bonding function turned off and the dielectric constant set in the range of 4 to 5. The rms value for (+)-catechin is 0.82 Hz, for example. A recent study on the temperature dependence of the pyran ring vicinal coupling constants for (+)-catechin found that the GMMX conformer search ensemble also gives a reasonable fit to the temperature coefficients of the proton-proton coupling constants [21]. This gives support to the distribution of energy found in the GMMX search conformer ensemble. Even the (-)-epicatechin-4β-phloroglucinol with the bulky group in the C(4β) position fit very well with the hydrogen bonding off [20].

3.2 D-GLUCITOL

Another example of the capability of the GMMX searching program is illustrated with the flexible carbohydrate, D-glucitol. Table 2 gives the variation of search parameters for D-glucitol with dielectric constant and hydrogen bonding. Table 3 shows the proton coupling constants computed with the GMMX protocol as compared with several experimental studies and a molecular dynamics modeling study. Several things become immediately clear from these results. First, the GMMX values are far superior when taken with the hydrogen bonding function off and when the dielectric constant is in the range of 4 to 5. This fits with the experimental data taken in water where hydrogen bonding would be disrupted. The higher dielectric constant is consistent with the highly polar solvent. In addition, the GMMX coupling constants appear to fit better than those found from the molecular dynamics ensemble. The rms deviation of the coupling constants as determined by GMMX compared to the average experimental NMR values is 1.3 Hz [19] compared with the molecular dynamics study at an rms of 1.9 Hz [15].

In Table 2 it is noted that there are a large number of "unique" conformers obtained especially as one goes to a higher dielectric constant. Some of this arises from the width of

TABLE 2. Comparison of the GMMX search parameters for D-glucitol as a function of dielectric constant and hydrogen bonding.

Conformation Searching Parameter	Calculated					
	H-bond Off			H-bond On		
	1.5	4.0	10.0	1.5	4.0	10.0
No. of conformers searched[1]	12,143	15,173	13,741	11,045	11,861	13,806
Unique conformers found[2]	1,763	2,056	3,661	671	671	2,693
Conformers in final ensemble[3]	784	770	2,202	352	333	1,704
$E_{minimum}$, Kcal/mol	3.40	3.39	7.25	1.82	1.82	6.79
E_{min} structure, % of ensemble	4.66	4.07	1.10	6.98	6.98	1.46
Average Boltzmann energy	5.16	5.20	8.75	3.58	3.26	8.25

(1) For dielectric constant 1.5 to 10.0 with dipole/dipole interactions.
(2) Number of conformers found with a 3.5 Kcal/mol window.
(3) Final number of conformers found within a 3.0 Kcal/mol window after successive reiteration with hydrogen atom remove/add command.

TABLE 3. Comparison of GMMX computed and experimental proton coupling constants in Hz for D-glucitol in water as a function of dielectric constant and hydrogen bonding.

Coupling Constant J	Exp.,Hz A[1]	B[2]	Calculated GMMX H-bond Off 1.5	4.0	10.0	H-bond On 1.5	4.0	10.0	Calc[3]
H_1H_2	6.55	6.55	7.49	7.64	7.21	0.65	2.52	3.70	5.08
$H_{1'}H_2$	3.55	4.25	4.34	3.49	3.83	3.24	3.39	5.02	4.24
H_2H_3	5.90	6.00	3.12	4.56	4.41	0.20	0.21	1.29	4.68
H_3H_4	1.70	2.47	5.44	3.44	3.50	0.46	0.61	0.82	1.56
H_4H_5	8.25	7.70	5.05	6.02	5.63	3.80	4.04	4.37	9.86
H_5H_6	2.95	3.33	3.70	3.26	3.36	6.38.	5.73	4.99	1.62
$H_5H_{6'}$	6.3	6.24	9.18	8.25	7.77	5.02	5.84	6.68	9.76

(1) Ref. 22.
(2) Ref. 23.
(3) Ref. 15, molecular dynamics with an aqueous shell.

the structure rms comparison window that is allowed but is also related to lowering of the energy of the electrostatic interactions for the high dielectric constant.

It is especially noteworthy that the very flexible end hydroxymethyl groups are predicted with fair success for D-glucitol, see Table 3. The J_{12}, $J_{1'2}$, J_{56} and $J_{56'}$ proton coupling to the hydroxymethyl groups in D-glucopyranose are also predicted very well [19].

4. Other Extended Karplus Methods

4.1. VICINAL PROTON-PROTON COUPLING EQUATIONS

There are several points to keep in mind when using any program with the Altona equation. One is that the constants for the equation were determined with a limited number of vicinal coupling constants, actually just those taken from 109 compounds [5]. In this study the rms deviation between the experimental and the fit was approximately 0.48 Hz. Consequently, it would not be unreasonable to have an rms value in the range of 0.4-0.8 Hz unless specific parameters developed for a set of compounds were used. In addition, it is important to understand which set of equations is being used in a computer program. For example, the Altona equation has been reparameterized [8] to give an rms value of 0.36 Hz, and other empirical $^3J_{HH}$ proton-proton coupling equations have been developed, e.g., that by Imai and Ōsawa [24]. This latter study suggests that their equation gives an rms fit of 0.33 Hz which is considerably better than that of the original Altona equation. The Imai/Ōsawa equation takes into account changes in C-C bond distances and the C-C-H bond angles. Bond angles have been shown to play an important role in vicinal coupling constants. A recent treatment by Barfield and Smith [25] shows that the C-C-H bond angle effect increased as the dihedral increased or decreased from 90° with the largest effect being in the direction for $\Phi > 90°$. Considering many of these factors, the rms fit found for (+)-catechin, e.g., 0.82 Hz, seems reasonable when using the Altona equation [5]. However, neither comparisons of fits with different empirical vicinal equations nor an examination of the bond angle factor on flavan-3-ols or carbohydrates has been done. However, Aliev and Sinitsyna [26] have made another modification of the Altona equation and have made comparisons for a series of γ-piperidones.

Other considerations are the method of searching and the relative energies or free energies (preferred if obtainable) one obtains for the conformers in the ensemble. This can vary greatly with different programs currently in use and can easily change the magnitude obtained with the Boltzmann summing equation (3).

4.2. EQUATIONS SPECIFICALLY FOR SIX-MEMBERED RINGS

An equation for treating six-membered rings has also been suggested by Hassnoot [7], and a specific method for the additive characteristics of carbohydrate ring proton-proton coupling constants has been developed [27]. In this latter case, the authors attempted to treat the pyranose methylol group, too.

4.3. CARBON(13) COUPLED TO PROTON KARPLUS EQUATIONS

Finally, it should be pointed out that these ideas have been extended to many forms of coupling constants and conformational analysis. For example, Tvaroška et. al. [28,29,30] extended studies with considerable success to $^3J^{13}_{CH}$ constants in carbohydrates across the glycosidic bond, $^1J^{13}_{CH}$ at the glycosidic bond and for other types of compounds, e.g., in 1-thioglycosides [31]. So although there is much work needed to refine many of these equations, and especially the parameter used within the equations, it appears that the aid rendered to conformational analysis warrants the research efforts being made.

5. Conclusions

The prediction of NMR coupling constants in flexible molecules depends upon both the searching capability of the computer program in use and the quality of the parameters used with the extended Karplus equation employed. The GMMX protocol is one method that can give, in many cases, very good estimates of average NMR proton-proton coupling constants that would be expected for a flexible molecule in solution.

6. Acknowledgments

FLT wishes to thank the University of Science and Technology, Lille 1 for support through a Visiting Research Professorship, 1996, and the NATO committee for support in attending the NATO Advanced Research Study Workshop in Greece, 1996.

7. References

1. Bovey, F.A. (1969) *Nuclear Magnetic Resonance Spectroscopy*, Academic Press, New York.
2. Bystrov, V.F. (1976) *Prog. Nucl. Magn. Reson. Spectrosc.* **10**, 41.; Bystrov, V.F. (1972) Spin-spin interactions between geminal and vicinal protons, *Usp. Khim.* **41**, 512-553; Abraham, R.J. and Gatti, G. (1969) Rotational isomerism VII. Effect of substituents on vicinal coupling constants in XCH2CH2Y fragments, *J. Chem Soc. B* 961-968.

3. Karplus, M. (1959) Contact electron-spin coupling of nuclear magnetic moments, *Chem Phys.* **30**, 11-15.

4. Karplus, M. (1963) Vicinal proton coupling in Nuclear Magnetic Resonance, *J. Am. Chem. Soc.* **85**, 2870-2871.

5. Haasnoot, C.A.G., DeLeeuw, F.A.A.M., and Altona, C. (1980) The relationship between proton-proton NMR coupling constants and substituent electronegativities-I, *Tetrahedron* **36**, 2783-2792.

6. Altona, C., Ippel, J.H., Westra Hoekzema, A.J.A., Erketlens, C., Groesbeek, M., and Donders, L.A. (1989) Relationship between proton-proton NMR coupling constants and substituent electronegativities, *Magnet. Reson. Chem.* **27**, 564-576.

7. Haasnoot, C.A.G. (1993) Conformational analysis of six-membered rings in solution: Ring puckering coordinates derived from vicinal NMR proton-proton coupling constants, *J. Am. Chem. Soc.* **115**, 1460-1468.

8. Altona, C., Francke, R., de Haan, R., Ippel, J.H., Daalmans, G.J., Westra Hoekzema, A.J.A., and Van Wijk, J. (1994) Empirical group electronegativities for vicinal NMR proton-proton couplings along a C-C bond: Solvent effects and reparameterization of the Haasnoot equation, *Magnet. Reson. Chem.* **32**, 670-678.

9. MacroModel, Stille, C. (Version 5.0, 1996), Department of Chemistry, Columbia University, New York, NY 10027.

10. PCMODEL (Version 5.0), Serena Software, P.O. Box 3076, Bloomington, IN 47402-3076.

11. GMMX (Version 1.0, modified), Serena Software, P.O. Box 3076, Bloomington, IN 47402.

12. INSIGHT II, NMR_Refine, NMRchitect User Guide, October 1985, San Diago, Molecular Simulations, Inc., 1996, pp 2-26.

13. Cumming, D.A. and Carver, J.P. (1987) Reevaluation of rotamer populations for 1,6 linkage: reconciliation with potential energy calculations, *Biochemistry* **26**, 6676-6683.

14. Tobiason, F.L. and Hemingway, R.W. (1994) Predicting heterocyclic ring coupling constants through a conformatinal search of tetra-O-methyl-(+)-catechin., *Tetrahedron Lett.* **35**, 2137-2140.

15. Tvaroška, I., Kozár, T. and Hricovíni, M. (1990) Oligosaccharides in solution in *Computer Modeling of Carbohydrate Molecules*, French, A.D. and Brady, J.W. (eds.) ACS, Wash., D.C., p. 162.

16. Hemingway, R.W. and Karchesy, J.J. (eds.), (1989) *Chemistry and Significance of Condensed Tannins*, Plenum Press, New York.

17. Hemingway, R.W. and Laks, P.E. (eds.), (1992) *"Plant Polyphenols: Synthesis, Properties, Significance,"* Plenum Press, New York.

18. French, A.D. and Brady, J.W. (eds.), (1990) *Computer Modeling of Carbohydrate Molecules*, ACS, Wash., D.C.

19. Tobiason, F.L., Vergoten, G., and Mazurier, J. (1997) Predicting carbohydrate chain and heterocyclic ring coupling constants in monosaccharides using GMMX conformational searching, *Theochem* (special issue, to be published).

20. Hemingway, R.W., Tobiason, F.L., Mc Graw, G.W., and Steynberg, J.P. (1996) Conformation and complexation of tannins: NMR spectra and molecular search modeling of flavan-3-ols, *Magn. Reson. Chem.* **34**, 424-433.

21. Tobiason, F.L., Kelley, S.S., Midland, M.M., and Hemingway, R.W. (1997) Temperature dependence of (+)-catechin pyran ring coupling constants as measured by NMR and modeled using GMMX search methodology, *Tetrahedron Lett.* (to be published).

22. Hawkes, G.E. and Lewis, D. (1984) Proton nuclear magnetic resonance spectra and conformations of alditols in deuterium oxide, *J. Chem. Soc., Perkin Trans. 2*, 2073-78.

23. Hoffman, R.E., Rutherford, T.J., Mulloy, B. and Davies, D.B. (1990) 1H NMR assignment and conformational analysis of glucitol: Comparison with maltitol, *Magnet. Reson. Chem.* **28**, 458-464.

24. Imai, K. and Ōsawa, E. (1990) An empirical extension of the Karplus equation, *Magnet. Reson. Chem.* **28**, 668-674.

25 Barfield, M. and Smith, W.B. (1992) Internal H-C-C angle dependence of vicinal ^1H-^1H coupling constants, *J. Am. Chem. Soc.* **114**, 1574-1581.

26. Aliev, A.É. and Sinitsyna, A.A. (1993) Empirical analysis of vicinal proton-proton spin-spin coupling constants from the classic and modified Karplus equations in the γ-piperidone series, *Proc. of Russian Academy of Sciences*, 1207-1213.

27. Altona, C. and Haasnoot, C.A.G. (1980) Prediction of anti and gauche vicinal proton-proton coupling constants in carbohydrates: A simple additivity rule for pyranose rings, *Org. Magnet. Reson.* **13**, 417-429.

28. Tvaroška, I. (1990) Dependence on saccharide conformation of the one-bond and three-bond carbon-proton coupling constants, *Carbohydr. Res.* **206**, 55-64.

29. Tvaroška, I. (1991) One bond carbon-proton coupling constants: Angular dependence in α-linked oligosaccharides, *Carbohydr. Res.* **221**, 83-94.

30. Tvaroška, I. (1992) One-bond carbon-proton coupling constants: Angular dependence in β-linked oligosaccharides, *J. Biolmol. NMR* **2**, 421-430.

31. Tvaroška, I., Mazeau, K., Blanc-Muesser, M., Lavaitte, S., Driguez, H. and Taravel, F.R. (1992) Karplus-type equation for vicinal carbon-proton coupling constants for the C-S-C-H pathway in 1-thioglycosides, *Carbohydr. Res.* **229**, 225-231.

PROTEINS AND LIPIDS

BIOMOLECULAR STRUCTURE AND DYNAMICS:
Recent Experimental and Theoretical Advances

R. KAPTEIN, A.M.J.J. BONVIN[#] and R. BOELENS,

Bijvoet Center for Biomolecular Research, Utrecht University,

Padualaan 8, 3584 CH Utrecht, The Netherlands

Present address: *Laboratorium für Physikalische Chemie, ETH-Zentrum,*

CH-8092 Zürich, Switzerland

1. Introduction

In the last decade NMR spectroscopy has become an important alternative to X-ray crystallography. The principal advantage of NMR is that it allows the study of biomolecules in solution, thus closer to their real physiological environment. The method is limited, however, to biomolecules of relatively small size due to limitations in spectral resolution and line broadening effects. The upper limit is presently situated around molecular weights of 30 kD. Knowledge of three-dimensional structures might lead, for example, to a better understanding of structure-function relationships or recognition processes. For this purpose, accurate structures are required and, as we shall see, the use of so-called relaxation matrix approaches provides a means to improve the accuracy of the method.

The process of structure determination based on NMR data can be divided into four stages:

- assignments of the proton resonances,
- determination of structural constraints from NOE and *J*-coupling data,
- generation of structures,
- refinement of these structures.

The first and laborious task of assigning proton resonances is still the main bottle-neck of the structure determination, but progress is made in this area by the development of new multidimensional NMR

G. Vergoten and T. Theophanides (eds.), Biomolecular Structure and Dynamics, 189–209.

techniques (3D-4D) [1-8] and the use of labelled samples (^{15}N, ^{13}C). We will focus here on the last three stages involving the determination of structural constraints (distances, dihedral angles) and the generation and refinement of the structures.

Structure determination by NMR is based primarily on the nuclear Overhauser effect (NOE) [9,10]. Its origin is dipolar cross relaxation between protons, which is a function of distances and molecular motions. Because of the approximate r^{-6} dependence of the NOEs on the interproton distances, these can only be measured between protons relatively close in space ($\sim< 5$Å). NOEs are generally transformed into distance constraints which are then used, together with dihedral angle constraints obtained from J-coupling data, for the generation of structures using Distance Geometry (DG) [11-14] or Simulated Annealing (SA) techniques [15-17]. These structures are then usually refined with a combination of restrained Energy Minimisation (EM) and Molecular Dynamics (MD) simulations [18-20]. Several approaches have been proposed for the interpretation of NOEs in term of distances. The simplest one classifies the NOEs in strong, medium and weak peaks and attributes corresponding distance ranges to these three categories. However, more accurate distances can be obtained from relaxation matrix calculations which take into account all indirect magnetisation transfers ("spin diffusion") [21-31]. A more direct method for refinement was proposed by Yip and Case [32], in which NOE intensities are directly used as structural constraints, thus avoiding the transformation into distances. In this method, the structures are directly refined against experimental NMR data by comparison of theoretical and experimental NOE intensities.

Here, we will first discuss the underlying theory of the nuclear Overhauser effect, including a description of intramolecular motions. The use of NOEs as structural constraints, indirectly by conversion into distances or directly by comparison of experimental and theoretical intensities, will be presented together with the most common methods used to generate and refine structures. This will be illustrated with results from our laboratory for the structure determination and refinement of the Arc repressor of phage P22, a dimeric DNA-binding protein (2x53 residues).

2. The Nuclear Overhauser Effect

2.1. RELAXATION MATRIX THEORY

Multispin relaxation in a biomolecule in solution can be approximately described by the generalised Bloch equations. The evolution of the longitudinal magnetisation components in a system of N spins is governed by a set of coupled differential equations of the form [33]

$$\frac{d}{dt}\, \Delta M(t) = -\, R\, \Delta M(t) \tag{1}$$

where the vector $\Delta M(t)$ comprises the deviations from thermal equilibrium for all spins i

$$\Delta M(t)_i = [M_z(t) - M_0]_i \tag{2}$$

and R is the relaxation matrix. It comprises contributions from cross relaxation and from external relaxation. For a system of dipolar-coupled spins, neglecting cross-correlation, R has the dimension of the number of spins N in the systems. Since the dipolar interaction is a function of time, the relaxation rates are intimately connected with the molecular motion. The elements of R can be expressed as [33,34,10]

$$R_{ii} = \rho_{ii} = K \sum_{j \neq i} \left(J_{ij}^{\,0}(0) + 3J_{ij}^{\,1}(\omega_0) + 6J_{ij}^{\,2}(2\omega_0) \right) + R_{leak}$$

$$\tag{3}$$

$$R_{ij} = \sigma_{ij} = K \left(-J_{ij}^{\,0}(0) + 6J_{ij}^{\,2}(2\omega_0) \right)$$

where ω_0 is the Larmor frequency and $K = (2\pi/5)\,\gamma^4\, \hbar^2$; ρ_{ii} is the auto relaxation rate, with R_{leak} as additional term to account for all non-dipolar relaxation mechanisms, and σ_{ij} is the cross relaxation rate between spin i and j. $J_{ij}^{\,m}(\omega)$ denotes the spectral densities at multiple values of the spin Larmor frequency $((m \times \omega_0)$ with $m = 0, 1, 2)$. Cross relaxation is normally measured by observing intensity changes upon saturation or inversion of spins. Two-dimensional NMR techniques have been developed to investigate the resulting nuclear Overhauser effects (NOE) [35]. The integrated intensities I_{ij} in a 2D NOE spectrum recorded with mixing time τ_m, are given by the matrix equation [35]:

$$I(\tau_m) = \exp(-R\tau_m)\, M_0 = A\, M_0 \tag{4}$$

which is the formal solution of the differential Eq. 1 for N start conditions. This defines the exponential matrix of normalised NOE intensities A

$$A = \exp(-R\tau_m) \tag{5}$$

that will be referred to as "the NOE matrix".

The spectral densities $J_{ij}{}^m(\omega)$ in the definitions of the relaxation rates ρ_{ii} and σ_{ij} are cosine Fourier transforms

$$J_{ij}{}^m(\omega) = \int_0^\infty C_{ij}{}^m(t)\, \cos(\omega t)\, dt \tag{6}$$

where the $C(t)$ are correlation functions describing the time evolution of the dipolar interaction. Assuming that the molecule is a rigid body moving isotropically in solution, the correlation functions in Eq. 6 can be described by a simple exponential of the form $\exp(-t/\tau_c)$, τ_c being the correlation time for the overall rotation of the molecule. With this simple definition, the spectral density $J_{ij}{}^m(\omega)$ becomes:

$$J_{ij}{}^m(\omega) = \frac{1}{4\pi r_{ij}^6} \left(\frac{\tau_c}{1+(\omega\tau_c)^2} \right) \tag{7}$$

If the assumption of a rigid body is not valid, which is usually the case for biomolecules, the description of $C(t)$ by a simple exponential is no more correct. The correlation function should then be defined in a more general form as

$$C_{ij}{}^m(t) = \left\langle \frac{Y_{2m}(\Phi_{ij}^{lab}(t_0+t))\, Y_{2m}^*(\Phi_{ij}^{lab}(t_0))}{r_{ij}^3(t_0+t)\, r_{ij}^3(t_0)} \right\rangle \tag{8}$$

Here the angular brackets indicate a time average over t_0 (which is equivalent to an ensemble average); Y_{2m} are the second order spherical harmonics and r and Φ^{lab} denote the length and polar angles of the interproton vector in the laboratory frame of coordinates. The form of the spectral density functions $J_{ij}{}^m(\omega)$ could be obtained experimentally from relaxation measurements. The feasibility of this approach has been recently demonstrated by Peng and Wagner [36,37] who mapped the spectral densities of N-H vectors using heteronuclear relaxation experiments. In theory, this function can also be computed from a long MD trajectory. With present day computational facilities $C(t)$ can only be computed with sufficient statistical accuracy for t-values of the order of 0.1 to 1.0 ns. Fortunately, for many interproton vectors $C(t)$ is observed to reach a plateau value after a few ps, indicating that fast picosecond motions are well separated from slower processes [38,39].

Following the approach developed by Lipari and Szabo [40,41], a "model-free" description of $C(t)$ can be set up in terms of two characteristic times, τ_p, the time in which $C(t)$ decays to the initial plateau value due to the fast internal motions, and τ_c, the correlation time for the overall rotation of the molecule. Assuming isotropic tumbling, and transforming to molecule fixed coordinates Φ^{mol} one has [38]

$$C_{ij}^{m}(t) = \frac{\exp(-t/\tau_c)}{4\pi} C_{ij}^{int}(t) \tag{9}$$

where the internal motion correlation functions $C^{int}(t)$ are defined as

$$C_{ij}^{int}(t) = \frac{4\pi}{5} \sum_{n=-2}^{2} \left\langle \frac{Y_{2n}(\Phi_{ij}^{mol}(t_0+t)) Y_{2n}^{*}(\Phi_{ij}^{mol}(t_0))}{r_{ij}^{3}(t_0+t) r_{ij}^{3}(t_0)} \right\rangle \tag{10}$$

According to the addition theorem $C_{ij}^{int}(0) = \langle r_{ij}^{-6} \rangle$. The plateau value of the correlation function, $C_{ij}^{int}(t)$, can thus be defined quite generally as $S_{ij}^{2} \langle r_{ij}^{-6} \rangle$, where S_{ij}^{2} is a generalised order parameter. It has a value between 0 and 1, and can be calculated from an MD trajectory by using Eq. 10 and estimating the plateau value for each interproton vector. Within this simplified model the functions $C_{ij}^{int}(t)$ can be written as

$$C_{ij}^{int}(t) = \left\langle \frac{1}{r_{ij}^{6}} \right\rangle \left[S_{ij}^{2} + (1-S_{ij}^{2}) \exp(-t/\tau_p) \right] \tag{11}$$

where the initial decay has been written as an exponential. Combining Eqs 6 and 8-11, one arrives at [38,39]

$$J_{ij}^{m}(\omega) = \frac{1}{4\pi} \left\langle \frac{1}{r_{ij}^{6}} \right\rangle \left[\frac{S_{ij}^{2} \tau_c}{1+(\omega\tau_c)^2} + \frac{(1-S_{ij}^{2}) \tau_{cp}}{1+(\omega\tau_{cp})^2} \right] \tag{12}$$

with $\tau_{cp}^{-1} = \tau_c^{-1} + \tau_p^{-1}$. Assuming that $\tau_p \ll \tau_c$, the second term in Eq. 12, related to the initial decay, vanishes and $J_{ij}^{m}(\omega)$ becomes similar to the definition of Eq. 7, with S_{ij}^{2} as additional scaling factor and r_{ij}^{-6} replaced by the average $\langle r_{ij}^{-6} \rangle$.

Besides the effects of fast local motions, other dynamic processes like, for example, methyl group rotation and aromatic ring flips can affect the relaxation behaviour of biomolecules. Their inclusion in the calculation of the theoretical NOE intensities has been described in detail by Koning *et al.* [42].

2.2   COMPUTATIONAL ASPECTS

Basically two methods are available for the computation of theoretical NOE intensities. These have been recently reviewed by Forster [43]. The first involves finding the eigenvectors and eigenvalues of the real symmetrical matrix $\mathbf{R}$ by standard matrix techniques [21]:

$$\Lambda = \mathbf{X}^{-1}\,\mathbf{R}\,\mathbf{X} \tag{13}$$

Combining this expression with Eq. 5 leads to

$$\mathbf{A} = \exp(-\mathbf{R}\tau_m) = \mathbf{X}\,\exp(-\Lambda\tau_m)\,\mathbf{X}^{-1} \tag{14}$$

The matrix diagonalisation scales as N^3 where N is the dimension of the matrix $\mathbf{R}$ and can become extremely time-consuming when N increases. The calculations can be speeded up by making use of the sparseness of the relaxation matrix; due to the r^{-6} dependence on the distances of the cross relaxation rates, only a few elements of $\mathbf{R}$ are significantly larger than zero. Thus, it is possible to apply a spherical cut-off around each proton pair defining a subset of NOEs with only the neighbours. In this way, for each peak, a small eigenvalue problem will then be solved. The second approach [44] for the calculation of theoretical intensities is based on a numerical integration of the differential equations (Eq. 1) describing the time dependence of the NOE intensities. Several integration schemes exist (Taylor series integration, Euler algorithm, various Runge-Kutta methods) and their accuracy and performance have been addressed by Forster [43]. A choice between the various methods has to be made depending on the dimension of the system and on the purpose of the NOE calculation; for example, numerical integration is no more suitable if eigenvalues and/or eigenvectors are required (*e.g.* for the analytical NOE gradient, see section (2.4).

2.3   FROM NOES TO DISTANCES

2.3.1. *Two-spin approximation*

Proton-proton distance constraints are most conveniently derived from cross-peak intensities in 2D NOE spectra taken at various short mixing times. The initial build-up rate of these peaks is proportional to the cross relaxation rates σ_{ij} between protons i and j [45]. Therefore, these cross relaxation rates can be measured either from a single 2D NOE spectrum taken with a sufficiently short mixing time or, more accurately, from a build-up series recorded with various mixing times [45-47]. For a large molecule, assuming isotropical tumbling and no internal motions, the $J_{ij}^{0}(0)$ term in Eq. 3 dominates the cross relaxation rate σ_{ij} which becomes simply related to the distance r_{ij} and the correlation time τ_c:

$$\sigma_{ij} \propto -\tau_c \, r_{ij}^{-6} \qquad (15)$$

Therefore, with a known calibration distance r_{cal} the proton-proton distances follow from the relation

$$r_{ij} = r_{cal} \, (\sigma_{cal} \, / \, \sigma_{ij} \,)^{1/6} \qquad (16)$$

In practice Eqs 15 and 16 are only approximately valid. There are two main problems associated with accurate determination of proton-proton distances. The first is that proteins are not rigid bodies, and intramolecular mobility leads to non-linear averaging of distances and to different effective correlation times for the different interproton vectors in the molecule. The second is that of indirect magnetisation transfer or "spin-diffusion" [48]. In reality the NOE cross-peaks are the result of multispin relaxation and only in the limit of extremely short mixing times (where the signal-to-noise ratio is poor) is the two-spin approximation of Eq. 15 valid. For these reasons, a common approach is that of translating the NOE intensities into distance ranges (e.g. 2-3, 2-4, 2-5 Å for strong, medium and weak NOEs, respectively) rather than attempting to obtain precise distances [49].

It has become possible to circumvent the spin diffusion problem by the use of relaxation matrix calculations which take into account all indirect magnetisation transfers. These methods will be developed in the next paragraph.

2.3.2. Relaxation Matrix Approaches

When a model for the structure and dynamics of a molecule is available, the NOEs can be calculated from the spectral densities as shown above (see Eqs 3, 5, 13, 14). The opposite route from experimental NOEs to relaxation parameters is also possible. The relaxation matrix **R** can be obtained after diagonalisation of the NOE matrix **A** [22]

$$\mathbf{X}^{-1}\,\mathbf{A}\,\mathbf{X} = \mathbf{D} = \exp(-\Lambda\tau_m) \tag{17}$$

$$\mathbf{R} = -\,\mathbf{X}\left[\frac{\ln\,\mathbf{D}}{\tau_m}\right]\mathbf{X}^{-1} \tag{18}$$

Thus, theoretically the matrix $\mathbf{R}$ can be obtained directly from 2D NOE experiments taken at suitably chosen mixing times. In practice, this is not possible for large molecules, since the experimental NOE matrix is incompletely known due to overlap and missing assignments. Hybrid matrix methods has been proposed to circumvent this problem, in which the experimental data are supplemented by theoretical NOEs calculated from a model structure as implemented in IRMA [23,25], MARDIGRAS [26], MORASS [50], FIRM [30] and CROSREL [31]. An overview of the Iterative Relaxation Matrix Approach (IRMA) is given in Figure 1.

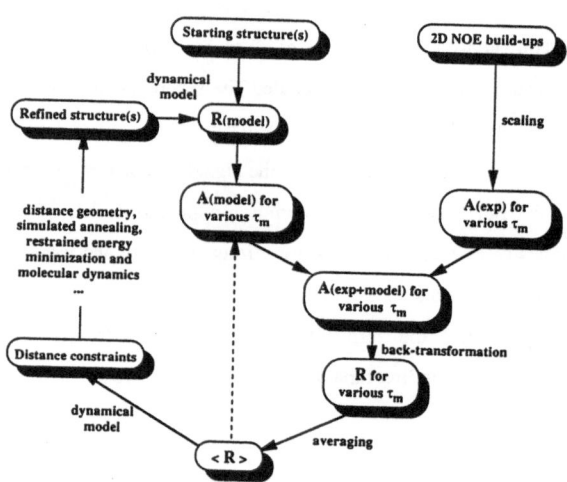

Figure 1: Overview of the IRMA scheme. (The stepped line indicates the possible direct use of the back-calculated relaxation matrix for the computation of new theoretical NOEs for a progressive mixing of theoretical and experimental NOE data as found in other implementations (see text)).

The starting point to the calculations can be a single structure or an ensemble of structures, the relaxation matrix being built from the distances of the corresponding contributions averaged as $<r^{-6}>$ in the latter case. A theoretical NOE matrix is then computed using standard matrix techniques (cf. Eqs 13-14) and combined with the experimental data. The resulting matrix can then be transformed back to a corrected relaxation matrix from which new distances are calculated. These distances are used to generate an improved model. The whole process is repeated until convergence is obtained. In other implementations (MARDIGRAS, MORASS, FIRM), the theoretical information is progressively introduced in the hybrid matrix and the back-calculated corrected relaxation matrix is directly used to calculate new NOE matrices (stepped path in Figure 1). Only after a few cycles are distances calculated and used to generate new structures. To increase the accuracy, the calculations can be performed for a series of mixing times; upper and lower bound margins can then be related to the precision with which the elements of $<R>$ can be calculated, *i.e.* their variation with the mixing time. It has been shown that distances obtained from these relaxation matrix procedures result in an increased accuracy of the structures [51,52].

Similar procedures have been developed that do not rely on a starting model [27-29]. We have shown, however, that the choice of the model structure in IRMA has little effect on the calculated distances [52]: for crambin a 46-amino acids protein, distances obtained from a linear chain of crambin and from the well-refined crystal structure [53] only differ in average by 0.15 Å with a maximum deviation of 0.65 Å for long range contact in a β-sheet. In general, however, relaxation matrix calculations can be started from low resolution model structures that have been obtained at previous structure determination stages from qualitative distance restraints.

2.4.   DIRECT USE OF NOES AS STRUCTURAL CONSTRAINTS

2.4.1. *2D NOE Intensities*

Yip and Case [32] proposed a method for structure refinement based on a direct comparison of experimental and theoretical NOE intensities, avoiding the conversion to distance constraints. The experimental NOE intensities are directly introduced as constraints in a refinement procedure using a penalty function of the form:

$$V_{NOE} = \sum w_{ij} \left[\left(f \, A_{ij}^{theo} \right)^x - \left(A_{ij}^{exp} \right)^x \right]^y \qquad (19)$$

where A_{ij}^{exp} and A_{ij}^{theo} represent the experimental and theoretical NOE intensities between protons i and j, respectively, f is a scaling factor and w_{ij} a weighting function. Often a quadratic potential is used (y=2), directly with the NOE intensities (x=1) [32,54-56] or with their sixth-root (x=1/6) [57] or inverse sixth-root (x=-1/6) [58], the last two definitions being closer to a standard distance based potential. To account for the fact that NMR experimental data are collected as time- and ensemble-averaged quantities, the theoretical intensities in Eq. 19 can be replaced by their time- or ensemble-averages denoted as $\overline{A_{ij}^{theo}}$ in a similar way as for distances [59,60] and J coupling restraints [61]. The averaged intensities, and no longer those calculated from a single static structure, are now required to satisfy the experimental constraints [62]

Calculation of the NOE forces. Calculation of the NOE forces requires the calculation of the gradient of the NOE potential function of Eq. 19 and therefore the calculation of the gradient of the NOE intensities ∇A^{theo}. An analytical solution of this gradient has been proposed by Yip and Case [32] which has been implemented since then by others as well [57,56,63]:

$$\nabla_\mu A_{ij}^{theo} = \sum_{rstu} X_{ir}\, X_{rs}^{-1}(\nabla\mu R)_{st} X_{tu}\, X_{uj}^{-1}\, f_{ru}$$

$$f_{ru} = \frac{\exp(-\Lambda_r\tau_m) - \exp(-\Lambda_u\tau_m)}{(\Lambda_r-\Lambda_u)} \quad (r{\neq}u) \; ; \qquad f_{rr} = -\tau_m \exp(-\Lambda_r\tau_m)$$

(20)

Another approach is based on a numerical evaluation of ∇A^{theo} [54,64a]. Both methods are time consuming. An approximation for the calculation of ∇A^{theo} can be based on the first two terms of the power series of the exponential of Eq. 5 [55]

$$A = \exp(-\tau_m R) = 1 - \tau_m R + \frac{1}{2}\,\tau_m^2\, R^2 \, ...$$

(21)

and the gradient becomes

$$\nabla A_{ij} = \nabla (1 - \tau_m R)_{ij} = -\tau_m \nabla R_{ij}$$

(22)

With this simple approximation, the NOE forces F_{NOE} can be computed rapidly, each NOE peak only giving rise to forces on the corresponding protons. The computational costs are considerably reduced in comparison with those for an analytical evaluation of the gradient, the time consuming part becoming the computation of the theoretical NOE intensities. The entire procedure scales as a N^3 problem if a diagonalisation method is used for calculating the NOEs. This can be reduced by the use of a spherical cut-off around each proton pair ij defining an experimental NOE peak. With this approximation the spin diffusion contributions of all neighbours within two spheres around the two protons defining a NOE peak are taken into account. For each NOE a small relaxation matrix is calculated.

3. Structure Calculations

3.1. DISTANCE GEOMETRY (DG)

The metric matrix distance-geometry (DG) algorithm [64b,11] was known well before structure determination by NMR became possible [12,13]. This method does not rely on a starting conformation, is in this respect free from operator bias and therefore very attractive for structure determination. The DG procedure amounts to the following. First, upper and lower bound matrices U and L are set up for all atom-atom distances of the molecule. Some of the elements u_{ij} and l_{ij} follow from standard bond lengths and bond angles of the covalent structure and from experimentally found distance ranges from NOEs and J-coupling constants. A bound smoothening procedure using triangle inequalities extends the constraints to all elements of U and L. This procedure can even be extended to satisfy tetrangle inequalities as for example implemented in the DG-II package [65]. Then, a distance matrix D is set up with distances chosen randomly between upper and lower bounds, $l_{ij} \leq d_{ij} \leq u_{ij}$. The elements of D can be chosen in such a way that they not only lie between their respective upper and lower bounds, but also themselves obey the triangle inequalities; this latter technique is known as *random metrisation* . The so-called "embedding" algorithm then finds the best 3D structure consistent with the distances d_{ij}. This structure must be optimised with an error function consisting of chirality and distance constraints. This forces the amino acids side chains to adopt the correct chirality and the distances to satisfy the upper and lower bound criteria, although usually the DG structures still contain a number of violations of the distance bounds. By repetition of the DG calculations with randomly chosen distance matrices D, families of structures can be obtained which allow to judge how uniquely the structure is determined by the constraints.

Another method, also termed distance geometry but using a quite different mathematical procedure, was suggested by Braun and Go ‾ [14]. Here, the biomolecular conformation is calculated by minimising a distance constraint error function. Special features of the method are that dihedral angles are used as independent variables rather than Cartesian coordinates and that it uses a variable target function, first satisfying local constraints (between amino acids nearby in the polypeptide chain) and later including long-range constraints. This approach avoids becoming trapped in a local minimum of the target function. Usually one starts with various initial conformations obtained by taking random values for the dihedral angles. This method, which has been implemented since in various programs, e.g. DISMAN [14] and more recently DIANA [66], has rather similar efficiency and convergence properties as the metric matrix distance-geometry algorithm.

3.2.   SIMULATED ANNEALING (SA)

Although the distance-geometry methods do not need starting structures and are therefore not subject to operator bias, this does not mean that they sample the allowed conformational space (consistent with the bounds) in a truly random fashion. Simplified molecular dynamics calculations with only geometric constraints were developed to increase the sampling of the DG methods. These are termed in general as simulated annealing (SA) procedures. Although these latter can be directly used to generate structures starting from random coordinates, they are most commonly combined with distance-geometry calculations, the starting structures for the simulated annealing stage being obtained by distance geometry embedding. Simulated annealing involves raising the temperature of the system followed by slow cooling in order to overcome local minima and locate the region of the global minimum of the target function V_{tot} [16]. This is achieved by solving Newton's equations of motion

$$m_i \, \mathbf{r}''_i = \mathbf{F}_i \tag{23}$$

with the forces given by

$$\mathbf{F}_i = -\delta V_{tot} / \delta \mathbf{r}_i \tag{24}$$

The potential V_{tot} usually contains the following terms:

$$V_{tot} = V_{covalent} + V_{repel} + V_{disre} + V_{dihre} + V_{NOE} \tag{25}$$

$V_{covalent}$ is a potential to maintain correct bond lengths, angles, planes and chirality, V_{repel} is a simple repulsion term to prevent unduly close non-bonded contact, V_{disre} and V_{dihre} represent the experimental

NOE distance and dihedral angles constraints, respectively, and V_{NOE} is an extra term for direct comparison of theoretical and experimental NOE intensities as defined in Eq. 19. Usually the NOE distance constraints are represented by a flat well potential of the form

$$V_{disre} = \begin{cases} \frac{1}{2}K_{disre}\,(d_{ij}-l_{ij})^2 & 0 \leq d_{ij} \leq l_{ij} \\\\ 0 & l_{ij} \leq d_{ij} \leq d_{ij} \\\\ \frac{1}{2}K_{disre}\,(d_{ij}-u_{ij})^2 & u_{ij} \leq d_{ij} \end{cases} \qquad (26)$$

where l_{ij} and u_{ij} are the value of lower and upper limits of the target distances, respectively. Other forms that contains a linear part at long distances are also used [19] to avoid too large energies which may give computational problems. A similar definition as in Eq. 26 can be used for dihedral angle constraints, but other forms have also been proposed as [67]:

$$V_{dihre} = K_{dihre}\,[1 - \cos(\phi - \phi_0)] \qquad (27)$$

where ϕ_0 represents the dihedral angle value obtained from J-coupling data. The various terms in Eq. 25 have variable force constants which can be progressively increased during the annealing.

The combination of distance geometry, for generating starting structures with an approximate folding by embedding a subset of atoms, and simulated annealing calculations provides a powerful and efficient tool for the structure determination of biomolecules [16,68]. This hybrid method should be particularly useful to solve the structure of large proteins for which computational costs may become a limiting factor.

3.3. RESTRAINED MOLECULAR DYNAMICS AND ENERGY MINIMISATION (RMD, REM)

The quality of biomolecular structures based on geometric constraints can be improved by taking energy considerations into account. For instance, in DG structures amino acids side chains often adopt eclipsed conformations. Also, hydrogen bonds and salt bridges may not be formed unless they are specifically introduced as constraints. In restrained molecular dynamics refinement structures are optimised simultaneously with respect to a potential energy function and a set of experimental restraints. Of course the success of this method now depends to a large extend on the quality of the force field used. It is therefore important to realise the limitations and approximate nature of this force field, especially when the calculations do not include solvent molecules.

In RMD calculations, Eqs 23 and 24 are integrated with the potential function given by

$$V_{tot} = V_{bond} + V_{angle} + V_{torsion} + V_{vdW} + V_{Coulomb} + V_{disre} + V_{dihre} + V_{NOE} \quad (28)$$

The first two terms tend to keep bond lengths and angles at their equilibrium values. $V_{torsion}$ is a sinusoidal potential describing rotations about bonds; for V_{vdW} (the van der Waals interaction) usually a Lennard-Jones potential is taken, and $V_{Coulomb}$ describes the electrostatic interactions. The three last terms, similar to those in Eq. 25, distinguishes RMD from more conventional MD simulations.

A RMD simulation is usually preceded and followed by restrained energy minimisation (REM) using steepest descent or conjugated gradient methods to bring the energy down to an acceptable level. REM using the same potential energy function as RMD (Eq. 28) usually induces only small changes in the structures by driving them to the closest minimum and is not able to take them out of this. By contrast, RMD is able to overcome barriers of the order of kT because of the kinetic energy in the system and therefore has a much larger radius of convergence. RMD works as an efficient minimiser, since excess potential energy, converted to kinetic energy, is drained off by coupling the system to a thermal bath of constant temperature. The sampling of the conformational space consistent with the experimental constraints can be increased by using high temperature simulations [69] or potential energy annealing conformational search methods (PEACS) [70] in which the potential energy is coupled to a reference energy bath which energy level is progressively decreased over the run, or by introducing a fourth dimension [71] as recently implemented in GROMOS (4D-MD) [72], which, by increasing the number of degrees of freedom in the system, allows to circumvent energy barriers present in the 3D potential energy hyper-surface.

Further extensions of the method, especially useful for highly mobile structures, include time-averaging [59,73] or ensemble-averaging [60,74] of the experimental restraints. In these methods, V_{disre} is calculated from time- or ensemble-averaged distances instead of instantaneous ones. Forces are thus applied on the corresponding protons only if the averaged distance violates the restraint. These methods allow a larger conformational freedom consistent with the NOE restraints. The application of time-averaged RMD to Tendamistat [73] suggested that some side-chains are more flexible than what conventional refinement had indicated. The use of time-averaged restraints has recently been described for direct structure refinement with J-coupling [61] and NOE [62] data.

4. Application: The Arc Repressor

A general view of the structure refinement process with use of relaxation matrix calculations can be best illustrated at the hand of a specific example: the Arc repressor. The solution structure of the Arc repressor of *Salmonella* bacteriophage P22, a dimeric sequence-specific DNA binding protein [75], has been determined from 2D NMR data using an "ensemble" Iterative Relaxation Matrix Approach (IRMA) followed by direct NOE refinement with DINOSAUR [76].

824 NOE build-up series in H_2O and D_2O and 130 additional qualitative distance constraints were collected for Arc. This resulted in a set of 954 constraints consisting of 350 intraresidue, 215 sequential, 181 medium range and 208 long range constraints; 144 of the total set correspond to intermonomer constraints. Twenty-four hydrogen bonds were also explicitly defined. The constraints were duplicated to account for the dimeric nature of Arc giving a total set of 2004 constraints (NOE's + hydrogen bonds). Of the set of 954 constraints per monomer, 824 NOEs with good quality build-up curves were used in the IRMA and DINOSAUR calculations, the others being introduced as qualitative constraints. From the stereospecific assignment procedure, which resulted in the assignment of 16 β-methylene diastereotopic proton pairs (40%) and of the methyl groups of 4 valine residues, constraints could be imposed on 21 χ_1 dihedral angles.

In the initial stage, a set of 51 structures was generated with Distance Geometry and further refined with a combination of restrained energy minimisation and restrained molecular dynamics using the GROMOS force field [77] in a parallel refinement protocol. Distance constraints were obtained from the set of 854 NOE build-ups *via* relaxation matrix calculations from the ensemble of structures except for the initial cycle in which a structure obtained by Breg *et al.* [78] was chosen as model. Methyl group rotation, aromatic ring flips and internal mobility effects (*via* order parameters obtained from a free MD run in water) were included in these calculations. Convergence was obtained after three IRMA cycles. The best 16 structures were finally refined with direct NOE constraints following a slow-cooling simulated annealing protocol from 600 to 1 K in 1.5 ps followed by restraint energy minimisation [79]. In this final refinement stage, theoretical NOE intensities were directly compared to the experimental data using a simple quadratic restraining potential (x=1, y=2 in Eq. 19) with as weighting factor an experimental error defined as $w_{ij} = (N + \varepsilon A_{ij}^{exp})^{-2}$ [55] N corresponds to a noise level and ε to a relative error on the experimental intensities. The forces were derived using the simple two-spin approximation for the gradient of the NOE function (Eq. 22). Dynamic assignment was applied to the peaks involving unassigned diastereotopic groups.

A view of the backbone of the 16 best Arc structures at various stages during the refinement (DG/SA, IRMA 1st, 2nd and 3rd cycles and DINOSAUR) is presented in Figure 2, which clearly shows the improvement of the structures. The precision of the well defined region of Arc including the β-sheet and the two α-helices (residues 8 to 48) increases during the IRMA refinement as confirmed by the backbone r.m.s.d.which decreases from 1.0 Å for the DG/SA structures to 0.50 Å for the structures obtained after the third IRMA cycle. This value increases slightly to 0.55 Å after the direct NOE refinement indicating that the use of direct NOE constraints might introduce more variability in the structures (due to spin diffusion effects there are more ways to fulfil a NOE constraint than a distance constraint).

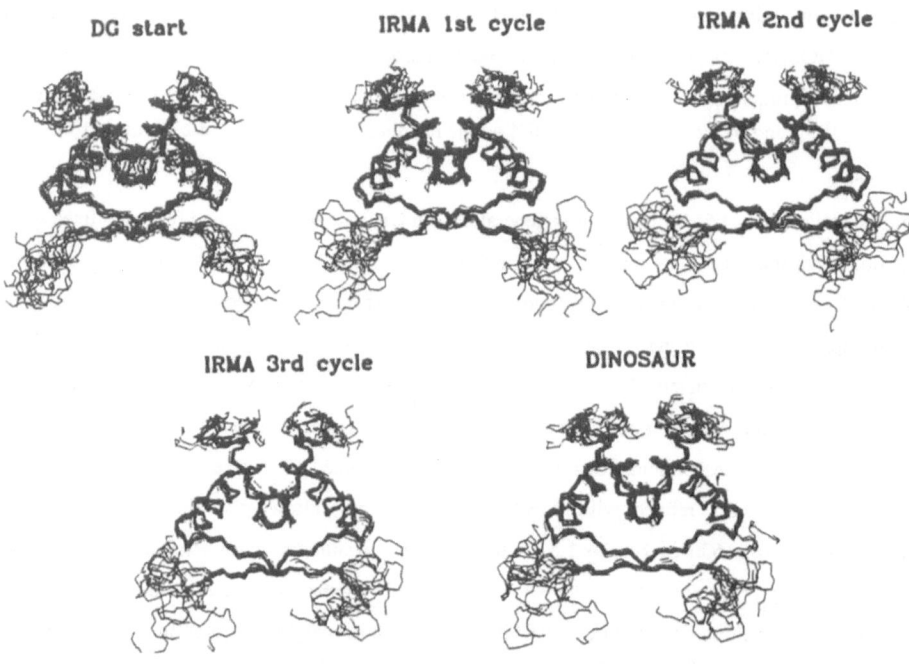

Figure 2: View of the Arc repressor structures during the refinement with NMR constraints. The structures corresponding to the sixteen best final conformations are shown directly after the DG/SA calculations in the first IRMA cycle, after each IRMA cycle and after direct NOE refinement with DINOSAUR. The structures were superimposed on C_α,C,N atoms of the well defined region of Arc (residues 8-48 of each monomer).

The R factors ($Q^{1/6}$ definition [76]), however, show a substantial decrease during the DINOSAUR refinement step as can be seen from Figure 3. The final structures satisfy the NMR constraints with no significant deviations from dihedral angle restraints obtained from J coupling data and very good agreement with the experimental NOE data. The distribution of the experimental constraints as function of the residue sequence is plotted in Figure 3, together with the precision (r.m.s.d. on C_α and all heavy atoms), the backbone angular order parameters and the $Q^{1/6}$ factors [51,80].

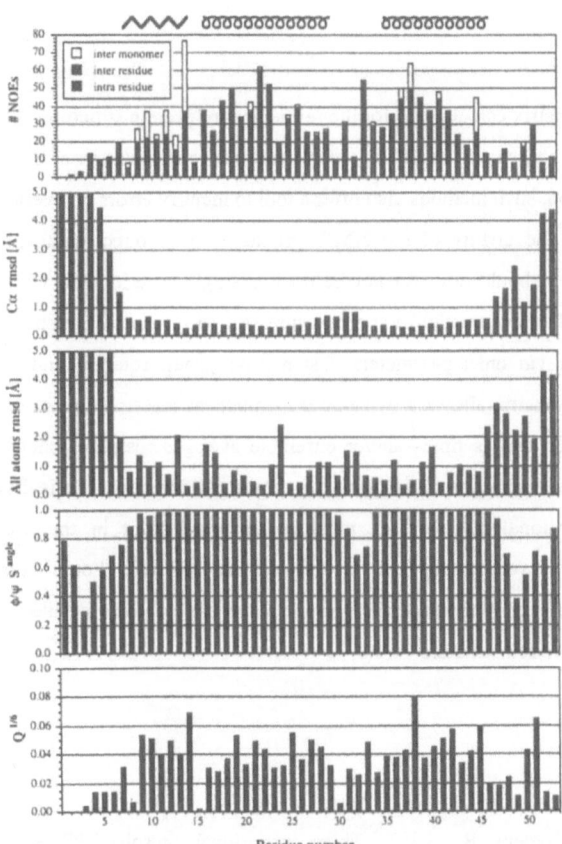

Figure 3: Number of constraints, r.m.s.d. (from the average) on C_α atoms and on all heavy atoms (the structures were superimposed on C_α,C,N atoms of the well defined region of Arc (residues 8-48 of each monomer)), backbone angular order parameters (averages from ϕ and ψ angular order parameters) (Eq. 41) and $Q^{1/6}$ factors as function of the residue sequence for the final Arc structures. Secondary structure

206

A correlation can be noticed between the number of constraints and the precision per residue, but no such relation can be found for the $Q^{1/6}$ factors which all represent low values (~0.05) (for comparison, the $Q^{1/6}$ factors for a linear chain of Arc have values around 0.55). Regions of high variability are found in the N- and C-terminal parts for which few constraints are available. The structures, which have been deposited in the Brookhaven Protein Data Bank (entry 1ARR), have good stereochemical qualities and present an extensive pattern of hydrogen bonds and a good packing with a very well defined core.

5. Conclusion

The use of relaxation matrix calculations for the refinement of protein structures from NMR, as illustrated here for the Arc repressor, should result in a more accurate representation of the solution structure of a biomolecule. In addition, such methods also offer a tool to identify errors or overinterpretation in distance constraints and assess the quality of the NMR structures, by a direct comparison of theoretical and experimental data. They should however not be used as single criterion for accuracy but together with others like the stereochemical quality of the structure. An adequate description of motional effects like internal mobility (*e.g. via* order parameters, fast methyl group rotation and aromatic ring flips) or conformational heterogeneity, allows a dynamic description of the structure in solution. A treatment of the experimental NMR data as time- and/or ensemble averaged restraints should even lead us further toward a more realistic description of solution structures. Additional work is however required in this direction since the motional model is the most controversial point in the use of relaxation matrix calculations. The new developments both in software and hardware should extend the range of such computations to proteins of increasing size and allow the use of more detailed motional models for a better description of solution structure.

6. References

1. Vuister, G.W., Boelens, R. (1987). Three-dimensional *J*-resolved NMR spectroscopy. *J. Magn. Reson.* **73**, 328-333.
2. Oschkinat, H. , Griesinger, C., Kraulis, P.J., Sørensen, O.W., Ernst, R.R., Gronenborn, A.M., Clore, G.M. (1988). Three-dimensional NMR spectroscopy of a protein in solution. *Nature* **332**, 374-376.
3. Griesinger, C., Sørensen, O.W., Ernst, R.R. (1989). Three-dimensional Fourier spectroscopy. Application to high-resolution NMR. *J. Magn. Reson.* **84**, 14-63.
4. Kay, L.E., Marion, D., Bax, A. (1989). Practical aspects of 3D heteronuclear NMR of proteins. *J. Magn. Reson.* **84**, 72-84.

5. Fesik, S.W. , Zuiderweg, E.R.P. (1990). Heteronuclear three-dimensional NMR spectra of isotopically labelled biological macromolecules. Q. Rev. of Biophys. **23**, 97-131.

6. Boelens, R. , Vuister, G.W., Padilla, A., Kleywegt, G.J., de Waard, P., Koning, T.M.G., Kaptein, R. (1990). Three-dimensional NMR spectroscopy of biomolecules. In: Structure & Methods, Volume 2: DNA Protein Complexes & Proteins, Adenine Press, 63-81.

7. Clore, G.M., Gronenborn, A.M. (1991a). Application of three- and four-dimensional heteronuclear NMR spectroscopy to protein structure determination. Prog. N.M.R. Spectr. **23**, 43-92.

8. Bax, A., Grzesiek, C. (1993). Methodological advances in protein NMR. Acc. Chem. Res. **26**, 131-138.

9. Noggle, J.H., Schirmer, R.E. (1971). The nuclear Overhauser effect-chemical applications, Academic, New York.

10. Neuhaus, D., Williamson, M. (1989). The nuclear Overhauser effect in structural and conformational analysis. VHC Publisher, New York.

11. Crippen, G.M., Havel, T.F. (1978). Stable calculation of coordinates from distance information. Acta Crystallogr. **A34**, 282-284.

12. Havel, T.F., Kuntz, I.D., Crippen, G.M. (1983). The theory and practice of distance geometry. Bull. Math. Biol. **45**, 665-720.

13. Havel, T.F., Wüthrich, K. (1984). A distance geometry program for determining the structures of small proteins and other macromolecules from nuclear magnetic resonance measurement of intramolecular ^{1}H-^{1}H proximities in solution. Bull. Math. Biol. **46**, 673-698.

14. Braun, W., Go ⁻, N. (1985). Calculation of protein conformations by proton-proton distance constraints. A new efficient algorithm. J. Mol. Biol. **186**, 611-626.

15. Clore, G.M., Brünger, A.T., Karplus, M., Gronenborn , A.M. (1986a). Application of molecular dynamics with interproton distance restraints to three-dimensional protein structure determination: a model study of crambin. J. Mol. Biol. **191**, 523-551.

16. Nilges, M., Clore, G.M., Gronenborn, A.M. (1988). Determination of three-dimensional structures of proteins from interproton distance data by hybrid distance geometry-dynamical simulated annealing calculations. F.E.B.S. Lett. **229**, 317-324.

17. Scheek, R.M., van Gunsteren, W.F., Kaptein, R. (1989) Molecular Dynamics techniques for determination of molecular structures from Nuclear Magnetic Resonance data. In: Oppenheimer NJ, James TL (eds) Nuclear Magnetic Resonance, Methods in enzymology. Vol. 177. Academic, New York, 204-218.

18. Van Gunsteren, W.F., Kaptein, R., Zuiderweg, E.R.P. (1983). Use of Molecular Dynamics computer simulations when determining protein structure by 2D-NMR. In: Olson, W.K. ed. Nucleic acid conformation and dynamics, report of the NATO/CECAM workshop, Orsay, 79-92.

19. Kaptein, R., Zuiderweg, E.R.P., Scheek, R.M., Boelens, R., van Gunsteren, W.F. (1985). A protein structure from nuclear magnetic resonance data. Lac repressor headpiece. J. Mol. Biol. **182**, 179-182.

20. Clore, M.G., Gronenborn, A.M., Brünger, A.T., Karplus, M. (1985). Solution conformation of a heptadecapeptide comprising the DNA binding helix F of the cyclic AMP receptor protein of Escherichia coli. Combined used of ^{1}H nuclear magnetic resonance and restrained molecular dynamics. J. Mol. Biol. **186**, 435-455.

21. Keepers, J.W., James, T.L.(1984).A theoretical study of distance determination by NMR two-dimensional nuclear Overhauser effect spectra. J. Magn. Reson. **57**, 404-426.

22. Olejniczak, E.T., Gampe, R.T. Jr, Fesik, S.W. (1986). Accounting for spin diffusion in the analysis of 2D NOE data. J. Magn. Reson. **67**, 28-41.

23. Boelens, R., Koning, T.M.G., Kaptein, R. (1988). Determination of biomolecular structures from proton-proton NOEs using a relaxation matrix approach. J. Mol. Struct. **173**, 299-311.

24. Boelens, R., Koning, T.M.G., van der Marel, G.A., van Boom, J.H., Kaptein, R. (1989a). Iterative procedure for structure determination from proton-proton NOEs using a full relaxation matrix approach. Application to a DNA octamer. J. Magn. Reson. **82**, 290-308.

25. Koning, T.M.G. (1990a). IRMA: Iterative Relaxation Matrix Approach for NMR structure determination. Application to DNA fragments. PhD thesis, Utrecht University, the Netherlands.

26. Borgias, B.A., Cochin, M., Kerwood, D.J., James, T.L. (1990). Relaxation matrix analysis of 2D NMR data. Progress in NMR Spectroscopy **22**, 83-100.

208

27. Koehl, P. Lefèvre, J.-P. (1990). The reconstruction of the relaxation matrix from an incomplete set of nuclear Overhauser effects. *J. Magn. Reson.* **86**, 565-583.
28. Madrid, M., Llinás, E., Llinás, M. (1991). Model-independent refinement of interproton distances generated from ^{1}H-NMR Overhauser intensities. *J. Magn. Reson.* **93**, 329-346.
29. Van de Ven, F.J.M., Blommers, M.J.J., Schouten, R.E., Hilbers, C.W. (1991). Calculation of interprotons distances from NOE intensities. A relaxation matrix approach without requirement of a molecular model. *J. Magn. Reson.* **94**, 140-151.
30. Edmonson, S. (1992). NOE R-factors and structural refinement using FIRM, an iterative relaxation matrix program. *J. Magn. Reson.* **98**, 283-298.
31. Leeflang, B.R., Kroon-Batenburg, L.M.J. (1992). CROSREL: Full relaxation matrix analysis for NOESY and ROESY NMR spectroscopy. *J. Biomol. NMR* **2**, 495-518.
32. Yip, P., Case D.A. (1989). A new method for refinement of macromolecular structures based on nuclear Overhauser effect spectra. *J. Magn. Reson.* **83**, 643-648.
33. Solomon, I. (1955). Relaxation processes in a system of two spins. *Phys. Rev.* **99**, 559-565.
34. Tropp, J. (1980). Dipolar relaxation and nuclear Overhauser effects in nonrigid molecules: The effect of fluctuating internuclear distances. *J. Chem. Phys.* **72**, 6035-6043.
35. Macura, S., Ernst, R.R. (1980). Elucidation of cross relaxation in liquids by two-dimensional NMR spectroscopy. *Molec. Phys.* **41**, 95-117.
36. Peng, J., Wagner, G. (1992a). Mapping of spectral density functions using heteronuclear NMR relaxation measurements. *J. Magn. Reson.* **98**, 308-332.
37. Peng, J., Wagner, G. (1992b). Mapping of the spectral densities of N-H bond motions in Eglin C using heteronuclear relaxation experiments. *Biochemistry* **31**, 8571-8586.
38. Olejniczak, E.T., Dobson, C.M., Karplus, M., Levy, R.M. (1984). Motional averaging of proton nuclear Overhauser effects in proteins. Prediction from a molecular dynamics simulation of lysozyme. *J. Am. Chem. Soc.* **106**, 1923-1930.
39. Koning, T.M.G., Boelens, R., van der Marel, G.A., van Boom, J.H., Kaptein, R. (1991). Structure determination of a DNA octamer in solution by NMR spectroscopy. Effect of fast local motions. *Biochemistry* **30**, 3787-3797.
40. Lipari, G., Szabo, A. (1982a). Model-free approach to the interpretation of nuclear magnetic resonance relaxation in macromolecules. 1. Theory and range of validity. *J. Am. Chem. Soc.* **104**, 4546-4559.
41. Lipari, G., Szabo, A. (1982b). Model-free approach to the interpretation of nuclear magnetic resonance relaxation in macromolecules. 2. Analysis of experimental results. *J. Am. Chem. Soc.* **104**, 4559-4570.
42. Koning, T.M.G., Boelens, R., Kaptein, R. (1990b). Calculation of the nuclear Overhauser effect and the determination of proton-proton distances in the presence of internal motions. *J. Magn. Reson.* **90**, 111-123.
43. Forster, M.J. (1990). Comparison of computational methods for simulating nuclear Overhauser effects in NMR spectroscopy. *J. Comp. Chem.* **12**, 292-300.
44. Marion, D., Genest,. M., Ptak, M. (1987). Reconstruction of NOESY maps. A requirement for a reliable conformation analysis of biomolecules using 2D NMR. *Biophys. Chem.* **28**, 235-244.
45. Kumar, A., Wagner G., Ernst, R.R., Wüthrich K. (1981). Buildup rates of the nuclear Overhauser effect measured by two-dimensional proton magnetic resonance spectroscopy: implications for studies of protein conformation. *J. Am. Chem. Soc.* **103**, 3654-3658.
46. Wagner, G., Wüthrich, K. (1979). Truncated driven nuclear Overhauser effect (TOE). A new technique for studies of selective proton-proton Overhauser effects in the presence of spin diffusion. *J. Magn. Reson.* **33**, 675-680.
47. Dobson, C.M., Olejniczak, E.T., Poulsen, F.M., Ratcliffe, R.G. (1982). Time development of proton nuclear Overhauser effects in proteins. *J. Magn. Reson.* **48**, 97-110.
48. Kalk, A., Berendsen H.J.C. (1976). Proton magnetic relaxation and spin diffusion in proteins. *J. Magn. Reson.* **25**, 343-366.
49. Wüthrich, K. (1986). NMR of proteins and nucleic acids. Wiley, New York.
50. Post, C.B., Meadows, R.P., Gorenstein, D.G. (1990). On the evaluation of interproton distances for three-dimensional structure determination by NMR using a relaxation rate matrix analysis. *J. Am. Chem. Soc.* **112**, 6796-6803.

51. Thomas, P.D., Basus, V.J., James, T.L. (1991). Protein solution structure determination using distances from 2D NOE experiments: effect of approximations on the accuracy of derived structures. *Proc. Natl. Acad. Sci. USA* **88**, 1237-1241.
52. Bonvin, A.M.J.J., Rullmann, J.A.C., Boelens, R., Kaptein, R. (1993a). "Ensemble" iterative relaxation matrix approach: a new NMR refinement protocol applied to the solution structure of crambin. *PROTEINS: Struct., Funct. & Genetics* **15**, 385-400.
53. Hendrickson, W.A., Teeter, M.M. (1981). Structure of the hydrophobic protein crambin determined directly from the anomalous scattering of sulphur. *Nature* **220**, 107-113.
54. Baleja, J.D., Moult, J., Sykes, B.D. J. (1990). Distance measurement and structure refinement with NOE data. *J. Magn. Reson.* **87**, 375-384.
55. Bonvin, A.M.J.J., Boelens, R., Kaptein, R. (1991a). Direct NOE refinement of biomolecular structures using 2D NMR data. *J. Biomol. NMR* **1**, 305-309.
56. Mertz, J.E., Güntert, P., Wüthrich, K., Braun, W. (1991). Complete relaxation matrix refinement of NMR structures of proteins using analytically calculated dihedral angle derivatives of NOE intensities. *J. Biomol. NMR* **1**, 257-269.
57. Nilges, M., Habazettl, J., Brünger, A.T., Holak, T.A. (1991a). Relaxation matrix refinement of the solution structure of Squash Trypsin Inhibitor. *J. Mol. Biol.* **219**, 499-510.
58. Stawarz, B., Genest, M., Genest, D. (1992). A new constraint potential for the structure refinement of biomolecules in solution using experimental NOE intensity. *Biopolymers* **32**, 633-642.
59. Torda, A.E., Scheek, R.M., van Gunsteren, W.F. (1989). Time-dependent distance restraints in Molecular Dynamics simulations. *Chem. Phys. Lett.* **157**, 289-294.
60. Scheek, R.M., Torda, A.E., Kemmink, J., van Gunsteren, W.F. (1991). Structure determination by NMR: the modeling of NMR parameters as ensemble averages. In: "Computational aspects of the study of biological macromolecules by NMR". Hoch, J. C. *et al.* (eds), Plenum Press, New York, 209- 217.
61. Torda, A.E., Brunne, R.M., Huber, T., Kessler, H., van Gunsteren, W.F. (1993). Structure refinement using time-averaged *J*-coupling constant restraints. *J. Biomol. NMR* **3**, 55-66.
62. Bonvin, A.M.J.J., Boelens, R., Kaptein, R. (1994). Time- and ensemble-averaged direct NOE restraints. *J. Biomol. NMR* , **4**, 142 149.
63. Pothier, J., Gabarro-Arpa, J., Le Bret, M. (1993). MORNIN: A quasi-Newtonian energy minimizer fitting the nuclear Overhauser data. *J. Comp. Chem.* **14**, 226-236.
64a. Baleja, J. (1992). NOE-based structure refinement without the use of NMR-derived interproton distances. *J. Magn. Reson.* **96**, 619-623.
64b. Blumenthal, L.M. (1970). Theory and application of distance geometry. Chelsea, New York.
65. Havel, T.F. (1991). An evaluation of computational strategies for use in the determination of protein structures from distance constraints obtained by nuclear magnetic resonance. *Prog. Biophysics Molec. Biol.* **56**, 43-78.
66. Güntert, P., Braun, W., Wüthrich, K. (1991). Efficient computation of three-dimensional protein structures in solution from nuclear magnetic resonance data using the program DIANA and the supporting programs CALIBA, HABAS and GLOMSA. *J. Mol. Biol.* **217**, 517-530.
67. De Vlieg, J., Boelens, R., Scheek, R.M., Kaptein, R., van Gunsteren, W.F. (1986). Restrained Molecular Dynamics procedure for protein tertiary structure determination from NMR data: a Lac repressor headpiece structure based on information on *J*-couplings and from presence and absence of NOE's. *Isr. J. Chem.* **27**, 181-188.
68. Holak, T.A., Gondol, D., Otlewski, J., Wilusz, T. (1989). Determination of the complete three-dimensional structure of the trypsin inhibitor from squasl seeds in aqueous solution by nuclear magnetic resonance and a combination of distance geometry and dynamical simulated annealing. *J. Mol. Biol.* **210**, 635-648.
69. Bruccoleri, R.E., Karplus, M. (1990). Conformational sampling using high-temperature molecular dynamics. *Biopolymers* **29**, 1847-1862.
70. Van Schaik, R.C., van Gunsteren, W.F., Berendsen H.J.C. (1992). Conformational search by potential energy annealing: algorithm and application to cyclosporin A. *J. Computer-Aided Mol. Design* **6**, 97-112.
71. Crippen, G.M. (1982). Conformational analysis by conformational embedding. *J. Comp. Chem.* **3**, 471-476.

WHAT DRIVES ASSOCIATION OF α-HELICAL PEPTIDES IN MEMBRANE DOMAINS OF PROTEINS? ROLE OF HYDROPHOBIC INTERACTIONS.

R. G. EFREMOV[1,2]* and G. VERGOTEN[1]

[1]*Université des Sciences et Technologies de Lille, Centre de Recherches et d'Etudes en Simulations et Modélisation Moléculaires (CRESIMM), Bâtiment C8, 59655 Villeneuve d'Ascq, Cedex, France;*
[2]*Shemyakin and Ovchinnikov Institute of Bioorganic Chemistry, Russian Academy of Sciences, Ul. Miklukho-Maklaya, 16/10, Moscow V-437, 117871 GSP, Russia;*

ABSTRACT

In this study we assessed the factors maintaining spatial structure of the α-helix bundles in membrane protein domains as well as in several globular pore-forming proteins. Their hydrophobic and packing properties were analyzed using Monte Carlo simulations in solvents of different polarity, molecular hydrophobicity potential calculations and environmental profiles method. The results obtained in independent techniques are in reasonable agreement, complement each other, and provide a detailed presentation of the spatial hydrophobic nature of membrane helix domains. Despite the lack of sequence homology between the individual α-helices constituing the bundles, their hydrophobic characteristics are very similar and enable a common polarity template to be delineated. The hydrophobic template method was shown to be powerful in recognition and characterization of membrane segments forming the bundles with five and more helices. It efficiently complements environmental profiles methods and threading techniques.

KEYWORDS: structure prediction / hydrophobic organization / membrane-spanning helices / molecular hydrophobicity potential / helix bundles / ion channels

ABBREVIATIONS: FHB - five α-helix bundle; TM - transmembrane; MHP - molecular hydrophobicity potential; MC - Monte Carlo method; ESS - energy of solute-solvent interactions; 2D, 3D - two- and three-dimensional; BRh - bacteriorhodopsin; RC - photoreaction center *Rhodopseudomonas viridis*;

*Author to whom correspondence should be addressed:
Present address: Université des Sciences et Technologies de Lille, Centre de Recherches et d'Etudes en Simulations et Modélisation Moléculaires (CRESIMM), UFR de Chimie, Bâtiment C8, 59655 Villeneuve d'Ascq, Cedex, France;

G. Vergoten and T. Theophanides (eds.), Biomolecular Structure and Dynamics, 211–228.
© *1997 Kluwer Academic Publishers.*

212

1. INTRODUCTION.

The prediction of the spatial organization of membrane domains in proteins from sequence remains one of the most challenging problems in the field of structural biology. Membrane-embedded parts in proteins are of crucial importance for numerous processes in the cell, like transmembrane (TM) transport and signalling, intercellular interactions, and maintenance of the integrity of the cell. An understanding of the structure-activity relationships for membrane proteins is of great fundamental and practical interest. The very nature of membrane proteins (high hydrophobicity, insolubility in nondetergent solutions) makes difficult their structural analysis using experimental techniques. Thus, despite the large number of membrane protein sequences gathered to date, only a few three-dimensional (3D) structures are known to an atomic resolution [for recent reviews see, for example, 1, 2]. This provokes interest in the computer-aided molecular modeling of membrane domains.

Among the membrane-spanning peptides α-helices represent the most frequently occurring conformation. They were shown to exist in the membrane as single autonomous motifs as well as arrangements of helix bundles, although membrane domains in porins and some other proteins reveal β-barrel architecture or mixed helix/sheet folds [2]. Therefore, understanding of the factors determining assembling of the membrane protein parts as well as structure prediction of α-helical membrane moieties are considered as important tasks towards solution of the folding problem for integral membrane proteins. Detailed revision of structures of known membrane-embedded parts in proteins is among the approaches destined to solve these problems. It makes possible formulation of common principles of their structural organization. Thus, analysis of the first 3D structures of bacterial photoreaction centers (RC) [3] led to a conclusion that association of TM helices in the bundles is driven mainly by hydrophobic interactions, while specific salt bridges and hydrogen bonds were also observed. In addition, the membrane-exposed surface of these proteins is believed to be more hydrophobic than that on the helix-helix interfaces and interior of the bundle whereas atomic packing and surface area are similar in both cases. Another interesting observation is that the lipid-exposed residues are less conserved among closely related proteins than those buried in the interior.

Important data on the factors driving helix association can be obtained also from analysis of several known spatial structures of globular proteins revealing pore-forming α-helix bundles. Thus, a promising structural motif is found in several cytosolic enzyme toxins [4]. The family includes cholera toxin and shiga toxin families, and pertussis toxin. Although their sequences and pathogenic mechanisms differ, oligomers of B-subunits of the toxins share the same folding type: it is a pentamer in which each monomer comprises two three-stranded antiparallel β-sheets disposed at the outer surface of the complex, and an α-helix lining a central "pore". While the helix bundles in the toxin molecules do not serve to pump ions, analysis of such spatial motif is very useful in understanding of its packing. The reason is that the similar folds were proposed for some membrane ion transporters. Thus, 9 Å electron microscopic projection maps of the nicotinic achetylcholine receptor (AChR) show that the membrane domain is formed by five α-helix bundle (FHB) surrounded by putative β-sheet layer [5]. The FHB-fold has also been proposed as a model of Ca^{2+} ion channel constitued by TM peptides of phospholamban [6]. Experimental and theoretical studies of several other membrane proteins and channel-forming peptides have led to molecular models with five TM helices lining the ion pore [7].

Dominant role of hydrophobic forces in folding of membrane protein domains stimulates an interest in assessment of their hydrophobic organization. Knowledge of spatial polarity properties of membrane-spanning bundles imposes significant constraints on possible arrangements of TM segments. This allows to reduce considerably a number of possible orientations of helices and construct restricted number of 3D models which can be used as a basis for subsequent molecular modeling. Previously, the hydrophobic characteristics of membrane α-helical moieties were studied essentially by hydrophobic moment analysis [8]. This method employs Fourier transform to calculate periodic differences of hydrophobic and hydrophilic residues in order to identify amphipathic helices from sequences and sequence alignments. Thus, inspection of 3D structures of membrane domains in the RCs revealed that the most hydrophobic sides of TM helices are exposed to lipids, whereas the most hydrophilic ones to the protein interior [3]. The hydrophobic properties of each TM segment were measured in terms of magnitude of the hydrophobic moment and its orientation about the helix axis.

Similar procedures were applied to the analysis of sequences of TM peptides in other membrane proteins [9, 10]. Generally, they were employed to detect the most prominent stretches of hydrophobic/hydrophilic and/or variable residues. Subsequently, these data were used as constraints in algorithms destined for prediction of spatial arrangement of TM helices. Despite the fact that such analysis can delineate the most prominent motif of polarity in the sequence of TM peptide, often more detailed information is needed to assess properly environmental characteristics of residues in the membrane bundles. For example, orientation of hydrophobic moment vectors calculated for bacteriorhodopsin (BRh) [11, 12] reveals no obvious correlation with the orientation of helices with respect to the lipid. Moreover, many hydrophobic residues are observed in the interior of the bundle and at the helix-helix interfaces as well as some polar residues are pointing toward the lipid.

Packing of parallel and antiparallel helix bundles in globular proteins was studied in numerous works (see e.g., 13, and references therein). In addition, a number of attempts were made to investigate principles of spatial α-helix arrangement in membrane protein domains [12, 14-17]. These studies provide a considerable insight into the problem but atomic-scale *ab initio* prediction of the spatial structure of multi-helix bundles is still a challenge task. The difficulty is caused by the necessity to account for an important contribution of hydrophobic effect into assembling of the bundle. That is why, detailed analysis of the high-resolution structures of the α-helix domains in membrane proteins and bacterial toxins is important to delineate the major factors driving helix association.

In the present study we pursued the following purposes: (i) to assess detailed spatial hydrophobic properties of TM helices in several membrane proteins with known structure and those for "pore-lining" α-helices in bacterial toxins; (ii) to compare characteristics of these helices and to reveal common (if any) hydrophobic templates peculiar to the helix bundles; (iii) to check how the polarity properties thus calculated, could be employed to predict helix-helix packing in membrane domains. The data will be used to recognize a folding type for membrane helix bundles in proteins with unknown 3D structure, and to delineate the most probable lipid-, protein-, and channel-exposure of TM peptides.

2. CHARACTERIZATION OF SPATIAL HYDROPHOBIC PROPERTIES OF α-HELICES.

Recently we have developed a method for analysis of hydrophobic properties of spatial molecular models of TM helices [18, 19]. The computational procedure employs two independent techniques - statistical mechanics Monte Carlo (MC) simulations of nonpolar (usually propane) and polar (water) solvents around the peptides, and 3D molecular hydrophobicity potential (MHP) calculations. In the first approach, polarity of a helix exposure is analyzed in terms of the average peptide-solvent interaction energies (ESS), whereas in the second one, it is assessed using the 3D MHP distribution on the helix surface. Earlier it was shown [19] that the results obtained in the frameworks of both formalisms are in reasonable agreement, complement each other, and provide a detailed presentation of the spatial hydrophobic nature of the peptides. The method was applied to several TM α-helical peptides from proteins with known structure. Resulting hydrophobic characteristics were compared with experimentally observed lipid- and protein-exposure of these segments in 3D structures of the membrane bundles. The approach was also employed in the hydrophobic mapping of several channel-forming α-helical peptides (in the rat epithelial Na channel (rENaC), segment M2 in AChR, α5 peptide in δ-endotoxin, and some others). The results obtained were used to predict the residues lining the pore as well as exposed to a nonpolar environment.

The formalism of 3D MHP utilizes a set of atomic physico-chemical parameters evaluated from octanol-water partition coefficients (log P) of numerous chemical compounds [20]. It permits detailed assessment of the hydrophobic and/or hydrophilic properties of various parts of the molecules. The MHP created by a group of N atoms (identified by index i) in any point j of space (e.g., on the protein surface), is calculated according to the formula:

$$MHP_j = \sum_{i=1}^{N} f_i \, e^{-r_{ij}} \qquad (1),$$

where f_i is the atomic hydrophobicity constant of atom i, r_{ij} is a distance between atom i and point j. Positive and negative signs of MHP correspond to hydrophobic (nonpolar) and hydrophilic (polar) environments of point j, respectively. The details of MHP calculations as well as MC simulations of TMS in solvents of different polarity can be found elsewhere [18, 19, 21]. This procedure provides pictorial one- (1D) and two-dimensional (2D) representations of nonpolar and polar patterns on the helix surfaces.

The approach has been applied to the analysis of hydrophobic organization of BRh which is one of the best-characterized membrane proteins whose model is available with atomic resolution [22]. The stucture consists of seven TM α-helices arranged in a kidney-shaped manner and contains only one cofactor - retinal attached to Lys216 in the

215

interior of the bundle. This is therefore a good system for assessment of its hydrophobic organization. In this work we calculated hydrophobic properties of individual α-helical segments in BRh and analyzed them with relation to spatial arrangement of the segments in the membrane moiety. This permits elucidation of the relationship between the polar properties of TM peptides and their orientation in the assembly. Approximations and limitations of the computational approach were discussed previously [19].

Fig. 1 (top) shows a 2D contour map of MHP on the surface of TM helix A in BRh. Only the hydrophobic areas with high values of MHP are indicated. It is seen that the most prominent stretch of hydrophobicity is observed in the range of angle α 135° - 270° and extends through the whole peptide length. It is formed by residues 11, 15, 19, 22, 26, 27. Another hydrophobic zones correspond to residues 10, 14, 25, 29 (310° < α < 360°) and 28 (0° < α < 50°). The strongest polar areas are attributed to residues 16, 21, 24. The plot of atomic peptide-propane interaction energies (ESS) summarized in the arc with angular size 90° versus rotation angle α of the helix is shown in the bottom part of Fig. 1 with a solid line. The prominent minimum corresponds to the helical face with 160° < α < 260°. Low energies are also observed for the regions 0° - 20° and 300° - 360°. The curve characterizing the angular distribution of surface MHP of this segment is presented with a dotted line. These plots demonstrate strong anticorrelation, confirming that the most prominent hydrophobic surface regions (high values of MHP) strongly interact with nonpolar solvent revealing the lowest energies of peptide-propane interaction.

Obviously, angular distribution plots give a picture of polar and nonpolar sides averaged over the helix length. In order to compare in more detail these measures of hydrophobicity, we superimposed the peptide atoms with the lowest energies of interaction with propane and with water (averaged in a set of MC configurations) upon 2D MHP map (Fig. 1, top). It is seen that almost all the atoms strongly interacting with nonpolar solvent fall into the areas with large positive values of MHP or are located close to their boundaries. On the contrary, the atoms revealing low energies of interaction with water are mainly distributed outside these areas. The exceptions are provided by the tryptophan residues 10 and 12 and methionine 20: some of their atoms with low energies of peptide-propane interactions lay outside the areas of high MHP.

Analysis of the experimentally determined structure of BRh [22] reveals that the lipid-exposed face of the segment contains residues 10, 11, 14, 15, 18, 19, 22, 25, 26, 29. The interface with helices B and G is formed by residues 12, 13, 16, 20, 24, 27, 28. The first region corresponds to the intervals of α 170°-360° and 0°-30°, whereas the second one corresponds to α = 30 - 180°. Therefore, that side of the helix which was predicted to be the most hydrophobic is exposed to the bilayer in the BRh bundle. On the contrary, the most polar side is in contact with the other protein segments. Note, that the most variable side of this TM segment (it is marked with 'v' in Fig. 1, top) is assigned as the strongest hydrophobic one on 2D MHP map and on the angular distribution plots. Although this is believed to be a common feature for the membrane α-helix bundles [3], the exceptions are not rare [12].

Analogous MHP- and ESS- maps were also analyzed for the other six TM segments of BRh as well as for numerous other membrane helices [12, 19, 23]. The main inference made from such analysis is that all the helices reveal a good accord between

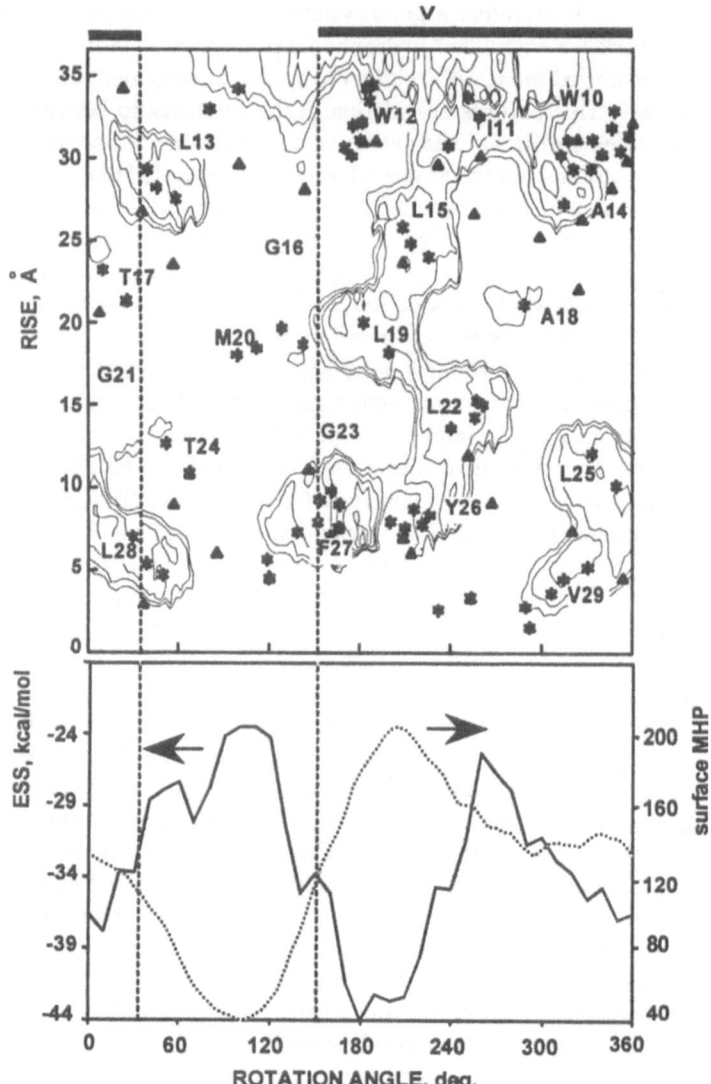

Figure 1. Spatial hydrophobic properties of the membrane-spanning α-helical segment A in BRh. **(top)** Two-dimensional isopotential map of the molecular hydrophobicity potential (MHP) on the peptide surface. The value on the X-axis corresponds to the rotation angle about the helix axis; the parameter on Y-axis is the distance along the helix axis. Only the areas with MHP > 0.1 are shown. Contour intervals are 0.02. The positions of residues are indicated by letters. The symbols * and △ represent projections of peptide atoms revealing the lowest energies of peptide-propane (ESS < -0.9 kcal/mol) and peptide-water (ESS < -30 kcal/mol) interactions, respectively. The most variable side of the segment is labeled with symbol "v". Filled bars and dashed lines represent lipid-exposure of the segment in the experimental structure [22]. **(bottom)** Solid line, angular distribution of peptide-propane interaction energy (ESS); dotted line, angular distribution of MHP on the peptide surface. ESS and MHP are summarized inside the sectors 90°-width.

MHP- and ESS- polarity characteristics. This confirms that the energies of peptide-solvent interaction and surface 3D MHP provide a qualitatively similar description of spatial hydrophobic properties of helical segments. We should point out that the two formalisms applied here for description of peptide hydrophobicity are independent and based on different principles. Thus, the OPLS force field parameters used in MC calculations were derived in the result of computer simulations of pure liquids and peptides in crystalline state [24], while atomic hydrophobicity constants implemented in the 3D MHP approach, were extracted from the analysis of experimentally observed partition coefficients of different organic compounds in octanol-water mixtures [20]. Therefore, a good accordance between these two formalisms, gives a strong argument that both approaches are able to provide a realistic description of hydrophobic nature of TM peptides.

The knowledge of spatial hydrophobic/ hydrophilic properties is important for modeling of membrane domains, especially of ion channels which often are constructed as bundles of parallel or antiparallel helices forming a pore. The challenge in modeling of such bundles resides in construction of a number of initial conformations for further refinement and analysis. In these assemblies, the hydrophilic side chains of the helices line the central pore, whilst the hydrophobic residues interact favourably with the fatty acyl chains of lipids. Therefore, the hydrophobicity mapping of channel-forming peptides like that described above, may provide important constraints on their mutual arrangement in the membrane domain.

3. HYDROPHOBIC ORGANIZATION OF MEMBRANE DOMAINS. BACTERIORHODOPSIN.

Angular 1D and 2D hydrophobicity plots of all TM helices of BRh calculated as that shown in Fig. 1, being superimposed on the experimental structure [22], illustrate the hydrophobic organization of the membrane bundle of BRh. For the most pictorial presentation of hydrophobic characteristics of each TM helix, we plot its angular distribution of surface MHP (or ESS) (Fig. 1, bottom) in polar coordinates (α, ρ), where α - rotation angle, ρ - corresponding value of MHP (or ESS). Because the plots obtained in both, MHP- and MC-approaches provide principally the same hydrophobic/ hydrophilic nature of the helices, we will consider polarity of TM peptides expressed only in terms of MHP on their surface. Fig. 2 shows angular MHP-plots for TM helices arranged as in the membrane bundle of BRh. Analysis of this hydrophobicity map permits the following conclusions:

1. There is no prominent tendency in orientation of hydrophobic/hydrophilic sides of helices with respect to the bilayer. Thus, the helices expose their strong hydrophobic surfaces only to the bilayer (helix A), only inside the protein (D, F), or both, to the membrane and to the interior of the bundle (B, C, E, G);

2. The neighbouring helices, except A, B and F in pairs A-G, B-C and E-F, respectively, are turned to each other with their sufficiently hydrophobic sides thus providing close packing of the segments and screening of the protein interior from surrounding lipids;

218

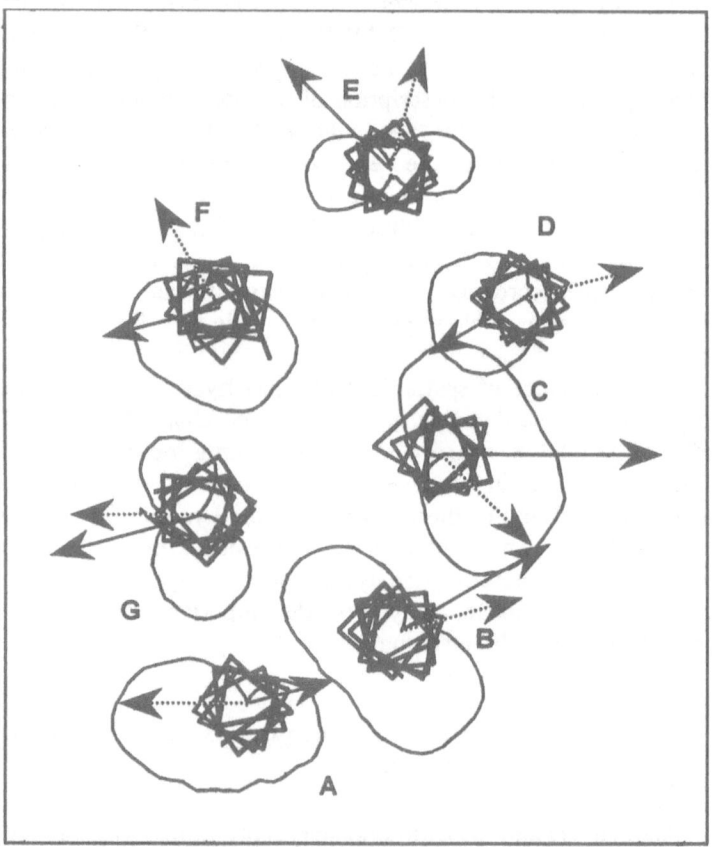

Figure 2. Extracellular view of the membrane bundle in BRh. TM helices (labelled A-G) are taken from the experimental structure [22] and oriented perpendicular to the figure plane. Hydrophobic properties of individual helices are shown in terms of their surface molecular hydrophobicity potential (MHP) plotted in polar coordinates (α, ρ), where α - rotation angle, ρ - corresponding value of MHP. Variability and hydrophobic moment vectors of TM segments are indicated with dashed and solid arrows, respectively.

3. The most nonpolar parts of the external surface of the bundle correspond to segments A, B, C, whereas relatively hydrophilic ones - to segments D, E, F and G.

4. While all the segments are composed mainly from nonpolar residues and, therefore, can be delineated in the sequence using hydropathy plots or hydrophobic moment analysis, the spatial distribution of their hydrophobic properties can be significantly different. For example, helices E and G reveal two pronounced hydrophobic and two hydrophilic sides. 1D-hydrophobicity plots of helices A, B and C also demonstrate that there is no clearly defined single hydrophobic side which can be assigned with the hydrophobic moment vector.

5. No correlation was observed between orientation of the most hydrophobic and the most variable sides of TM helices in BRh. Only for helices A and C the variability moment vectors (Fig. 2) point toward strongly nonpolar faces of the segments.

Otherwise, such vectors correspond to relatively polar sides of helices. Based on the analysis of the RCs structures, it was postulated earlier [3], that the most variable and the most hydrophilic surfaces tend to be on the opposite sides of membrane-spanning helices. Therefore, our results for BRh do not support this observation.

6. Orientation of hydrophobic moment vectors of TM peptides in BRh does not correlate with the most nonpolar sides of the helices (expressed in terms of MHP- and ESS-distributions) as well as with their most variable sides. Moreover, there is no obvious relationship between orientation of the hydrophobic moment vectors and orientation of the segments with respect to the lipid.

Detailed analysis of the hydrophobic maps and pairwise MHP-contacts of the membrane moiety [12] shows that the structure of the membrane bundle is stabilized to a large extent by hydrophobic interactions between neighbouring helices. Moreover, helix C is involved in strong hydrophobic contacts with segments E, F and G and, probably, serves as a main factor maintaining the double-layer shape of BRh. This is consistent with the experimentally observed smallest exposure of helix C to lipids. Nonpolar contacts of the retinal chromophore with helices C, F and, in less degree, D, E, also contribute to the stability of the complex. Basing on the topology of helix-helix nonpolar contacts, one can propose that the hydrophobic core of the bundle is formed by TM peptides C, E, F, G. Each of them is involved in hydrophobic interaction with 5, 3, 3, 3 other segments, respectively. On the other hand, helices A, B, and D reveal nonpolar contacts only with two neighbours in the bundle and, therefore, contribute somewhat less to the overall stability of the assembly.

The results described above are compatible with those obtained in complementary approaches. Thus, the energetic analysis of the experimental model of BRh [13] shows that the helix bundle is maintained sufficiently by hydrophobic helix-helix interactions. In addition, helices A and B were shown to be practically uninfluenced by the other TM segments, helix C was proposed to play a key role in determining the bundle conformation, strongly interacting with the other helices, whereas helix D gives rise to the weakest interactions. Such "virtual" independence of pair A-B in the bundle and tight association of the other helices conform with the experimental data on chymotryptic cleavage of BRh [reviewed e.g., in 16].

Additional data on the principles of helix association in a membrane could be also obtained from the analysis of spatial structures of some bacterial toxins revealing α-helix bundles with a central pore. While such proteins do not contain transmembrane domains, analysis of their pore-forming motifs is important because the similar folds were proposed for some membrane ion transporters like e.g., AChR, phospholamban, δ-endotoxin, and others.

4. HYDROPHOBIC TEMPLATE FOR MEMBRANE HELIX BUNDLES.

Spatial hydrophobic properties of pore-forming helices in B-pentamers of bacterial toxins - helices A in verotoxin-1 (VT-A) and D in cholera-toxin (CT-D) - are

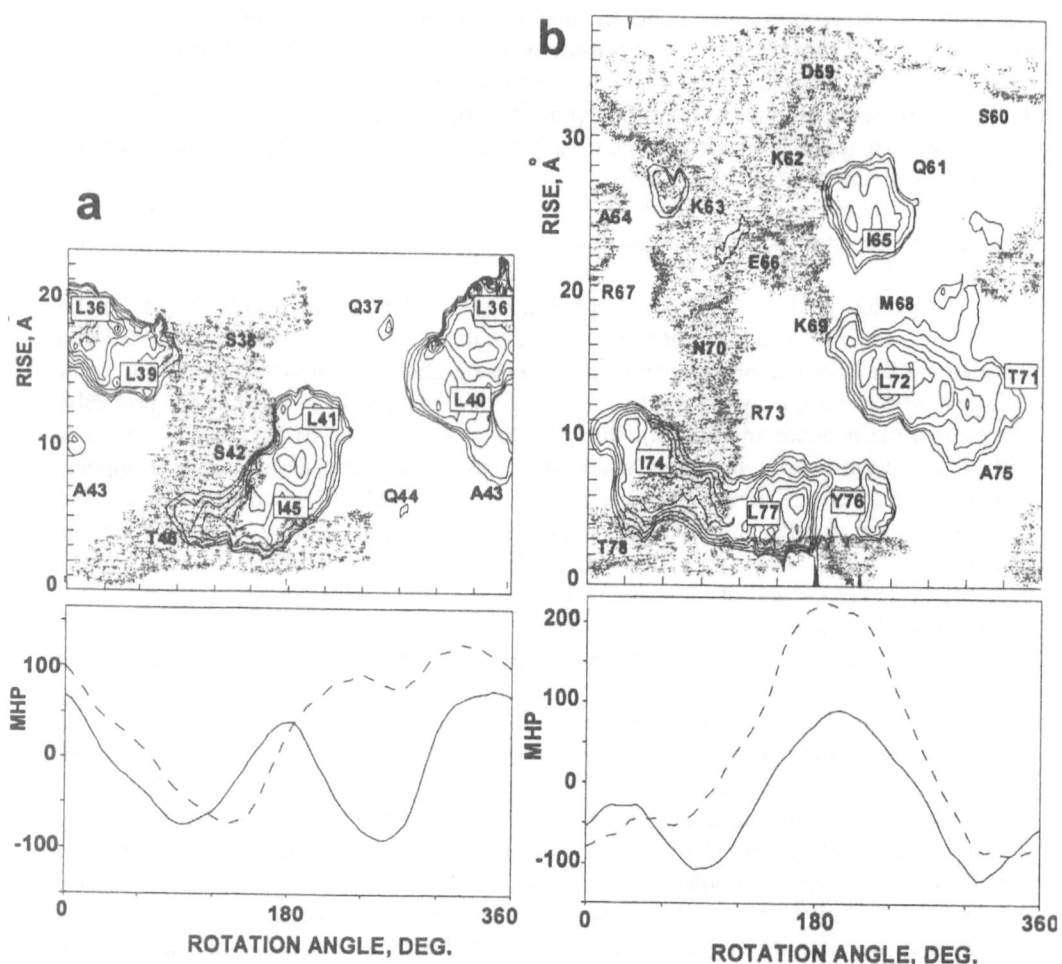

Figure 3. Spatial hydrophobic properties of pore-forming α-helices in B-pentamers of bacterial toxins. **a.** helix A of verotoxin-1; **b.** helix D of cholera-toxin; **(Top)** Two-dimensional isopotential map of the molecular hydrophobicity potential (MHP) on the peptide surface. Grey shadow represents pore-lining surface in the experimental structure. **(Bottom)** *Solid line*, angular distribution of MHP created by the peptide atoms on it's surface; *dotted line*: angular distribution of MHP created by the neighbouring protein parts of the pentamer on the peptide surface. Other details are the same as in the legend to Figure 1.

presented in Fig. 3 in terms of 2D MHP isopotential contour maps (top) and 1D MHP plots (bottom). The origin (with a rotation angle α = 0°) of the map for helix CT-D was shifted along X-axis to provide the best fit with 2D MHP map of helix VT-A [23]. The same kind of alignment has also been applied to corresponding 1D MHP plots. It is seen that polarity properties of the individual helices as well as the hydrophobic organization of the FHBs are quite similar. The peptide surfaces reveal two vertical polarity stretches spanning almost all the helix length and separated by the two nonpolar regions. The polar patterns (and the hydrophobic ones) lie on the opposite sides of the peptides. The strongly hydrophobic surface areas are disposed at the interfaces between neighbouring helices in the bundle. One of the hydrophilic stretches corresponds well to the pore-lining surface, whereas the second one is in contact with a β-sheet barrel surrounding the helix bundle. Quite similar features are also inherent in the polarity properties of pore-forming helices in another toxin - pertussis toxin (PT) [23] (data not shown).

To check, how the "own" polarity properties of the individual helices (calculated without taking into account the rest of the protein) agree with the "external" ones (those created by its environment in the 3D structure), the following protocol has been employed. We have assessed MHP on the van der Waals surface of the helices VT-A and CT-D induced by all atoms of the B-pentamer, except the helices A and D themselves. The resulting 1D MHP plot is shown in Fig. 3 (bottom) with a dotted line. It is seen that the "own" and the "external" plots correlate well, except the interfaces α-helix/β-sheet (e.g., region with $220° < α < 280°$ for VT-A). This means that the neigbouring helices are packed *via* their strongly hydrophobic sides, while the helix/sheet contacts correspond to rather less favorable interaction of the polar helix and nonpolar sheet surfaces.

Such conlusions are in a good agreement with the results of assessment of the toxins' spatial structure with the 3D profile method [25]. Thus, residues disposed in the helix/sheet contacts reveal the lowest scores (unfavorable environment), whereas residues at the hydrophobic helix-helix interfaces demonstrate the highest scores (favorable environment) [23]. Therefore, the helix bundle is packed rather better (in terms of consistency with the other high-resolution 3D protein structures) than the FHB/β-layer interface. Similar inferences have been also made from estimation of the "environment free energy", EFE, [26] of residues in the helices: the lowest EFE-values (favorable environment) are attributed to residues at the helix-helix interface, whereas the highest ones (unfavorable environment) - to residues in the helix/sheet contacts.

Interestingly, 2D maps and 1D-plots of MHP reveal very similar angular disposition of the pore as well as helix-helix and helix-sheet contacts. This means that the hydrophobic properties of individual α-helices, being superimposed on the 3D experimental models of the bundle, are very close in the toxins under study. The hydrophobic template found for the FHB-fold is quite different from those usually proposed for modeling of ion channels. In such models (which are often built based on the coordinates of these toxins), α-helices constituent the bundle are oriented to the central pore with their relatively polar sides, whereas the most hydrophobic regions are turned outside. But as follows from the data presented above, this does not correspond exactly to the hydrophobic organization of the FHB in known protein structures.

The FHB-fold is also a characteristic feature of the membrane ion channels [7]. Therefore, it is interesting to check, whether or not the FHB-polarity template found in bacterial toxins can be attributed to membrane-spanning peptides in these channels. We have calculated 1D MHP plots for several TM α-helices whose structure is known with atomic resolution or which exhibit a considerable body of structural information provided by mutagenesis, selective labeling, and other techniques. The following peptides were considered: membrane helices of BRh (BRh-A, B, ..., G) and RC *R. viridis* (PRC-M3, M4), helix M2 of AChR (AChR-M2), TM α-helix of phospholamban (PLP), membrane α5 segment of δ-endotoxin (DTOX), membrane-spanning peptides of α-, β-, and γ-subunits of rENaC (M1A, M1B, etc.), and TM peptide of influenza virus M2 protein (VMT-2).

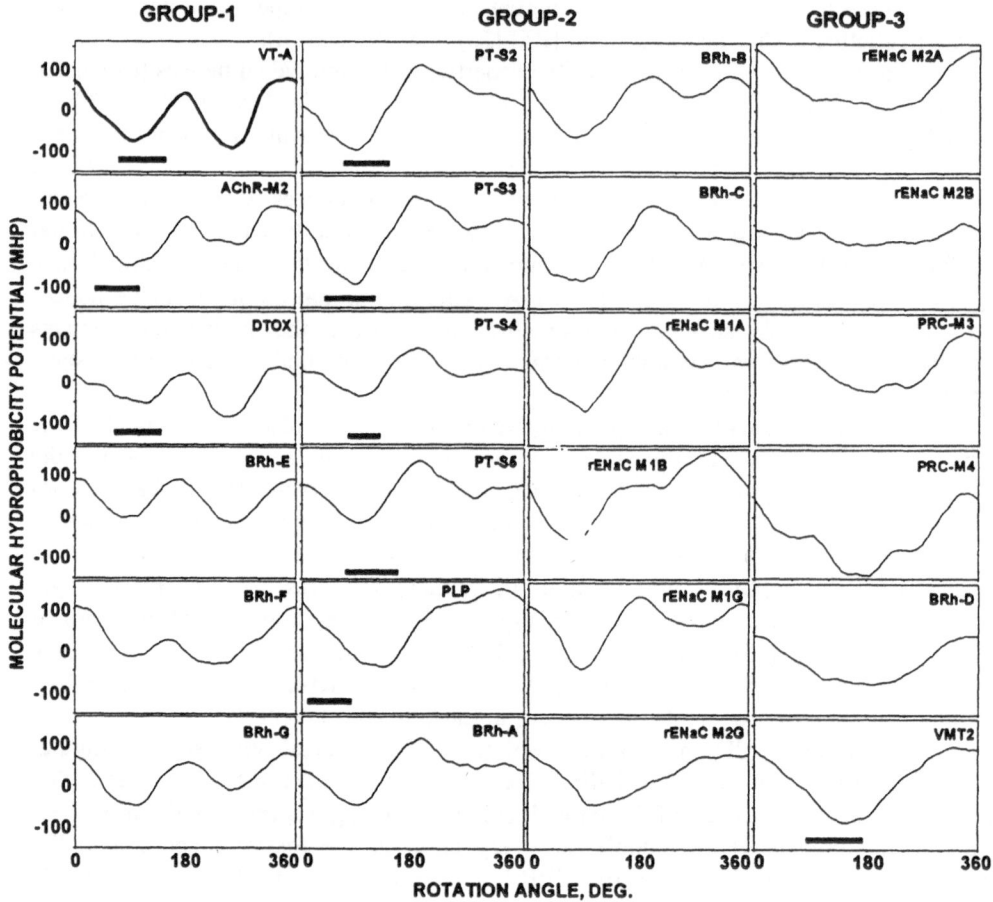

Figure 4. One-dimensional plots of the molecular hydrophobicity potential on the surface of pore-forming helices in bacterial toxins (helix A of verotoxin-1, VT-A, helices in S2-, S3-, S4-, S5-subunits of pertussis toxin, PT-S2, S3, etc.), and some transmembrane (TM) α-helices. Abbreviations used: BRh-A + BRh-G, helices A÷G of bacteriorhodopsin; PRC-M3, PRC-M4, helices M3 and M4 of the photoreaction center *R. viridis*; AChR-M2, helix M2 of the acetylcholine receptor (α-subunit); PLP, TM helix of phospholamban; VMT2, TM helix of influenza virus M2 coat protein; DTOX, α5 segment of *Bacillus thuringiensis* δ-endotoxin; rENaC-M1A, M2A, M1B, M2B, M1G, M2G, putative TM helices M1 and M2 of α-, β-, and γ-subunits of the rat epithelial amiloride-sensitive Na$^+$ channel, respectively. All the plots are aligned relatively that of VT-A [23]. Known pore exposure of some segments is shown with filled bars.

Their 1D angular MHP plots together with the plots of pore-lining helices of verotoxin, VT, and S2-, S3-, S4-, S5-subunits of pertussis toxin, PT, are presented in Fig. 4. It is seen that some of these plots correlate well with the 1D MHP template characteristic for the FHB. Analysis of pairwise correlation coefficients between the profiles ([23], data not shown) reveals the highest degree of correspondence inside the following groups of helices. Group 1: VT-A, CT-D (Fig. 3b), AChR-M2, DTOX, BRh-E,

BRh-F, BRh-G; group 2: PT-S2, S3, S4, S5, BRh-A, BRh-B, BRh-C, PLP, rENaC M1A, M1B, M1G, M2G; group 3: BRh-D, PRC-M3, M4, rENaC M2A, M2B. For example, pairwise correlation coefficients inside the group 1 are all higher than 0.82. Therefore, pore-forming helices in the toxins were subdivided into two groups: one containing verotoxin and cholera-toxin, and the second one including different subunits of pertussis toxin. As seen in Fig. 4, the difference between the 1D MHP plots characteristic of the two groups is the depth of the minimum corresponding to the external helix side in the FHB. Interestingly, both MHP-templates of the FHBs (groups 1 and 2) fit well to the MHP-profiles of almost all helices in BRh (except BRh-D) and to TM segments of channel proteins known to form α-helix bundles in the membrane like AChR-M2, DTOX, PLP.

Besides the hydrophobic effect, electrostatic interactions are believed to play an important role in protein binding on the membrane surface and insertion. To check the influence of electrostatic interactions on assembly of the FHBs, we assessed electrostatic properties of their pore-forming helices in terms of electrostatic potential (ELP) on the surface [23]. The ELP values were represented by 1D and 2D plots using the same methodology as was employed in the MHP-approach described above. As follows from this analysis (data not shown), no prominent correlation was found between the ELP-distributions and orientation of helices in the bundles. Thus, in some segments the regions with highest ELP values line the pore, while in others they correspond to helix-helix or helix-sheet interfaces. An inference was made that the electrostatic organizations of the FHB-proteins differ significantly and do not permit delineation of a common ELP-template. That is why, we propose that the electrostatic interactions do not play a dominant role in assembling and stability of the helix bundles. The same observation was also made from the analysis of ELP-characteristics of membrane helices in the membrane channel proteins revealing polarity properties mathing the FHB-hydrophobicity template (see above).

The following inferences can be made in the result of our analysis of the FHBs:

1. Despite a lack of sequence homology, individual α-helices in the FHB reveal very similar hydrophobic properties on their surfaces: two relatively polar sides separated by strongly hydrophobic stretches. This permits delineation of a common polarity template for the FHB which can efficiently recognize α-helices involved in such helix bundles. In the FHB, α-helices are tightly packed *via* their nonpolar faces, whereas the contacts with surrounding β-layer are rather weaker. One of the hydrophilic sides of the helices is exposed to the central pore and another one is turned to the exterior of the bundle.

2. By contrast, ELP-distributions on the peptide surfaces are rather different for these helices, they do not correlate with orientation of helices in the bundle, and, therefore, do not reveal a common electrostatic template for the FHB.

3. Spatial structure of the FHB is maintained primarily by strong hydrophobic interactions between neighbouring α-helices, while the role of the helix contacts with surrounding β-sheet layer and importance of electrostatic interactions are significantly weaker.

4. Hydrophobic properties of several channel-forming TM α-helices (AChR-M2, DTOX, PLP) were found to be very similar to those inherent in the FHB polarity template.

5. According to their polarity properties, the membrane α-helices can be subdivided into three groups: two of them are characteristic for the bundles of five or more helices, whereas the third one corresponds either to four-helix folds or to non-bundle architectures.

6. The FHB-polarity template (expressed in terms of 1D-MHP plot) efficiently recognizes in the database of polarity motifs those belonging to α-helices constituting helix bundles in membrane proteins.

The results presented here emphasize that the FHB-hydrophobic template derived from the analysis of 3D structures of bacterial A-B$_5$ toxins can be successfully employed in identification of FHB-forming peptides in membrane ion channels with unknown spatial structure. It complements environmental profiles methods and threading techniques.

5. HELIX-HELIX PACKING IN MEMBRANE FROM A POINT OF VIEW OF HYDROPHOBIC INTERACTIONS.

The data obtained in the result of such a detailed assessment of spatial hydrophobic properties of individual TM helical peptides as well as inspection of hydrophobic organization of α-helix bundles, could be employed to predict helix-helix packing in the membrane moieties. Known types of interactions that have been shown to occur between protein TM helices, include: close packing van der Waals/hydrophobic interactions, charged-pair mediated interactions, intermolecular H-bonds, and covalent disulfide linkages between neigbouring TM peptides. As it seems from the analysis of known membrane protein structures, the first type of interactions occur most frequently, probably, because of its nonspecific character, whereas the other types of interactions are quite specific and depend on the sequence of TM peptide. In the subsequent discussion we will consider two methods accounting for hydrophobic interactions in the membrane protein domains. The other types of protein-protein interactions in the lipid bilayer were recently considered elsewhere [16, 27].

5.1. Optimization of MHP-contacts.

Earlier we have shown [17] that the spatial hydrophobic properties of TM segments calculated using the MHP method (see above), could be useful in prediction of packing of helical hairpins (two helices linked with short extramembrane loop). The main idea of such an approach is that mutual association of α-helices in the membrane domain is driven primarily by the hydrophobic/hydrophilic protein-protein interactions, whilst the role of surrounding lipids is rather weaker. The computational procedure is based on the optimization of hydrophobic (MHP) contacts between helices. Their optimal arrangement was proposed to correspond to minimum of the function:

$$U = -\sum_i \sum_j f_i\, f_j\, e^{-r_{ij}} \qquad (2),$$

where f_i and f_j are the atomic hydrophobicity parameters of atoms i and j of the first and the second peptides separated by distance r_{ij}. Validity of the computational scheme was tested in calculations of relative orientations of several pairs of TM helices in the RC *R. viridis* [17]. It was shown that, in spite of serious simplifications inherent in the method, the results of computer modeling agree fairly well with those obtained directly from the experimental X-ray structure of the RC.

In addition, the approach was tested by prediction of helix-helix interface in the TM domain of leader peptidase of *E. coli* which contains two TM-helices. Simultaneously, the residues lying on the interface were identified using Cys-scanning mutagenesis [28]. It is important to note that the experimental results were not yet known during the modeling (this work was carried out in collaboration with the group of Dr. G. von Heijne). The results obtained are quite encouraging: for one of the TMS position of the interface was predicted exactly, whereas for the second helix a deviation for the contact region in comparison with the experimentally-determined one was less than 80°. Discrepancy obtained for the second TMS could be explained either by influence of other factors (e.g., protein-lipid interactions), not accounted in this method or by distortion of conformational/hydrophobic properties of the helices in the result of mutagenesis and formation of disulfides (the calculations were done on unmodified peptides).

It seems that the computational procedure just described, is suitable mainly for multihelix bundles, where each TM-helix interacts with two or more neighbouring peptides. This is supported by the analysis of hydrophobic organization of such helix bundles (see below, sections 3 and 4). At the same time, spatial hydrophobic properties of TM helices in small membrane domains (e.g., helix dimers), often are quite different. In this case, the role of the lipid environment should be seriously taken into account. To check the factors influencing on the assembling of two α-helices in the membrane, an attempt was made to optimize protein-lipid interactions.

5.2. Optimization of protein-lipid contacts.

Quantitatively, the strength of protein-lipid interactions was approximated by (i) hydrophobicity (expressed in terms of the MHP) of protein parts on the outer surface of a helix dimer or (ii) with average energies of peptide-propane interactions (ESS) attributed to each atom of the helices. As it was mentioned before, these values were obtained in the result of MC simulations of TM peptides in solvents of different polarity [19]. A computational scheme employed is based on an assumption that the optimal mutual orientation of helices in dimer reveals the strongest interaction with surrounding nonpolar media mimicking a hydrophobic core of a membrane. In accordance with such a proposal, helices were rotated around their principal axes, and for each orientation the following function was evaluated:

$$F_{protein-lipid} = \sum_{i=1}^{M} F_i \qquad (3),$$

226

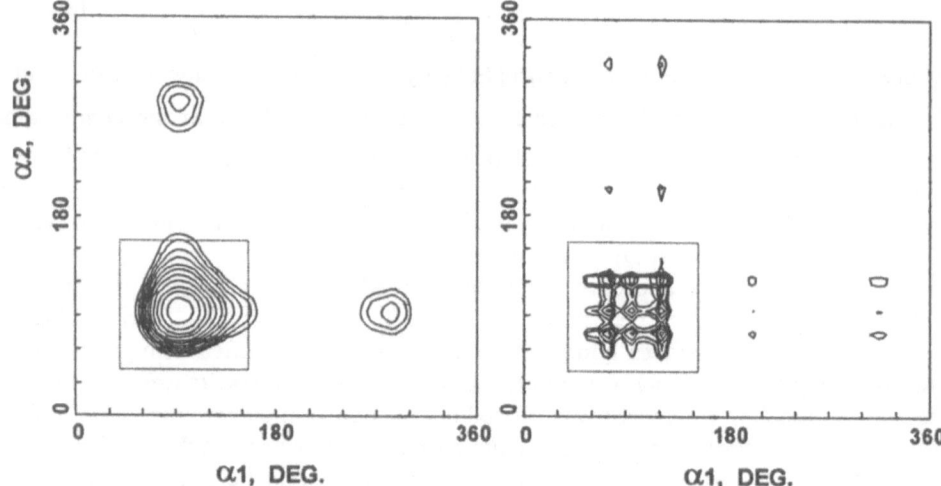

Figure 5. Packing of membrane α-helices in the dimer of glycophorin A by optimization of protein-lipid contacts. 2D isopotential contour maps $F_{protein-lipid}$ (α_1, α_2) (Eq. 3), expressed in terms of MHP (left) and ESS (right). α_1 and α_2 - rotation angles for the two helices. Only regions with the lowest values of $F_{protein-lipid}$ are shown. Region of the helix-helix interface determined from the mutagenesis data [16] is indicated with dotted square.

where F_i is either MHP in point i lying on the external surface of the dimer, or average ESS-value for atom i accessible at the external surface of the dimer. M - total number of surface points or atoms in the dimer, respectively.

The method was checked on several TM helix-helix pairs whose dimerization interface is known from the experiment. Thus, Fig. 5 shows 2D isopotential contour map ESS(α_1, α_2) (α_1 and α_2 are rotation angles for the two helices) obtained for the dimer of TM α-helices of glycophorin A. The interface region known from the mutagenesis data [16] is also indicated. It is seen that the results of the calculations in the frames of our simple model agree fairly well with the experimentally-derived geometry of the complex.

As a resume, we can conclude that in both cases considered above for multihelix membrane bundles and for dimers of TM segments, the methods based on the account of hydrophobic (MHP) or van der Waals (expressed in terms of the OPLS [24] force field) interactions are able to provide quite reasonable mutual orientations of helices. As discussed elsewhere [17, 19], there is a number of strong approximations made. For example, the helix tilt and extramembrane regions were not taken into consideration, etc. On the other hand, comparison with the experimental data shows that at least for the systems which were analyzed, the residues lying on the helix-helix interfaces are predicted correctly. This provides a good starting geometry for subsequent precise refinement of the model using minimization, molecular dynamics, and other techniques. We should note that employment of the empiric functions of type 2 or 3 not only leads to sufficient reducing of computational resources used to find an optimal geometry (in comparison with standard force-field based methods exploring all the conformational space available for the helix pair), but also gives an additional insight into our understanding of the forces driving helix-helix association in membrane helix bundles.

6. PERSPECTIVES.

Further applications of the MHP/ESS-mapping and the method of hydrophobicity template could be envisaged. Thus, polarity properties of putative membrane-spanning segments in proteins with unknown 3D structure could be calculated and compared with the FHB-motifs. Close correspondence between them will provide a sufficiently strong ground to believe that the peptides under study participate in formation of α-helix bundle in the membrane. Moreover, the approach enables the disposition of these peptides in the bundle to be determined. This information is necessary in building atomic-scale molecular models of the intramembrane protein domains because it imposes stringent constraints on the helix orientation in the assembly. In turn, the models provide a basis for rationalization of known structural and functional data as well as for design of further experiments. Our current work is also being pursued to extend this approach to membrane proteins with a β-barrel architecture of their membrane parts. Finally, recent progress in site-directed mutagenesis of TM segments calls for estimation of the effects induced upon amino acid substitution on the structural and hydrophobic properties of the peptides. The technique proposed in this study may provide an essential help in such analysis.

ACKNOWLEDGEMENTS.

E.R.G. was a recipient of an invited professor fellowship from the University of Science and Technology of Lille. This work was supported in part by the NATO Linkage grant HTECH.LG.951401.

REFERENCES.

1. von Heijne, G. (1994) Membrane proteins: from sequence to structure, *Ann. Rev. Biophys. Biomol. Struct.* **23**, 167-192.
2. Walker, J.E. and Saraste, M. (1996) Membrane protein structure, *Curr. Opin. Struct. Biol.* **6**, 457-459.
3. Rees, D.C., DeAntonio, L. and Eisenberg, D. (1989) Hydrophobic organization of membrane proteins, *Science*, **245**, 510-513.
4. Li, J. (1992) Bacterial toxins, *Curr. Opin. Struct. Biol.* **2**, 545-556.
5. Unwin, N. (1993) Nicotinic acetylcholine receptor at 9 Å resolution, *J. Mol. Biol.* **229**, 1101-1124.
6. Arkin, I.T., Adams, P.D., MacKenzie, K.R., Lemmon, M.A., Brunger, A.T. and Engelman, D.M. (1994) Structural organization of the pentameric transmembrane α-helices of phospholamban, a cardiac ion channel, *EMBO J.* **13**, 4757-4764.
7. Montal, M. (1996) Protein folds in channel structure, *Curr. Opin. Struct. Biol.* **6**, 499-510.
8. Eisenberg, D., Weiss, R.M. and Terwilliger, T.C. (1982) The helical hydrophobic moment: A measure of the amphiphilicity of a helix, *Nature (London)* **299**, 371-374.
9. Donnelly, D., Overington, J.P., Ruffle, S.V., Nugent, J.H.A. and Blundell, T.L. (1993) Modeling α-helical transmembrane domains: The calculation and use of substitution tables for lipid-facing residues, *Prot. Sci.* **2**, 55-70.
10. Du, P. and Alkorta, I. (1994) *Protein Eng.* **7**, 1221-1229.
11. Cronet, P., Sander, C. and Vriend, G. (1994) Modeling of transmembrane seven helix bundles, *Protein Eng.* **6**, 59-64.
12. Efremov, R.G. and Vergoten, G. (1996) Hydrophobic organization of α-helix membrane bundle in bacteriorhodopsin, *J. Prot. Chem.* **15**, 63-76.

228

13. Mumenthaler, C. and Braun, W. (1995) Predicting the helix packing of globular proteins by self-correcting distance geometry, *Prot. Sci.* **4**, 863-871.
14. Tuffery, P. Etchebest, C. Popot, J.-L. and Lavery, R. (1994) Prediction of the positioning of the seven transmembrane α-helices of bacteriorhodopsin - a molecular modeling study, *J. Mol. Biol.* **236**, 1105-1122.
15. Cramer, W.A., Engelman, D.M., von Heijne, G., and Rees, D.C. (1992) Forces involved in the assembly and stabilization of membrane proteins, *FASEB J.* **6**, 3397-3402.
16. Lemmon, M. and Engelman, D.M. (1994) Specificity and promiscuity in membrane helix interactions, *Quart. Rev. Biophys.* **27**, 157-218.
17. Efremov, R.G., Gulyaev, D.I. and Modyanov, N.N. (1992) Application of three-dimensional molecular hydrophobicity potential to the analysis of spatial organization of membrane protein domains. II. Optimization of hydrophobic contacts in transmembrane hairpin structures of Na,K-ATPase, *J. Prot. Chem.* **11**, 699-708.
18. Efremov, R.G., Gulyaev, D.I. and Modyanov, N.N. (1992) Application of three-dimensional molecular hydrophobicity potential to the analysis of spatial organization of membrane protein domains. I. Hydrophobic properties of transmembrane segments of Na,K-ATPase, *J. Prot. Chem.,* **11**, 665-675.
19. Efremov, R.G. and Vergoten, G. (1995) Hydrophobic nature of membrane-spanning α-helical peptides as revealed by Monte Carlo simulations and molecular hydrophobicity potential analysis, *J. Phys. Chem.* **99**, 10658-10666.
20. Ghose, A.K. and Crippen, G.M. (1986) Atomic physicochemical parameters for three-dimensional structure-directed quantitative structure-activity relationships. I. Partition coefficients as a measure of hydrophobicity, *J. Comput. Chem.* **7**, 565-577.
21. Efremov, R.G. and Alix, A.J.P. (1993) Environmental properties of residues in proteins: three-dimensional molecular hydrophobicity potential approach, *J. Biomol. Struct. & Dyn.* **11**, 483-507.
22. Grigorieff, N., Ceska, T.A., Downing, K.H., Baldwin, J.M. and Henderson, R. (1996) Electron-crystallographic refinement of the structure of bacteriorhodopsin, *J. Mol. Biol.* **259**, 393-421.
23. Efremov, R.G. and Vergoten, G. (1997) Pore-forming domain in pentameric bacterial toxins as a hydrophobic template for membrane ion channels, in press.
24. Jorgensen, W.L. and Tirado-Rives, J. (1988) The OPLS potential functions for proteins. Energy minimizations for crystals of cyclic peptides and crambin, *J. Amer. Chem. Soc.* **110**, 1657-1666.
25. Bowie, J.U., Luthy, R. and Eisenberg, D. (1991) A method to identify protein sequences that fold into a known three-dimensional structure. *Science,* **253**, 164-170
26. Koehl, P. and Delarue, M. (1994) Polar and nonpolar atomic environments in the protein core: implications for folding and binding. *Proteins: Struct. Funct. Genet.* **20**, 264-278.
27. Harrison, P.T. (1996) Protein:protein interactions in the lipid bilayer, *Mol. Membr. Biol.* **13**, 67-79.
28. Whitley, P., Nilsson, L. and von Heijne, G. (1993) A three-dimensional model for the membrane domain of Escherichia coli leader peptidase based on disulfide mapping, *Biochemistry* **32**, 8534-8539.

INFRARED SPECTROSCOPIC STUDIES OF MEMBRANE LIPIDS

José Luis R. Arrondo and Félix M. Goñi
Departamento de Bioquímica, Universidad del País Vasco
P.O. Box 644, 48080 Bilbao, Spain

1. INTRODUCTION.

Membranes constitute unique architectures in living cells. They isolate compartments inside the cell and the cell itself from the environment. Membranes are composed mainly of proteins and lipids, the latter being the components that confer them their properties of a permeability barrier. They are also involved in almost every aspect of cellular activity. The functional diversity is reflected in their structural diversity, with widely differing compositions of lipids and proteins. Lipids are hydrophobic molecules, sparingly or non-soluble in water. A membrane consists usually of a double leaflet of amphipathic lipids, mainly phospholipids, with a polar moiety in contact with the aqueous environment and an apolar moiety facing the other layer of phospholipid. Also, a lipid-water interfacial region can be defined. Figure 1 shows a membrane phospholipid bilayer and an outline of a single phospholipid molecule with its three regions depicted. The function of the membrane bilayer is not only to provide a barrier, instead the physico-chemical properties of membranes, such as the composition and fluidity of the lipid moiety, can affect a great variety of cell functions, including the modulation of the activity of membrane-associated enzymes and transport systems (1). Besides the phospholipids, other lipid molecules as well as proteins are present in the membranes conferring them their specific properties.

2. PHOSPHOLIPID BAND ASSIGNMENT.

The phospholipid shown in Figure 1 is a glycerophospholipid, in which the glycerol backbone is esterified to a polar molecule, and to two non-polar molecules, fatty acids. The spectrum of such phospholipid in aqueous suspension is shown in Fig. 2. Because of the strong water absorption, only the bands corresponding to the region 3000-1000 cm^{-1} can be seen. Phospholipid infrared bands in the spectrum can be divided into three different regions corresponding to the hydrophobic, interfacial and polar moieties of the lipid (for reviews, see (2-5)). The characteristic vibrations of covalently bonded atoms can be classified as "stretching", which involves changes in bond lengths, or "bending", corresponding to changes in bond angles. Table 1 shows the assignments of the phospholipid peaks in the spectrum of Figure 2.

G. Vergoten and T. Theophanides (eds.), Biomolecular Structure and Dynamics, 229–242.

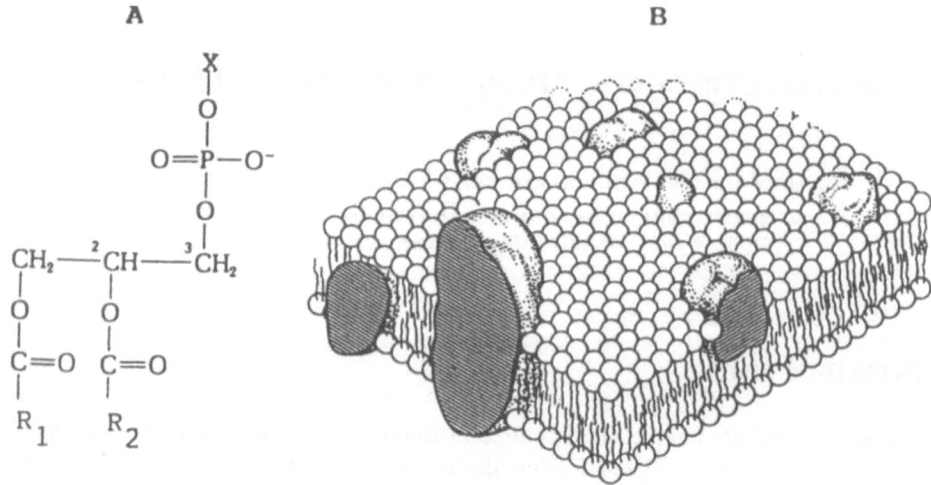

A   B

Figure 1. (A) A phospholipid molecule with the three regions that are detected in infrared spectra; the acyl polar chains, the interfacial C=O region and the headgroup moiety. X corresponds to the polar headgroup subtituent; R_1 and R_2 are the fatty acyl chains esterified to the glycerol backbone. (B) An outline of a membrane composed of phospholipids and proteins according to the fluid mosaic model.

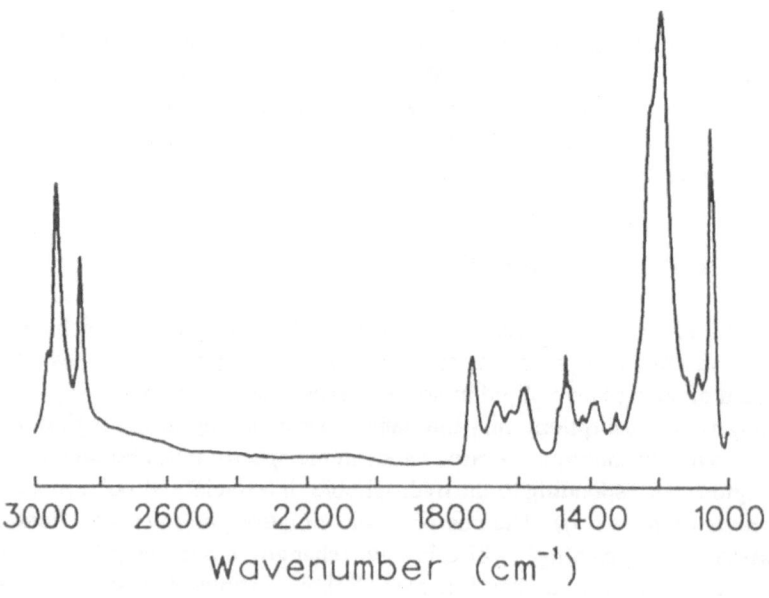

Wavenumber (cm⁻¹)

Figure 2. The infrared spectrum in the 3000-1000 cm⁻¹ region of a phospholipid in aqueous suspension.

TABLE 1.   Assignments of the lipid bands in the infrared spectrum of a membrane preparation.[a]

Wavenumber (cm⁻¹)	Assignment
2957	CH₃ stretching, asymmetric
2924	CH₂ stretching, asymmetric
2871	CH₃ stretching, symmetric
2853	CH₂ stretching, symmetric
1732	C=O stretching, esters
1467	CH₂ bending, scissoring
1456	(L_c + L_β gel phase)
1402	id.
1380	sn-1, α-CH₂ bending, scissoring
1233	CH₃ bending, deformation, symmetric
1171	PO₂⁻ stretching, asymmetric
1159	C-O stretching, single bond
1060	C-C stretching, skeletal
1082	R-O-P-O-R'
	PO₂⁻ stretching, symmetric

[a] Data corresponding to the spectrum of Fig. 2

2.1 ACYL CHAIN VIBRATIONAL MODES.

Bands arising from hydrophobic acyl residues have been easily assigned by comparing phospholipid bands to those of fatty acyl esters and other polymethylene chain compounds (6,7). C-H stretching vibrations give rise to bands in the 3100-2800 cm⁻¹ region. Asymmetric and symmetric CH₂ bands, at 2920 and 2850 cm⁻¹ respectively, are the strongest ones in a phospholipid spectrum. These bands are also most useful in the study of the physical properties of phospholipids. Vibrational bands corresponding to terminal methyl residues, CH₃, are found around 2956 cm⁻¹ (asymmetric stretching) and 2870 cm⁻¹ (symmetric stretching) appearing as shoulders of the stronger methylene bands. Olefinic group bands, =C-H, arising from unsaturated hydrophobic chains are usually located around 3010 cm⁻¹.

Methylene bending bands (scissoring) are located around 1470 cm⁻¹ and can split into two components in moti0on restricted chains. The CH₂ wagging progression is found between 1380 and 1180 cm⁻¹ in the gel phase of saturated phospholipids as a series of shoulders on the phosphate band (8). The methyl symmetric deformation mode around 1378 cm⁻¹, that was assumed to be insensitive to changes in lipid morphology, has been recently found as sensitive to cochleate phase formation in phosphatidylserine-containing model membranes (9). Full or partial isotopic substitution of the acyl chains can be used to avoid interferences with other components, i.e. in lipid mixtures or with proteins in lipid-protein interaction studies. In addition, selective deuteration can be used as an internal probe to study the characteristics of a specific methylene (10). Asymmetric and symmetric CD₂ vibrational bands are located around 2195 and 2090 cm⁻¹ respectively, whereas CD₃ bands are found at 2212, 2169 and 2070 cm⁻¹ (11).

2.2 THE INTERFACIAL REGION.

The most prominent bands arising from the interfacial region are due to stretching vibrations of the carbonyl group involved in ester bonds, in the region 1750-1700 cm^{-1}. These bands are responsive to changes in their environment, i.e. hydrogen bonding or polarity. Moreover, these vibrations occur in a region free of significant absorption by other infrared-active groups, mainly the amide I vibrational band of peptides and proteins, so they are of great interest. In dipalmitoylphosphatidylcholine (DPPC) this region consists of a broad band contour that appears to be the summation of two underlying components easily resolvable by deconvolution or derivation giving maxima near 1743 and 1728 cm^{-1}. In principle, these two bands were assigned to the C=O stretching vibrations of the *sn*-1 and *sn*-2 positions of the glycerolipid moiety, respectively (12). The assignment was based on the study of dry or poorly hydrated DPPC. However, with the use of fully hydrated, specifically ^{13}C=O labelled phospholipid bilayers, this assignment was questioned, since the difference between the *sn*-1 and *sn*-2 bands should be no more than 4 cm^{-1} and ^{13}C labelling of only one the positions still gave rise to two bands around 1743 and 1728 cm^{-1} (13,14). These studies concluded that the underlying components normally resolved in the stretching vibrations of C=O absorption bands are the summation of comparable contributions from both of the ester carbonyl groups and assignable to subpopulations of free and hydrogen-bonded ester carbonyl groups. More recently, a reevaluation of all the existing data has concluded that the values obtained for the dry or poorly hydrated samples cannot be translated to fully hydrated phospholipids and that instead of different bands arising from the *sn*-1 and *sn*-2 bonds, the bands are due to distinct hydrated subpopulations of both carbonyls (15).

2.3 HEADGROUP BANDS.

The most characteristic vibrational bands of the headgroup are those arising from the phosphate. Particularly, three bands appear in the 1300-1000 cm^{-1} region of a hydrated DPPC bilayer. As in the interfacial region, these bands are strongly dependent on the hydration state of the lipid and are sensitive to hydrogen bonding. The antisymmetric PO$_2^-$ stretching mode appears around 1240 cm^{-1} in "dry" phosphate and shifts to around 1220 cm^{-1} in fully hydrated DPPC. The symmetric PO$_2^-$ stretching mode appears around 1086 cm^{-1} in hydrated DPPC. A shoulder near 1060 cm^{-1} has been attributed to a R-O-P-O-R' stretching mode, which is equivalent to a P-O-C vibration with nonequivalent substituents (16). In choline and ethanolamine glycerophospholipids there are characteristic infrared bands due to these group vibrations, but they are of no significant diagnostic value (3).

3. PHYSICAL PROPERTIES OF MEMBRANE LIPIDS.

The structural and functional properties of biomembranes are highly dependent on the physical properties of their lipid matrix. Processes such as membrane protein activity, protein insertion or membrane fusion are highly dependent on the lipid polymorphism or lipid order. Infrared spectroscopy has been a technique of choice in the study of the physical properties of membrane lipids thanks to its applicability to the study of turbid suspensions and also because separate infrared bands are obtained from the different regions in the phospholipid molecule.

3.1 POLYMORPHIC PHASE TRANSITIONS.

3.1.1. *The gel to liquid-crystal thermotropic transition.*
The most common structure of aqueous lipid assemblies of biological origin is the bilayer or lamellar arrangement, which in turn can exist in various physical states. Transitions between two of these physical states can usually be induced by temperature. The thermotropic phase transition can be defined as a conformational change of the phospholipid. The gel phase (ordered) contains mainly *all trans* conformers, whereas the liquid crystal phase (unordered) is composed mainly of *gauche* conformers. Figure 3 shows how the gel-fluid transition is produced in a bilayer of dimirystoyl phosphatidyl serine (DMPS). The gel phase is highly ordered with hydrophobic interactions between the apolar fatty acyl chains and polar interactions in the headgroup region. The more disordered, fluid, liquid crystalline phase is associated to lateral diffusion and protein activity. The transition between these two states produces significant changes in the infrared spectrum, that are easily detected.

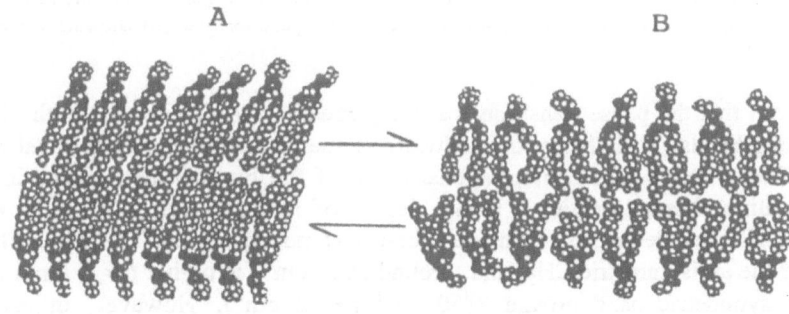

Figure 3. The gel (A) to liquid crystalline (B) phase transition of a bilayer, showing the conformational transition between all-trans and *gauche* conformers.

The vibrations of the non-polar chains are specially useful in characterizing the gel to liquid-crystalline phase transition of phospholipids. The bands corresponding to the symmetric and antisymmetric CH_2 vibrations undergo a shift in position associated to this thermotropic transition. Also, the transition is accompanied by an increase in bandwidth and a decrease in the intensity of the bands around 2850 and 2920 cm^{-1}.

Figure 4 shows the region 3000-2800 cm^{-1} below and above the main thermotropic transition and the plot of maximum wavenumber *vs* temperature for DMPS corresponding to the symmetric stretching vibration.

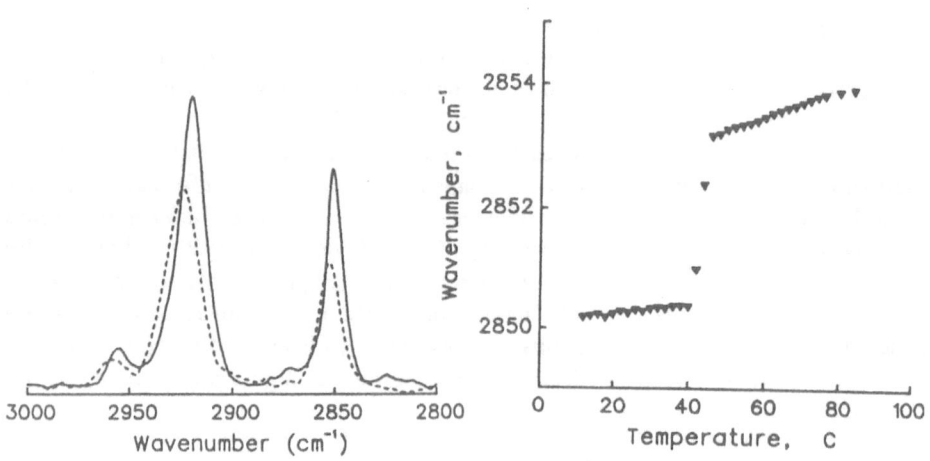

Figure 4. A plot of the CH$_2$ vibrational bands (A) above (solid line) and below (dashed line) the main phase transition of a phospholipid suspension. The wavenumber *vs.* temperature plot (B) showing the phase transition of DMPS.

It can be seen that the phase transition is accompanied by an increase in width of the CH$_2$ stretching bands. This change in bandwidth is due to the increased motional rates and to the larger number of conformational states of the hydrocarbon chains in the liquid crystalline phase, implying a higher degree of motional freedom of the CH$_2$ groups, that has been described as a higher librational movement (3). The width of the transition of the antisymmetric CH$_2$ band (around 2920 cm^{-1}) is higher ($\approx$ 4 cm^{-1}) than that of the symmetric band around 2850 cm^{-1} ($\approx$ 2 cm^{-1}). However, in natural membranes where proteins are present the vibrations of the aminoacid side chains can interfere with the antisymmetric stretching band, therefore the symmetric vibration band that is not affected by proteins is preferred in the study of the main thermotropic phase transition. This phase transition as monitored by the shift of the CH$_2$ stretching vibration is conformationally sensitive and is affected by the presence of other lipids or proteins, as will be shown later. The observed changes in bandwidth are sensitive to the phase transition, but they also reflect changes that do not result in an alteration of the proportion of *gauche* conformers.

The C=O stretching bands associated to the ester carbonyl groups of the interfacial region are sensitive to both conformational and enviromental factors and therefore are affected by the phase transition. In principle, a shift of 4 cm^{-1} to lower frequencies was observed in DPPC (3), but it was noted that the band is asymmetric since it arises from the superposition of two bands due to different hydration subpopulations of the interfacial region as stated above. Resolving the bands by deconvolution or derivation shows that the components are not much affected in band position by the phase transition and unexpectedly both components increase their frequencies by 2 cm^{-1}. The change observed in the overall band is then due to a change in the relative intensities of the components. The higher frequency component is attributed to the "free" (non-hydrogen bonded) carbonyl group and the lower frequency component band to the hydrogen bonded carbonyl group (4). Hydrogen bonding to both C=O groups increases in the liquid-crystalline phase as shown by the increase in the relative intensity of the low-frequency component bands.

The headgroup region of the phospholipids is characterized by the PO$_2^-$ stretching vibrations. From Fig. 2, it would be expected that the bands characteristic of this vibration would be affected by the phase transition. However, since the headgroup is in contact with water, the infrared bands are less sensitive to the phase transition because major changes in hydrogen-bonding to water do not occur. This is demonstrated because other phase transitions that involve dehydration of the headgroup, are accompanied by changes in the phosphate bands (17).

Besides the main transition, in saturated phosphatidylcholines and phosphatidyl glycerols, a pretransition has been described in the gel phase due to a transition from orthorhombic-like to hexagonal packing that is characterized by an increase of 2 cm^{-1} in the overall C=O stretching band and a collapse of the two bands near 1475 and 1465 cm^{-1} due to the CH$_2$ scissoring vibration into a single component at 1468 cm^{-1} .

3.1.2 The lamellar to hexagonal phase transition.

Phosphatidylethanolamine(PE), another major phospholipid class, can form non-bilayer structures at temperatures above the gel to liquid-crystal transition. This new transition corresponds in most cases to the conversion of the lamellar liquid-crystalline phase to an inverted hexagonal phase (i.e., a L$_\alpha$/H$_{II}$ transition), although in some circumstances some PEs may exhibit direct transitions from the gel phase to the H$_{II}$ phase. The different transitions of PEs have been studied by infrared spectroscopy by examining the three regions of the phospholipid (17-19). The differences observed with respect to DMPS or DPPC are due, as expected, to the different headgroup that not only allows in PE transition to non-lamellar phases but also produces in PE gel to liquid-crystalline transition temperatures higher than those exhibited by phosphatidylcholines of similar acyl chain composition. The infrared results that can be obtained in these systems are summarized in Fig. 5 where the band position vs temperature plots of the bands corresponding to the symmetric CH$_2$ stretching, the asymmetric PO$_2^-$ stretching and the asymmetric C=O stretching of hydrated egg yolk PE at pH 5 are depicted. The results corresponding to the CH$_2$ stretching around 2850 cm^{-1} show two transitions, one at $\approx$ 10 °C resulting from the gel to liquid-crystal transition and a second increase in wavenumber $\approx$35 °C that corresponds to the lamellar to hexagonal phase transition, due to an increased conformational disorder (19). The bands corresponding to PO$_2^-$ stretching vibration and C=O stretching vibration behave in the gel to liquid-crystal transition as in the phosphatidylcholines, but at 35 °C where the lamellar to hexagonal transition takes place, an increase in the frequency of the PO$_2^-$ stretching vibration and the C=O stretching vibration is seen.

236

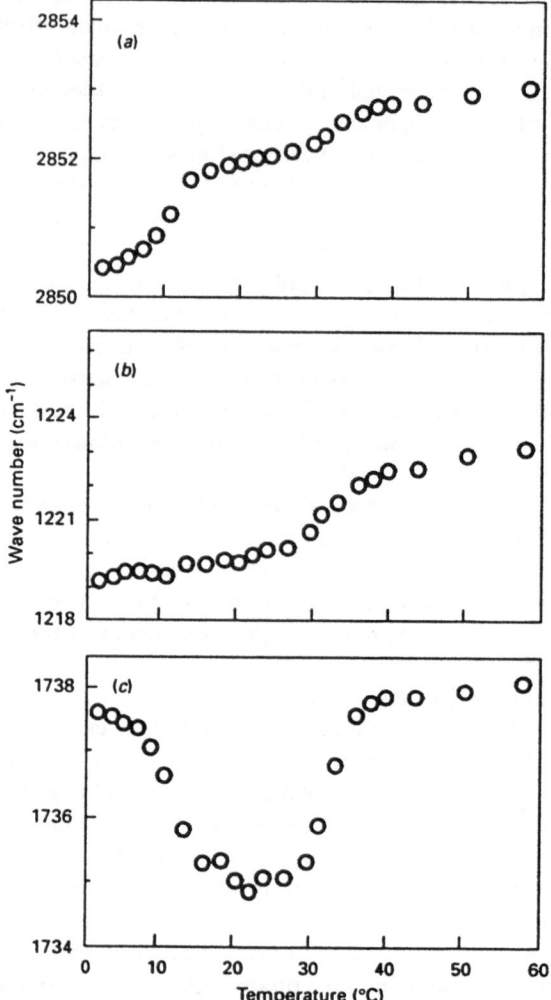

Figure 5. The wavenumber *vs.* temperature plot of the apolar region (A), the polar head (B) and the interfacial region (C) of a phosphatidylethanolamine at pH 5.0 in excess water (17).

These results of the three characteristic infrared bands indicate that the lamellar to hexagonal phase transition is accompanied by a dehydration of the headgroup and a lesser penetration of the water molecules in the phospholipid structure, together with stronger intermolecular interactions in the headgroup and interfacial regions (17).

3.2 MOLECULAR ORDER OF MEMBRANE LIPIDS.

An important physical parameter that is related to the biological activity of membranes is the molecular order and dynamics of their lipid constituents. Infrared studies of membrane lipids indicate that the increase in wavenumber observed in the main gel-fluid phase transition is indicative of the existence of a higher disorder (more *gauche* bonds) in the bilayer in the fluid state. In the studies of phospholipid gel to liquid-crystalline phase transition, the change in band position is not fully quantitatively related to the respective degrees of hydrocarbon chain order in the gel and liquid-crystalline state, although the change in frequency is related to the number of *gauche* bonds introduced at the T_m. A starting point for the quantitative studies was the proposal of a two-state model considering that the phospholipids existed in one of the two physical states (i.e. gel or liquid crystalline) and converting this "brute force"into thermodynamically useful quantities (8). Unfortunately, these correlations between the degree of hydrocarbon chain organization and the frequency of the CH_2 stretching bands does not hold when the CD_2 stretching vibrations of specifically deuterated lipids are measured as a function of chain composition and compared with the values of the order parameter observed by ^2H-NMR spectroscopy (20). Therefore, other acyl chain modes have been studied to quantify the proportion of *gauche* rotamers thus obtaining a molecular order parameter for the membrane lipids. Mendelsohn and coworkers (21) have analyzed the conformation-sensitive CD_2 rocking modes of specifically deuterated phospholipids. The approach is based on the finding that the frequency, in the rocking mode, of a CD_2 substituted into a hydrocarbon chain is sensitive to the conformation only in the immediate vicinity of the CD_2. Thus, a CD_2 adjoined by two *trans* bonds gives a band at 622 cm^{-1}, a CD_2 group adjoined by one *trans* and one *gauche* bond gives bands at 646 and 652 cm^{-1}, while a CD_2 group adjoined by two *gauche* bonds gives a band near 665 cm^{-1} (4). Integrated band intensities can then provide a quantitative measure of the fraction of *gauche* rotamers at a particular chain position. Such analysis revealed that in the gel phase few *gauche* rotamers are present, producing very weak bands. DPPC deuterated at positions 4, 6 and 10 and studied in the gel phase gave only *gauche* rotamers at position 4. In the liquid crystalline state the number of *gauche* rotamers is increased to 20-30 % depending on chain position of the CD_2. The method requires the use of specifically deuterated hydrocarbon chains, cannot differentiate between kink and single *gauche* conformers and must be used under conditions of poor hydration to avoid the strong water absorption in the CD_2 rocking region. Alternatively, Casal and McElhaney (22) have described a method based on the positions and intensities of isolated CH_2 wagging modes in disordered phases. In the 1300-1400 cm^{-1} region, these bands are specific for different types of *gauche* conformers. Thus, *gauche-trans-gauche* sequences (or kinks) give a band at 1367 cm^{-1}, end-*gauche* conformers a band at 1341 cm^{-1}, and double-*gauche* conformers a band at 1355 cm^{-1}. The integrated intensities of these bands are determined and normalized with respect to the symmetric methyl deformation mode at 1378 cm^{-1} to give a value which is directly related to the concentration of the different types of conformational species. Later, the relative intensities of the CH_2 wagging progressions have been studied and conversion factors determined for these relative intensities (23). Using the wagging progression approach and the conversion factors, Blume and co-workers (24) have studied the acyl chain conformational ordering in liquid crystalline bilayers in phospholipids differing in headgroup structure and chain length, and comparing the infrared results with the values obtained by ^2H-NMR. At 69 °C the highest number of *gauche* conformers are found for PCs, followed by DMPG$^-$, phosphatidylethanolamines, protonated DMPG, and phosphatidic acids.

4. INTERACTION OF LIPIDS WITH OTHER MEMBRANE COMPONENTS.

Membranes are a composite of different elements that interact with each other resulting in a final structure whose characteristics are influenced by all the molecules present. Infrared spectroscopy can distinguish among vibrations coming from different molecular species, such as lipids or proteins, but cannot distinguish in principle between vibrational bands arising from similar molecules, i.e. the CH_2 stretching vibrations from different phospholipid classes. Thus, in order to study the interaction of the various membrane components the use of specific vibrations or the deuteration of one of the molecules to be studied are common strategies in studying mixtures of membrane components.

4.1 LIPID-LIPID INTERACTIONS.

4.1.1 *Binary mixtures of phospholipids.*
The strategy to study mixtures of phospholipids is the use of isotopic substitution of one of the components of the mixture. Perdeuterated lipids were first used to separate the bands arising from each lipid component. Initially, phospholipids with the same headgroup, phosphatidylcholine, were used (25). Because of the isotopic shift the behaviour of both phospholipids can be easily distinguished. It is apparent from the results that the changes in frequency and width are not concerted. This result implies that the bandwidths in the coexistence of two phases are the superposition of two Lorentzian bands with different bandwidths. When the headgroups are different (26) coexistence of different phases can be seen with more complex plots; the IR data can be analyzed together with DSC results. More recently, besides deuteration of the acyl chain, isotopic substitution of the ester carbonyl groups has been performed (27), so that information from the interfacial region of each phospholipid can be separately retrieved. These more complex studies of e.g. mixtures of DPPC and dipalmitoylglycerol (DPG) allow the construction of dynamic phase diagrams, from which the biological significance of the presence of some minor lipids in biological membranes can be inferred.

4.1.2 *Cholesterol-phospholipid interactions.*
Cholesterol is a rigid, planar lipid that does not contain fatty acyl residues. Cholesterol or related molecules are found in eukaryotic membranes, but not in bacteria. Cholesterol has been recognized as a modulator of the membrane fluidity. Infrared studies showed that cholesterol in phospholipid bilayers causes an increase in the proportion of *gauche* conformers below the gel to liquid-crystalline phase transition temperature and a decrease in these conformers above the transition temperature as compared to a pure lipid bilayer. At equimolar cholesterol/DPPC mixtures, the main gel-to-fluid transitions disappears completely (5).

Studies of fatty acyl chain order through the CD_2 rocking in cholesterol-phospholipid mixtures (28) have shown that at positions 4 and 6 of the acyl chains of

DPPC an inhibition of *gauche* rotamers is produced, but the ability of the sterol to increase acyl chain order is much reduced at and beyond position 12.

4.1.3. *Interaction of membrane phospholipids with other lipids.*
Other lipids different than phospholipids or cholesterol can also be a part of the membrane influencing its structure and dynamics. Some of these lipids may take part in metabolic pathways, this is the case of ubiquinones. Ubiquinones are essential intermediates in the respiratory chains of mitochondria and bacteria; the structurally related plastoquinones play similar roles in photosynthetic electron transport chains. Ubiquinone-10 has been studied in pure form and incorporated into phospholipid bilayers (29). In pure form three distinct phases have been characterized: crystalline, isotropic liquid and liquid crystalline; when incorporated into phospholipid bilayers, ubiquinone-10 appears to be removed from the aqueous environment and is found to exist in the 4-70 °C range, according to the C=O stretching vibration, in an isotropic liquid phase in the form of small aggregates.

In eukaryotes, fatty acids are degraded in the mitochondrial matrix in the form of coenzyme A derivatives. Fatty acyl mitochondrial import involves their transient and reversible conversion into fatty acylcarnitines. Mixtures of perdeuterated DPPC with palmitoylcarnitine or palmitoyl-CoA in aqueous suspensions have been studied by looking at the CH_2 and CD_2 stretching vibrations (30). Palmitoylcarnitine mixes with DPPC without perturbing the gel-fluid transition of the phospholipid even at 1:2 molar ratios. However, at similar proportions palmitoyl-CoA smears out the transition as detected by the CD_2 stretching vibration.

The interaction of other lipid molecules of biological interest, such as α-tocopherol (31) or the platelet activating factor (32), with phospholipid bilayers has also been studied by infrared spectroscopy together with other techniques, such as DSC and NMR, showing the effect of the mixture of these lipids on the phase transition of phospholipids.

4.2 LIPID-PROTEIN INTERACTIONS.

Natural membranes are not only composed of lipids, but contain as well proteins that play an important role in their physiological function. Moreover, the physical state of the lipid influences protein activity. Infrared spectroscopy is an important technique in the study of protein-lipid interaction since in a single experiment information about both lipid conformation and protein structure can be obtained. In addition, no spectroscopic probes are needed, thus eliminating the problems associated to the presence of foreign molecules in the system.

A recent review on infrared spectroscopic studies of lipid-protein interactions in membranes (33) has extensively covered the field, including the effect of lipid on protein conformation and conversely, the effect of protein on lipid conformation. The reader is addressed to this paper for a detailed discussion of the problem.

5. BIOLOGICAL APPLICATIONS.

Most of the infrared studies of membrane lipids have been accomplished with model or isolated membranes because of the complexity of cells. However, with the introduction of deuterated lipids, has become possible discrimination between some phospholipids and other lipids or proteins. The first studies were performed on live *Acholeplasma laidlawii* cells grown on a deuterated fatty acid diet (34). The region 2300-2000 cm^{-1} contains only absorption bands originating from CD_2 stretching vibration. The response of live cells to temperature changes differs drastically from that of the isolated membranes. The fractional population of the liquid crystalline phase at any given temperature is always higher in membranes of live cells and the transition has a higher degree of cooperativity than in the lipids of isolated membranes. More recently, the conformational order of *A. laidlawii* has been studied looking at the CH_2 wagging progression (35). Conformational order in the live cells and in the isolated membranes was virtually identical over the range of cell viability (5-40 °C). In contrast, the membrane lipid extracts show much more conformational disorder from 5 to 25 °C than either the live cells or the membranes. This work yielded substantially different results than the previous one. The differences have been attributed either to differences in experimental conditions or to the use of perdeuterated fatty acids in the previous studies.

The state of order of Gram-negative bacterial membranes *in vivo* has also been studied by infrared spectroscopy monitoring the frequency shifts of the acyl chain methylene symmetric stretching band (36). Cells grown at different temperatures yielded distinct transition temperature profiles showing the adaptation of the "state of order" and "fluidity" of bacterial membranes to varying growth temperatures.

The study of the lipids of isolated native membranes and intact cells may also be carried out by chemically modifying their phospholipid environment using a water soluble hydrogenation catalyst in a deuterium-containing environment (37) for the *in situ* insertion of deuterium atoms into the fatty acyl chains of biological membranes.

6. ACKNOWLEDGMENTS.

This work has been supported in part by Grant PB91-0041 from DGICYT, Ministerio de Educación y Ciencia, Spain, and from the University of Basque Country (Grant No. 042-310-EB-219/95). The authors are members of "Grupo de Biomembranas, Unidad Asociada al C.S.I.C."

7. REFERENCES.

1. Farias, R.N., B. Bloj, R.D. Morero, and F. Siñeriz. (1975) Regulation of allosteric membrane-bound enzymes through changes in membrane lipid composition. *Biochim. Biophys. Acta* **415**, 211-225.
2. Fringeli, U.P. and H.H. Günthard. (1981) Infrared Membrane Spectroscopy. in E. Grell (ed.), *Membrane Spectroscopy*, Springer, Berlin. pp. 270-332.
3. Casal, H.L. and H.H. Mantsch. (1984) Polymorphic phase behaviour of phospholipid membranes studied by infrared spectroscopy. *Biochim. Biophys. Acta* **779**, 381-401.

4. Mantsch, H.H. and R.N. McElhaney. (1991) Phospholipid phase transitions in model and biological membranes as studied by infrared spectroscopy. *Chem. Phys. Lipids* 57, 213-226.

5. Lee, D.C. and D. Chapman. (1986) Infrared Spectroscopic Studies of Biomembranes and Model Membranes, *Bioscience Rep.* 6, 235-256.

6. Bellamy, L.J. 1980. The Infrared Spectra of Complex Molecules. Chapman and Hall, London.

7. Fischmeister, I. (1975) Infrared absorption spectroscopy of normal and substituted long-chain fatty acids and esters in the solid state. *Prog. Chem. Fats Other Lipids* 14, 91-162.

8. Mendelsohn, R. and H.H. Mantsch. (1986) Fourier transform infrared studies of lipid-protein interaction. in A. Watts and J.J.H.H.M. DePont (eds.), *Progress in Protein-Lipid Interactions*. vol.2, Elsevier, Amsterdam. pp. 103-146.

9. Flach, C.R. and R. Mendelsohn. (1993) A new infrared spectroscopic marker for cochleate phases in phosphatidylserine-containing model membranes, *Biophys. J.* 64, 1113-1121.

10. Sunder, S., D.G. Cameron, H.L. Casal, Y. Boulanger, and H.H. Mantsch. (1981) Infrared and raman spectra of specifically deuterated 1,2-dipalmitoyl-*sn*-glycero-3-phosphocholines, *Chem. Phys. Lipids* 28, 137-148.

11. Castresana, J., J.M. Valpuesta, J.L.R. Arrondo, and F.M. Goñi. (1991) An infrared spectroscopic study of specifically deuterated fatty-acyl methyl groups in phosphatidylcholine liposomes, *Biochim. Biophys. Acta* 1065, 29-34.

12. Levin, I.W., E. Mushayakarara, and R. Bittman. (1982) Vibrational assignment of the *sn1* and *sn2* carbonyl stretching modes of membrane phospholipids. *J. Raman Spectroscop.* 13, 231-234.

13. Blume, A., W. Hübner, and G. Messner. (1988) Fourier Transform Infrared Spectroscopy of $^{13}C{=}O$-Labeled Phospholipids Hydrogen Bonding to Carbonyl Groups. *Biochemistry* 27, 8239-8249.

14. Lewis, R.N.A.H. and R.N. McElhaney. (1992) Structures of the subgel phases of *n*-saturated diacyl phosphatidylcholine bilayers: FTIR spectroscopic studies of $^{13}C{=}O$ and ^2H labeled lipids, *Biophys. J.* 61, 63-77.

15. Lewis, R.N.A.H., R.N. McElhaney, W. Pohle, and H.H. Mantsch. (1994) Components of the carbonyl stretching band in the infrared spectra of hydrated 1,2-diacylglycerolipid bilayers: A reevaluation, *Biophys. J.* 67, 2367-2375.

16. Arrondo, J.L.R., F.M. Goñi, and J.M. Macarulla. (1984) Infrared Spectroscopy of Phosphatidylcholines in Aqueous Suspension. A Study of the Phosphate Group Vibrations. *Biochim. Biophys. Acta* 794, 165-168.

17. Castresana, J., J.L. Nieva, E. Rivas, and A. Alonso. (1992) Partial Dehydration of Phosphatidylethanolamine Phosphate Groups during Hexagonal Phase Formation, as seen by i.r. Spectroscopy, *Biochem. J.* 282, 467-470.

18. Lewis, R.N.A.H. and R.N. McElhaney. (1993) Calorimetric and spectroscopic studies of the polymorphic phase behavior of a homologous series of n-saturated 1,2-diacyl phosphatidylethanolamines, *Biophys. J.* 64, 1081-1096.

19. Mantsch, H.H., A. Martin, and D.G. Cameron. (1981) Characterization by infrared spectroscopy of the bilayer to nonbilayer phase transition of phosphatidylethanolamines, *Biochemistry* 20, 3138-3145.

20. Blume, A., W. Hübner, M. Müller, and H.D. Bäuerle. (1988) Structure and dynamics of lipid model membranes. FT-IR and ^2H NMR spectroscopic studies. *Ber. Bunsenges Phys. Che.* 92, 964-973.

21. Mendelsohn, R., M.A. Davies, J.W. Brauner, H.F. Schuster, and R.A. Dluhy. (1989) Quantitative determination of conformational disorder in the acyl chains of phospholipid bilayers by infrared spectroscopy. *Biochemistry* 28, 8934-8939.

22. Casal, H.L. and R.N. McElhaney. (1990) Quantitative Determination of Hydrocarbon Chain Conformational Order in Bilayers of Saturated Phosphatidylcholines of Various Chain Lengths by Fourier Transform Infrared Spectroscopy. *Biochemistry* 29, 5423-5427.

23. Senak, L., D. Moore, and R. Mendelsohn. (1992) CH_2 wagging progressions as IR probes of slightly disordered phospholipid acyl chain states, *J. Phys. Chem.* 96, 2749-2754.

24. Tuchtenhagen, J., W. Ziegler, and A. Blume. (1994) Acyl chain conformational ordering in liquid-crystalline bilayers: Comparative FT-IR and ^2H-NMR studies of phospholipids differing in headgroup structure and chain length, *Eur. Biophys. J.* 23, 323-335.

25. Dluhy, R.A., R. Mendelsohn, H.L. Casal, and H.H. Mantsch. (1983) Interaction of Dipalmitoylphosphatidylcholine and Dimyristoylphosphatidylcholine-d54 Mixtures with Glycophorin. A Fourier Transform Infrared Investigation. *Biochemistry* 22, 1170-1177.

242

26. Brauner, J.W. and R. Mendelsohn. (1986) A comparison of differential scanning calorimetric and Fourier transform infrared spectroscopic determination of mixing behavior in binary phospholipid systems, *Biochim. Biophys. Acta* **861**, 16-24.

27. López-García, F., J. Villalaín, J.C. Gómez-Fernández, and P.J. Quinn. (1994) The phase behavior of mixed aqueous dispersions of dipalmitoyl derivatives of phosphatidylcholine and diacylglycerol, *Biophys. J.* **66**, 1991-2004.

28. Davies, M.A., H.F. Schuster, J.W. Brauner, and R. Mendelsohn. (1990) Effects of cholesterol on conformational disorder in dipalmitoylphosphatidylcholine bilayers. A quantitative IR study of the depth dependence. *Biochemistry* **29**, 4368-4373.

29. Castresana, J., A. Alonso, J.L.R. Arrondo, F.M. Goñi, and H.L. Casal. (1992) The physical state of ubiquinone-10, in pure form and incorporated into phospholipid bilayers. A Fourier-transform infrared spectroscopic study, *Eur. J. Biochem.* **204**, 1125-1130.

30. Echabe, I., M.A. Requero, F.M. Goñi, J.L.R. Arrondo, and A. Alonso. (1995) An infrared investigation of palmitoyl-coenzime A and palmitoylcarnitine interaction with perdeuterated-chain phospholipid bilayers. *Eur. J. Biochem.* **231**, 199-203.

31. Villalaín, J., F.J. Aranda, and J.C. Gómez-Fernández. (1986) Calorimetric and infrared spectroscopic studies of the interaction of α-tocopherol and α-tocopheryl acetate with phospholipid vesicles, *Eur. J. Biochem.* **158**, 141-147.

32. Salgado, J., J. Villalaín, and J.C. Gómez-Fernández. (1993) Effects of platelet-activating factor and related lipids on dielaidoylphosphatidylethanolamine by DSC, FTIR and NMR, *Biochim. Biophys. Acta Bio-Membr.* **1145**, 284-292.

33. Arrondo, J.L.R. and F.M. Goñi. (1993) Infrared Spectroscopic Studies of Lipid-Protein Interactions in Membranes. in A. Watts (ed.), *Protein-Lipid Interactions.* Elsevier Science Publishers, Amsterdam. pp. 321-349.

34. Cameron, D.G., A. Martin, D.J. Moffat, and H.H. Mantsch. (1985) Infrared Spectroscopic Study of the Gel to Liquid-Crystal Phase Transition in Live Acholeplasma laidlawii Cells, *Biochemistry* **24**, 4355-4359.

35. Moore, D.J., M. Wyrwa, C.P. Reboulleau, and R. Mendelsohn. (1993) Quantitative IR studies of acyl chain conformational order in fatty acid homogeneous membranes of live cells of *Acholeplasma laidlawii* B, *Biochemistry* **32**, 6281-6287.

36. Schultz, C. and D. Naumann. (1991) In vivo study of the state of order of the membranes of Gram-negative bacteria by Fourier-transform infrared spectroscopy (FT-IR), *FEBS Lett.* **294**, 43-46.

37. Szalontai, B., L. Vigh, F. Joó, L. Senak, and R. Mendelsohn. (1994) In situ modification of the phospholipid environment of native rabbit sarcoplasmic reticulum membranes, *Biochem. Biophys. Res. Commun.* **200**, 246-252.

TIME-RESOLVED INFRARED SPECTROSCOPY OF BIOMOLECULES

H. GEORG, K. HAUSER, C. RÖDIG, O. WEIDLICH AND F. SIEBERT
Institut für Biophysik und Strahlenbiologie, Albert-Ludwigs-Universität
Albertstr. 23, D-79104 Freiburg, Germany

1. Introduction

The elucidation of the molecular mechanism of biological and biochemical systems represents still a challenge in the field of biochemical/biophysical research. The knowledge of the structure of these systems at high resolution, determined by X-ray cystallography or NMR spectroscopy (see contributions in this volume) will certainly help in the understanding of the mechanism. However, although much progress has been made in recent years with both methods of structure determination, there are still many biologically important systems of which the structure is unknown. This applies especially to membrane systems whose structures are still very difficult to determine. Therefore, it is necessary to have alternative methods providing molecular information on structure, interaction and their changes during the reaction. Infrared spectroscopy is a classical tool in Physical and Analytical Chemistry sensitive to such molecular properties. However, for many years it has mainly been applied to relatively small molecules, because it had been thought that the infrared spectrum of a protein is too complex. The numereous overlapping bands would inhibit the deduction of molecular details necessary for deriving the mechanism of biochemical reactions. However, it has been shown that during the reaction of enzyme usually only a small part of the complex system is actually involved, the remainder serving mainly as an appropriate matrix. Therefore, if it were possible to derive the infrared spectra of only those parts actually involved, the evaluation would provide important molecular details on structure and interaction from which the mechanism could be deduced. Here, the idea of infrared difference spectroscopy comes up: by forming the difference spectra between different states of a reaction, only those groups will be reflected that are involved, i.e. that undergo molecular changes. The absorption bands of the large rest cancel each other. There is, however, one requirement for this method to work: it has to be sensitive enough to detect the infrared bands of a single residue (amino acid, ligand, co-factor etc.) against the background absorption of the large system. This means that one has to be able to detect absorbance changes as small as 10^{-4} to 10^{-5} against an average background absorbance of 0.5. We and others have shown in recent years that, using the modern technology of infrared spectroscopy such as Fourier transform spectroscopy, sensitive semiconductive detectors, devices with high energy through-put, the the goals can be attained. Furthermotre, since infrared spectroscopy is

G. Vergoten and T. Theophanides (eds.), Biomolecular Structure and Dynamics, 243–261.

an optical technique, it can be extended to time-resolved studies. This is an important aspect, since biochemical reactions are processes in time, and the determination of the temporal evolution of a reaction provides important information on its mechanism. We as also others have demonstrated, that, by exploiting the new developments in infrared technology, it is nowadays possible to extend the method of infrared spectroscopy to time-resolved studies. In recent years, several reviews have appeared on this subject [1-7]. In the following, the basic methods of time-resolved infrared spectroscopy as applied to biological systems will be described and typical applications presented. In addition, some new results from my group demonstrating the strength of this technique will be preseneted. However, before going into the details of the description of the different instruments and experiments, some general comments on infrared spectroscopy as applied to biological systems may be appropriate.

Most biological systems require for proper function the presence of water. Therefore, if the experiments are aimed at the mechanism or at the active state of the enzyme, the measurements have to be performed under the presence of water. However, water is a strong infrared absorber, limiting the sample thickness to 4 to 10 μm. This has the consequence that the concentration of the biological sample must be very high. In special cases, this may cause some problems due to protein aggregation which inhibits enzyme function. In addition, the strong absorbance of water decreases the intensity of the infrared beam almost over the entire spectral range of interest, making it mandatory to use the most sensitive equipment. However, as has been demonstrated in many practical cases, the presence of water does not cause serious technical problems. In addition, the use of ^2H_2O, having lower absorbance and shifted bands may resolve part of the difficulties. Since several groups in a protein undergo $^1H/^2H$ exchange upon exposure to ^2H_2O, this technique may provide additional information for band assignment (see below) and accessibility of groups to the aqueous phase. Also, another aspect of time-resolved infrared spectroscopy as applied to biological samples is, at least partially, related to the requirements of high sample concentration. In order to follow the time course of a reaction, it has to be triggered at a precise instant. In the more common case of time-resolved UV-vis spectroscopy, this is realized with the well-known stopped-flow or rapid-mixture techniques, by which the reaction partners are mixed in a very short time (approx. 1 ms) and the reaction susequently followed in the mixture. It is obvious, that these mixture techniques are difficult to realize with the high sample concentrations. A further limitation caused by the small size of the absorbance changes has to be mentioned. The corresponding small signals at the infrared detector requires the application of signal averaging technique. Thus, biochemical reactions can be only followed by time-resolved infrared spectroscopy if it can be triggered or made triggerable many times. If a broad spectral range has to be covered, this property is even more stringent, since a correspondingly larger number of experiments have to be repeated (see below). Thus, photoreaction of photobiological systems which are irreversible are difficult to investigate, whereas systems with cyclic reactions or with photochromic properties are conveniently studied.

Other remarks relate to the interpretation of the difference spectra. It is important to realize what molecular information can be deduced. In principle, the following questions may be asked: 1. what are the groups contributing to the difference spectra; 2. what are their molecular changes; 3. are these groups (and their molecular changes) important for the function. It is obvious that groups located in the catalythic center or groups involved in binding of reaction partners will show up iń the spectra. These may be amino acid side chains or co-factors such as chromophores in photobiological systems. However, it is also possible that during the reaction the systen undergoes conformational changes extending over a somewhat larger range. In this case, groups not directly involved in the reaction will experience an altered interaction and therefore show up in the difference spectra. In addition, these conformational changes will cause alterations of vibrations of the peptide backbone such as amide A (NH stretching), amide-I (mainly CO stretching) and amide-II (CN stretching coupled to NH bending) vibrations [8]. Thus, although the number of residues contributing to the difference spectrum is greatly reduced as compared to that contributing to the absorption spectrum of the total system, the former is still often too complex to allow a straightforward interpretation. Therefore, additional supporting techniques are generally required to deduce the molecular content and some of them will now be presented. The infrared spectra of model compounds is always an important start for the interpretation. The spectra of amino acid side chains are documented in the literature [9,10], but it is often advantageous to repeat corresponding measurements. Key residues are imidazole, protonated and unprotonated carboxyl groups, OH groups, methyl-guanidinium, side-chains of aromatic residues. It would be advantageous to have the corresponding spectra mesured in different solvents from which the influence of polarity and hydrogen bonding can be deduced and similar considerations apply to the spectra of co-factors such as chromophores [11]. However, these data are not always sufficient for the interpretation. Here, Molecular Biology provides an unique solution: by replacing one particular amino acid by another one, one not only can test it for its functional importance, but the comparison of the respective difference spectra can also identify the bands caused by these amino acids. This technique is now widely applied and references can be found in the review articles mentioned above. Despite of the clear advantages, the application of mutation has also its caveat: the influence of the mutation on the difference spectra may not be a direct one but rather reflect indirect interaction and their alterations by the mutation. In addition, the mutation may represent just a new system with drastically altered properties. Under these conditions, it may be difficult to compare the spectra of the wild type system with those of the mutated one. The use of isotopic labelling is a classical tool for identifying absorption bands and assigning them to special molecular vibrations (there is an excellen book collecting older data [12]). However, it is not obvious how to incorporate isotope-labels into biological systems. If co-factors are to be labelled, this may often be achieved, depending on the sytem, by exchanging the natural compound for an artificial one. However, the incorporation of labelled amino acids into a protein is more complicated. If the protein is (or can be) expressed by bacteria, they can be grown in an artificial medium containing this labelled amino acid. Depending on its biosynthetic pathway in the bacterial strain, this amino acid will be more or less incorporated. A level of 20 to 30 % is usually sufficient. But, due to the biosynthetic pathway, the label may also scramble to other amino acids, complicating the interpretation. Some of the aspects of isotopic labelling can be found in the old investigations on the biosynthesis of amino acids [13], and a new collection of papers dealing with the application of stable isotope labelling in the study of structural and mechanistic aspects of biomolecules has appeared recently [14]. One has to keep in mind that by adding a labelled amino acid to the culture medium in which the bacteria are grown, all amino acids of this kind in the expressed protein are labelled. Therefore, additional techniques are required to discriminate between the different sites of this special residue. However, a very interesting development has been recently made using the modern techniques of Molecular Biology. It has been shown that it is possible to incorporate a labelled amino acid at one specific site only [15,16]. Although the method is still elaborate, the numereous advantages it offers for the interpretation of the difference spectra are obvious. In the reviews mentioned above, there are also examples on the application of isotope labelling of co-factors (chromophores) and amino acids to infrared spectroscopic studies. In addition to the assignemnt of bands to specific residues, isotope labelling also allows a more detailed description of band in terms of normal modes, and thus a much more detailed analysis at a molecular level.

In Fig. 1 an example is shown how isotope-labelling provides information on conformational changes of a specific region of a protein. The system investigated is the light-driven proton pump bacteriorhodopsin. Since some further experiments on this system will be shown below, some general remarks on the properties of this system are necessary. It is a membrane protein, being constituted by seven α-helices spanning the membrane (structurally it belongs to the class of seven-helix-receptors) The light-sensitive part is the chromophore all-*trans* retinal which is bound to a lysine of the protein via a protonated Schiff base ([7,17-21] for review). Upon light absorption, it undergoes a cyclic photoreaction involving several intermediates, which is completed in about 10 ms. The initial action of light is the isomerization of the chromophore from the all-*trans* to the 13-*cis* geometry. This event induces the other important molecular changes necessary for proton pumping across the membrane. One essential step occurring with the formation of the so-called M intermediate is deprotonation of the Schiff base, since it is assumed that it is actually this proton which is being pumped. The prior intermediate, called L, appears therefore to be crucial since the essential molecular conditions for Schiff base deprotonation must already be present here. This intermediate appears in the time range of 2 µs, and time-resolved techniques for monitoring it would be required. However, in many photobiological systems it is possible to freeze intermediates of the photoreaction at low temperature. The corresponding difference spectrum shown in Fig. 1 was therefore taken under static conditions, illuminating the sample at 170 K. It is beyond the scope of this contribution to discuss the large number of positive (photoproduct state) and negative (initial state BR) bands. We only focus now on the spectral range between 1680 and 1620 cm^{-1}, representing the spectral range where amide-I modes show up. As has been already mentioned, they are sensitive to protein conformations. Thus, site-directed isotopic labelling of one C=O group in the peptide backbone would provide information whether this group is involved in the conformational changes. Bacteriorhodopsin offers

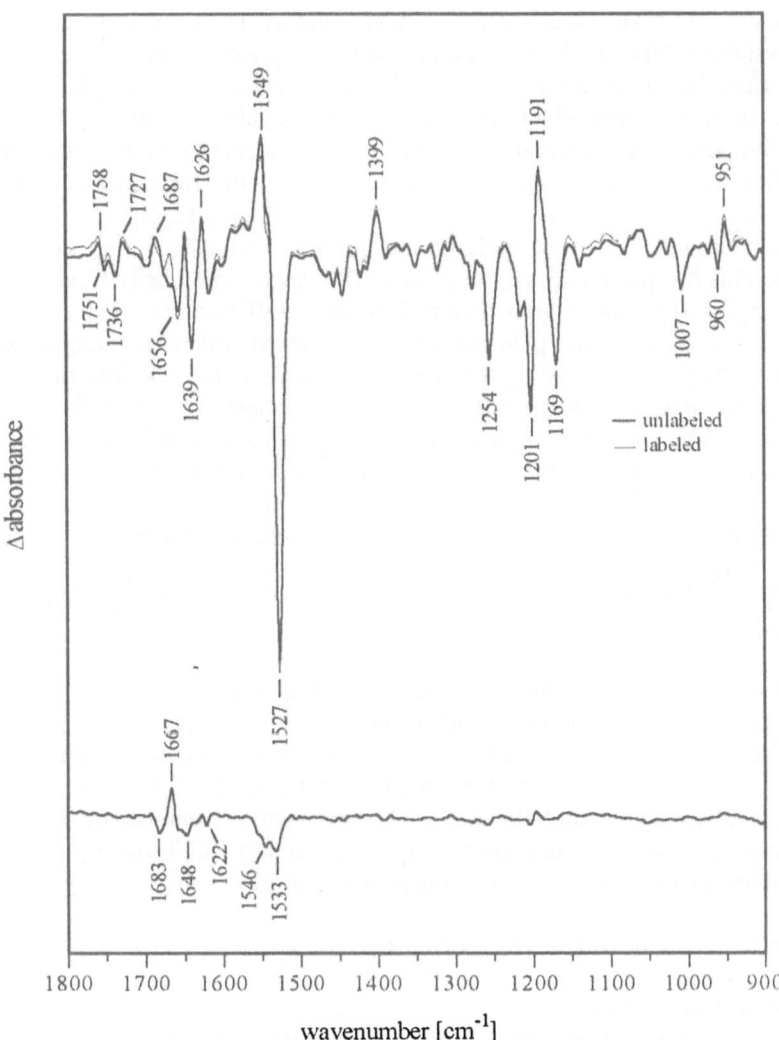

Fig. 1 BR→L difference spectrum of the T89C mutant of bacteriorhodopsin. Upper part: thick trace spectrum of unlabelled sample; thin trace: spectrum of sample containing ^{13}C at the C=O group on the carboxy-terminal side of T89. Lower part: difference between the two spectra, revealing clear differences caused by the labelling, indicating that the corresponding peptide group is involved in structural changes.

an unique way of site-directed labelling of such a group. The wild type system does not contain cysteines. Thus, replacing an amino acid at a position one is interested in by a cysteine offers the possibility to introduce also an isotopic label only at this position if a cysteine is incorporated of which the carboxyl group is labelled with ^{13}C. In Fig. 1, the mutant T89C has been investigated and the BR→L difference spectra of the unlabelled and labelled mutant are compared. The subtraction of the two spectra reveals clear differences in the amide-I range (1683 to 1648 cm^{-1}). Therefore, one can conclude that protein part around T89 must undergo some conformational change in the BR→L transition. This is especially interesting, since T89 is located on helix C, containing the proton acceptor D85 and proton donor D96 for Schiff base de- and reprotonation, respectively. Thus, it appears plausible that the observed structural changes located on helix C somehow regulate the proton transfer steps. (Amide-I changes involving T89 have also been observed in the BR→N transition, where the Schiff base is already reprotonated, data not shown here). Of course, one has to control that the T89→C mutation does not cause gross alterations in the properties of bacteriorhodopsin, and we have confirmed that they are reasonably small. This example demonstrates the advantages of isotopic labelling in general and of site-directed labelling in particular.

2. Description of Time-Resolved Infrared Techniques and their Applications to Biomolecules

This part is divided into a longer section dealing with techniques and applications in the time-range of 10 ns longer, and a shorter section describing techniques and experiments in the sub-nanosecond time range. The first section is further subdivided into a part containing narrow-band techniques, i.e. the spectral changes are monitored at a single wavenumber and the difference spectrum is obtained by repeating the experiments at all the wavenumbers of interest, and a part with broad-band techniques. The broad band techniques all employ the principle of FT-IR spectroscopy.

2.1 TIME RANGE OF 10 NS AN SLOWER

2.1.1 *Narrow band techniques*
The technique using a monochromatic monitoring beam is the straightforward extension of conventional time-resolved UV-vis spectroscopy into the infrared spectral range. Time-resolution is primarily limited by the rise-time of the detector and by the acquisition electronics, (transient recorder) and in practice also by the intensity of the monitoring beam. In its simplest implementation this technique uses as monitoring beam an infrared broad band source such as a globar or Nernst glower and a monochromator to select the wavenumbers. This is the version how time-resolved infrared measurements have first been performed with biological samples. A photoconductive mercury cadmium telluride (MCT) detector has been used. The photoreaction of rhodopsin [22] and bacteriorhodopsin [23] have been studied at a time-resolution of approx. 20 µs. Using the same technique, we have been later able to monitor the protonation of the acceptor of the proton from the Schiff base and assigned

it with isotopic labelling to an aspartic acid [24,25]. By extending the time-resolution with a faster transient recorder to 1 μs, we have been able to describe the photoreaction of bacteriorhodopsin from the L intermediate onwards, and we realized that the kinetics observed for absorption bands of the chromophore are also observed for bands of the protein. Thus it appears, that there is a tight coupling between the chromophore and the protein. By replacing the photoconductive by a photovoltaic MCT detector, the time resolution could be increased to approx. 20 ns and the KL intermediate (before L) could be measured [26,27].

In a more recent application the use of a monochromator has been crucial. We have studied the temperature induced unfolding and refolding process of proteins with time-resolved infrared spectroscopy, mainly monitoring the amide-I/II bands. The temperature increase was initiated by an infrared laser pulse of 200 μs width (Er:YAG-laser emitting at 2.9 μm) which is effectively absorbed by the aqueous solvent. Due to the sample geometry (thin layer, large area), the temperature decreases to the initial

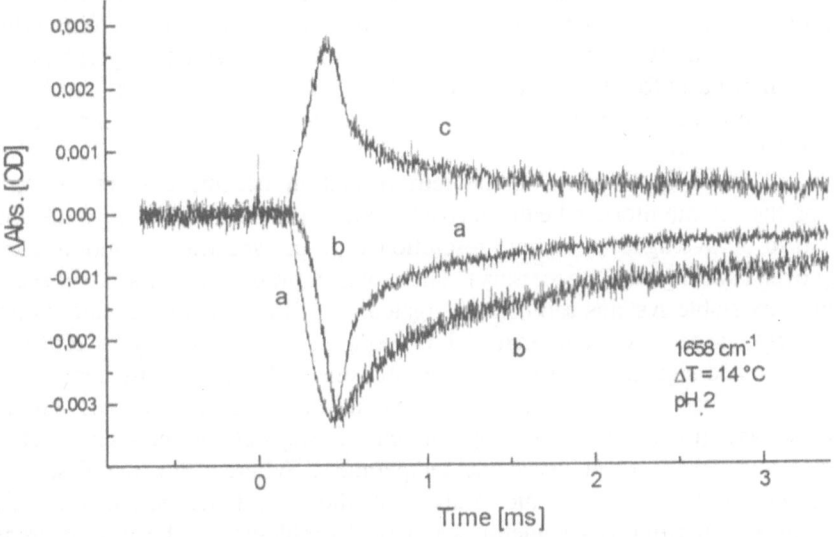

Fig. 2 Time-resolved absoprbance changes of unfolding in cytochrome c induced by a temperature pulse. Trace a: at 3200 cm^{-1}, providing information on the temperature profile; trace c: at 1658 cm^{-1}, starting temperature 20 °C., temperature does not reach unfolding transition; trace b: starting temperature 45 °C, here complete unfolding is induced. Rise in temperature deduced from trace a is 14 °C.

value within 4 ms. Thus, an observation window is available from 100 μs up to 2 ms for the unfolding process, and folding kinetics slower than 4 ms can be observed. A temperature increase of 5 to 15 °C can be easily obtained. The monochromator inserted between the sample and the detector not only blocks the exciting infrared laser pulse, but also the infrared signal emitted by the sample due to the temperature pulse. We have applied this method to the unfolding and refolding of cytochrome c at pH 2. In Fig. 2 typical absorbance changes are shown. Trace (a) represents the signal measured at 3200 cm^{-1}, reflecting the absorbance change of the water caused by the temperature pulse. A comparison with static measurements at different temperatures shows that the maximum temperature increase is 14 °C. Trace (c) represents the signal measured at 1658 cm^{-1} if the starting temperature is 20 °C. It shows mainly the temperature dependence of the absorbance of water and, with much smaller amplitude, that of the amide-I band of the protein. No unfolding occurs under this condition. Trace (b) represents the same measurement if the starting temperature is 45 °C. Here unfolding takes place. The signals descibes the disappearance of α-helical structure. The actual signal due to unfolding is much larger, since the signal of trace (a) is superimposed. The time-course shows that there are at least two steps in the unfolding process (maximum is delayed with respect to the temperature). In addition, the refolding lacks behind the temperature decrease. The comparison with static experiments on temperature induced unfolding show that complete unfolding is reached after approx. 250 μs. These preliminary results already demonstrate the feasability and scientific potential of the method.

In all these applications an obvious limitation became evident. Since the intensity of the IR monitoring beam derived from a thermal source is weak many signals had to be averaged (at a time-resolution of 1 μs. spectral resolution 6 cm^{-1}, averaging over 1000 signals is necessary). Therefore. in order to cover a broad spectral range, only very stable systems can be investigated, or, if the sample is replaced during a measurement, large amount of material is required. The only way to reduce the number of averaged signals is to increase the intensity of the monitoring beam, and in practice, this has been only realized by tunable infrared lasers. The most common lasers are CO laser (tunable from 2000 to 1540 cm^{-1} in steps of approx. 4 cm^{-1}) [28,29], and laser diodes operating at cryogenic temperature. Whereas for the first source tuning is achieved with an in-resonator grating, for the second one the temperature has to be controlled. This imposes sometimes practical problems for the tuning process. Nevertheless, useful information has been obtained on ligand transfer processes in cytochrome oxidase [30] and on protonation changes of carboxyl groups during the reduction of Q_B in the photosynthetic reaction center [31,32]. Due to the very slow recombination reaction which limits the repetition rate for signal averaging to 0.05 Hz the latter experiments would heve been difficult to perform with the weak intensity of a thermal infrared source. If a more practical high intensity source tunable over a large spectral range (1800 - 800 cm^{-1}) were available, the narrow-band technique would be a very convenient method of time-resolved infrared spectroscopy. With the lasers mentioned, it is now mainly applied for cases where it is sufficient to cover a small spectral range.

2.1.2 Broad-band techniques

In time-resolved UV-vis measurements it is common to use array detectors to simultaneously cover a broad spectral range. In principle, this technology is also available for time-resolved infrared spectroscopy. However, array detectors connected to appropriate electronics are still very expensive. In addition, the small pixel size limits the application (usual IR detectors have an area of 1 mm^2). Therrfore, the broad band techniques discussed here are onlyFourier transform infrared (FT-IR) spectroscopic methods. For more details of FT-IR spectroscopy the book by Griffiths and de Haseth is highly recommended [33]. For an understanding of the time-resolved implementations of FT-IR spectroscopy a few aspects of the principles of this method are given here. Instead of a monochromator, a Michelson interferometer is the main optical element and the intensity is measured as function of the path difference caused by the movable mirror. The Fourier transform of this intensity function (interferogram) yields the spectral intensity. In practice, the interferogram is digitized at fixed spacings determined by the highest wavenumber to be measured (closer spacing→higher wavenumber) and the length of the interferogram determines the spectral resolution (longer interferogram→higher resolution). The advantages of FT-IR spectroscopy are much higher sensitivity caused by the multiplex-advantage and the higher energy throughput since much larger apertures can be tolerated. Due to the small spectral changes occurring in IR-difference spectroscopy of biological systems (see above), this increased sensitivity is very important. The time needed to measure a single interferogram can be as short as a few milliseconds. Thus, if processes slower than 5 ms are to be followed. this so-called rapid-scan FT-IR spectroscopy is a very efficient method, since within the time needed to measure the interferogram a complete spectrum is finally obtained. In order to increase the signal/noise ratio, the experiments have to be repeated and the results averaged. This method has been successfully applied to the slow part of the photoreaction of bacteriorhodopsin [34,35] and the visual pigment rhodopsin [36]. We have used this method to study the influence of replacing the suggested proton acceptor of Schiff base deprotonation, D85, by a glutamic acid [37]. and, in combination with isotopically labelling of aspartic acids it was shown that now a glutamic acid becomes protonated, supporting the role of D85 as proton acceptor. Further, we have studied with this method the role of steric interactions between the 9-methyl group of retinal and the protein. It was shown that a special interaction between this group and W182 controls the rate of backisomerization of the chromophore. Here, a combination of chromophore modification (removal of the 9-methyl group) and of mutagenesis was used [38,39]. The obvious limitations in time-resolution are the velocity of the mirror and the rate at which the interferogram can be digitized. With present instrumentation the limitation is a few millisecond. Another disadvantage is the limited dynamic range. Since the spectral changes are very small, it would be advantageous to digitize the difference interferogram corresponding to the intensity changes directly. A method using such a principle is described later. It demonstrates that the increase in dynamic range of the digitization of the interferogram increases the signal/noise ratio. Nevertheless, the rapid-scan method is easy to apply, since manufacturers of modern FT-IR instruments provide the required software and hardware (triggering and synchronizing the reaction with the scan movement). In order to overcome the limitations determined by the velocity of the mirror, the stroboscopic rapid-scan FT-IR technique has been developed. Here, with one scan, only part of the interferogram is recorded, thereby shortening the time correspondingly (in the extreme limit only one interferogram point). With the next scans, the delay between the trigger of the reaction and the acquisition 'of the interferogram is varied, until the total interferogram is obtained. It is clear that the number of scans has to be increased correspondingly, making the method less efficient. Nevertheless, this technique has been successfully applied to the photoreaction of bacteriorhodopsin [40,41] and important informations on the proton pathway was obtained by studying special mutants [42]. Again, the application of the stroposcopic method has been facilitated since manufacturers offer the corresponding hard- and software. The maximum time-resolution is determined by the 16 bit AD converter of the interferogram which has a sampling rate ranging between 100 and 500 kHz. However, the remarks on the limited dynamic range of the digitization also apply here.

In order to overcome the disadvantages of the rapid-scan and stroboscopic methods (time-resolution, efficiency, digitization), we have developed the so-called step-scan technique. The idea is simple: if it were possible to move the mirror of the interferometer in a step-wise manner from one sampling point to the next one (like a monochromator from one wavenumber position to the next one), then it would be possible to record at each sampling position the transient change of the interferogram. By re-arranging the data in the computer, a complete set of interferograms is obtained, covering the time-range of the transient, and the Fourier transformation provides finally the difference spectra. It is obvious that the time-resolution is only limited by the rise time of the detector and by the acquisition electronics (like in the narrow-band methods). The main problem in the past was the availability of such an interferometer. This has changed since the late 80-ies, and we have shown that high quality time-resolved difference spectra can be obtained with this method [43]. In a later application, we have studied the $KL \rightarrow L$ transition of the bacteriorhodopsin photocycle with a time-resolution of approx. 400 ns [44]. We have been able to clearly identify the KL intermediate which is different from the picosecond K intermediate. In addition, we could confirm the existence of a KL/L equlibrium and important structural changes of the chromophore in the sequence $K \rightarrow KL \rightarrow L$ could be derived. We have shown that in K the chromophore is twisted all along the polyenen chain, and that this twist is later partially relaxed and progressively confined to the Schiff base region.

A more recent application is shown in Fig. 3. It relates to an important aspect of the function of bacteriorhodopsin. It has been realized that, in order to accomplish a vectorial transport across the membrane, two M states must exist [45,46] which differ in the accessibility of the Schiff base. In M_1, the Schiff base is still connected to the proton acceptor (D85) for Schiff base deprotonation, whereas in M_2 some conformational change must take place inhibiting the re-uptake from D85 and enabling Schiff base reprotonation from the donor, D96. In order to block the re-uptake of the proton from the same side to which it was ejected, the transition from M_1 to M_2

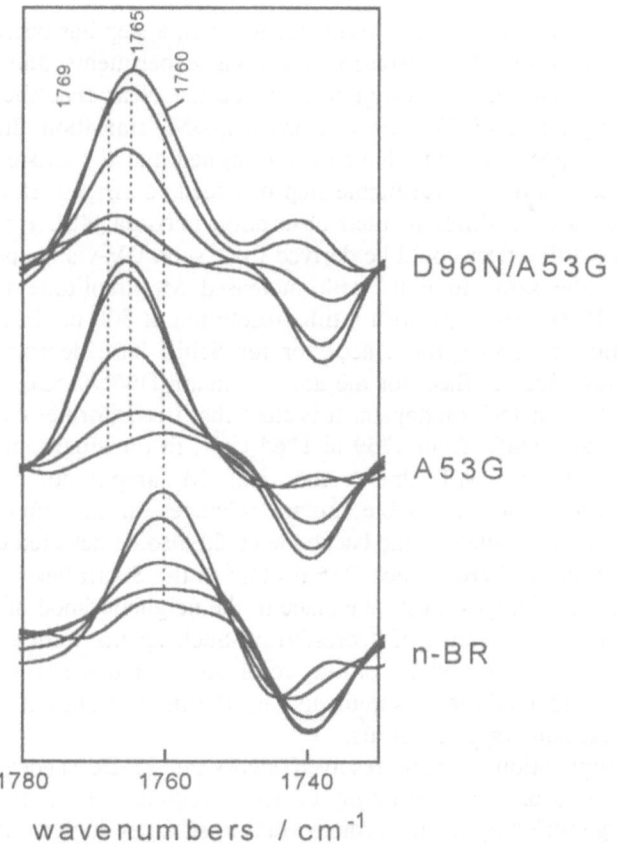

Fig. 3 Time-resolved step-scan measurements of wild type bacteriorhodopsin, the mutants A53G and A53G/D96N. The time evolution of the band due to protonation of D85 is shown for 2, 4, 8 and 20 μs after the flash. In the mutant, a clear $M_1 \rightarrow M_2$ transition can be seen, which is accompanied by a shift of the C=O band. In wild type BR, due to the small M_1 amplitude, no such trnasition can be detected.

has to be irreversible or quasi-irreversible. Evidence for such a step has been provided by a detailed kinetic analysis of time-resolved UV-vis experiments and electrical measurements [19,47]. From such investigations it became clear that the apparent amplitude of M_1 is very small (5 %), due to a fast $M_1 \rightarrow M_2$ transition. In order to increase this amplitude, special mutants have been designed and the existence of the two M states connected by a quasi-irreversible step has been confirmed. In one of the mutant, the two M states even differ in their absorption maxima [48]. However, no molecular cause for the differences could be derived from such UV-vis measurements. Therefore, we studied the same mutants with increased M_1 amplitude with time-resolved step-scan FT-IR spectroscopy with a time-resolution of 500 ns. In Fig. 3, the band due to protonation of D85 (proton acceptor for Schiff base deprotonation) is shown at different times after the flash for the double mutant D96N/A53G, the single mutant A53G and for WT bacteriorhodopsin. It is clear that in the former case, during the evolution of the band it shifts from 1769 to 1765 cm^{-1}, in the single mutant from 1760 to 1765, whereas for the WT, due to the small M_1 amplitude, no $M_1 \rightarrow M_2$ transition can be detected. In the double mutant, changes in the amide-I range reflecting structural changes of the peptide backbone could also be detected during the $M_1 \rightarrow M_2$ transition (not shown here). Since D85 is close to the Schiff base, the results demonstrate that structural changes must take place in the neighbourhood of the Schiff base, which could explain the change of accessibility. Such results would have been difficult to obtain with any other methods of time-resolved infrared spectroscopy so far discussed, since the accuracy of the mesurements and the time-resolution have to be very high to detetct these small spectral shifts.

In another application of time-resolved step-scan FT-IR spectroscopy we studied the photodissociation and rebinding of CO-myoglobin. Like hemoglobin, myoglobin is an oxygen-binding heme protein, but it storages oxygen in skeletal muscles. It binds with high affinity CO to the iron of the heme ring. Upon absorption of light by the heme, the CO-Fe bond is photolyzed and at room temperature the CO molecule is driven out of the protein. Subsequently, rebinding occurs. This system is the classical model system for studying the interaction of small ligands to proteins and the involved molecular changes. It has also been taken to elucidate the role of structural substates and their fluctuations (see [49] for review). Time-resolved infrared spectroscopy can provide information on structural changes occurring with the rebinding of the CO molecule. Fig. 4 shows corresponding results, representing a difference spectrum taken 4.4 µs after the flash. We have used wild type recombinant human myoglobin and the H64I mutant. We have chosen this mutant, since histidine 64 is located on the distal site of the heme and can interact with the bound ligand. The upper part of Fig. 4 shows the CO stretching band of bound CO. It is evident that the mutation causes an upshift of this band to a position, where in WT myoglobin only a shoulder is present. This result could also be obtained with static measurements. They suggest, that the shoulder of the CO band is due to a geometry of H64 where it cannot interact with CO. The lower part, showing spectra in the amide-I/II range, demonstrates that by the photolysis structural changes take place involving the peptide backbone. Since these structural changes are different for the mutant, it is clear that the

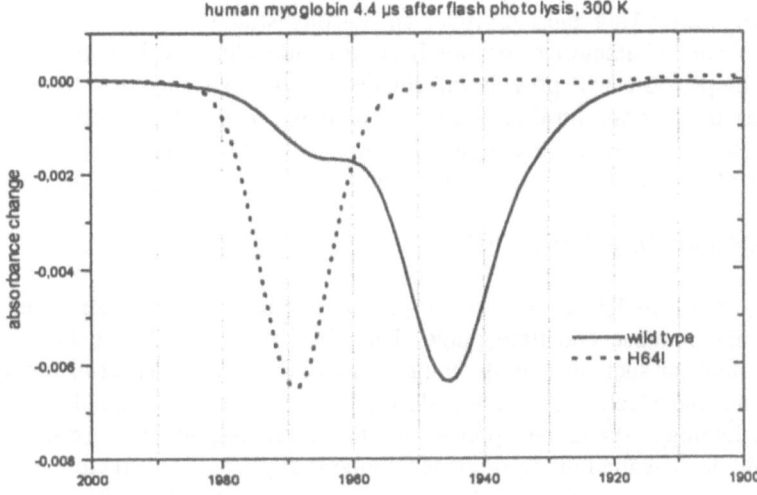

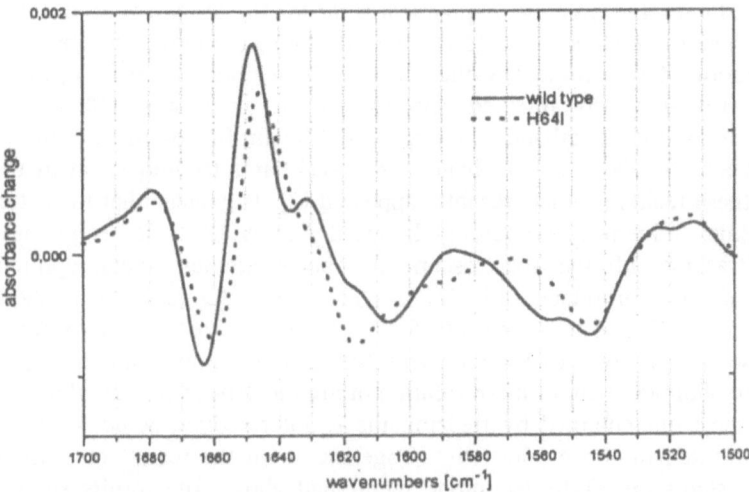

Fig. 4 Time-resolved step-scan measurements of the photolysis of human CO-myoglobin and of the mutant H64I. Solid trace: wild type, dotted trace: mutant. Mutation not only alters the position of the CO band of bound CO (upper panel), but also amide-I/II bands (lower panel), indicating that this residue is involved in structural changes of the peptide backbone, which occur binding of CO to the heme.

backbone containing H64 must be involved. Further studies with additional mutants will help to better characterize the involved structural changes. Experiments carried out at low temperature will provide information on the existence of intermediates in the rebinding process and on their molecular idendity. The step-scan technique is the ideal method for studying the dynamics in myoglobin. Receently, by using a photovoltaic MCT detector, the time-resolution has been increased to 50 ns [50].

2.2 SUBNANOSECOND TECHNIQUES

Important molecular events in photobiological systems take place in the picosecond and femtosecond time range. Therefore, it is desirable to have methods allowing to monitor such structural changes also in the infrared spectral range. Since both detectors and electronics are too slow to capture such fast signals, a complete different technology has to be applied. As in the corresponding ultra-fast UV-vis spectroscopy, the so-called pump-probe technique is applied. The principle is simple: a short laser pulse excites the sample, and a second short pulse with known delay interrogates the absorption changes of the sample caused by the first pulse. In order to cover the time course, the delay has to be varied between negative values up to the longest time. The delay between the two pulses is realized by different pathlengths between excitation and probing. A practical limit for the longest time is around 1 ns, since otherwise the distances would become too long, rendering the set-up unstable. Time resolution is determined by the puls width of both the exciting and probing pulses. In time-resolved infrared spectroscopy it is around 100 fs. Usually, deconvolution of the absorbance changes with the profile of the exciting pulse is required. There are both narrow-band and broad-band techniques available and a review on these techniques has recently appeared [3]. It is clear that these techniques are not routine, and require a thourough knowledge of modern laser technology.

Ultrashort infrared spectroscopy has been, among others, applied to the photocycle of bacteriorhodopsin [51,52], to electron transfer reactions in the bacterial reaction center [53], and to ligand transfer reactions in cytochrome c oxidase [54]. In the first case, it was shown that the time-resolved difference spectrum at 10 ps of the K intermediate obtained with a time-resolution of around 1 ps is very similar to the static difference spectrum obtained by freezing the K intermediate at 80 K. The spectral range was limited to above 1560 cm^{-1}. Togeteher with the results obtained with the nanosecond step-scan FT-IR technique mentioned above, the results show that the protein undergoes very fast conformational changes, but that further changes up to the time range of 500 ns are mainly restricted to the chromophore. Thus, time-resolved infrared difference spectroscopy has provided a deeper understanding of the dynaimcs of the proton pump bacteriorhodopsin. In the case of the photosynthetic reaction center, good agreement is obtained between the kinetics observed in the visible and in the infrared. Only at very early times, fast vibrational relaxation processes take place which are monitored in the infrared. Time resolution in these experiemnts has been 300-400 fs. The application of ultrafast infrared spectroscopy to biological systems is still at its beginning. It can be expected that important information on their dynamics can be obtained, such as energy dissipation, vibrational relaxation, fast molecular movements, protein folding.

The application of time-resolved infrared spectroscopy has already provided detailed molecular information on biological processes and further advances can be expected. There appears to be, howerver, a misbalance between the quality of the spectra (very high) and the possibility of their interpretation (not satisfactory). Much more theoretical work is necessary such as normal mode analysis on the basis of quantumchemical calculations in order to provide the framework for the interpretation of the infrared data at a molecular level.

3. Acknowledgement

Part of unpublished and published work performed in my group is based on the collaboration with the following collegues: M. Engelhard (providing unlabelled an labelled cysteine mutants of bacteriorhodopsin); J.K. Lanyi (providing mutants for the M_1/M_2 measurements and providing the W182F mutant for the study of the steric interactions in bacteriorhodopsin); M. Sheves (providing 9-demethyl retinal for the study of steric interactions in bacteriorhodopsin); M. Ikeda Saito (providing recombinant wild type and mutated human myoglobin. Without the help of such experts in Biochemistry, Protein Chemistry, Molecular Biology, Organic Chemistry, the success of infrared spectroscopy as applied to biological systems would be much more limited. We thank all of them for their fruitful collaboration.

4. References

1. Siebert, F. (1993) Infrared Spectroscopic Investigations of Retinal Proteins. In *Biomolecular Spectroscopy Part A*, (Edited by R.J.H. Clark and R.E. Hester), pp. 1-54. John Wiley & Sons, Chichester.
2. Siebert, F. (1995) Application of modern infrared spectroscopy to biochemical and biological problems. In *Biochemical Spectroscopy*, (Edited by K. Sauer), pp. 501-526. Academic Press, Orlando, Florida.
3. Siebert. F. (1995) Time-resolved infrared spectroscopy. In *Encyclopedia of Analytical Science*. (Edited by A. Townshend), pp. 2225-2232. Academic Press, London.
4. Siebert, F. (1996) Equipment: slow and fast infrared kinetic studies. In *Infrared Spectroscopy of Biomolecules*. (Edited by H.H. Mantsch and D. Chapman), pp. 83-106. Wiley-Liss, Inc.
5. Siebert, F. (1996) Monitoring the mechanism of biological reactions by infrared spectroscopy. *Mikrochim.Acta* (In Press).
6. Braiman, M.S. and K.J. Rothschild. (1988) Fourier transform infrared techniques for probing membrane protein structure. *Ann.Rev.Biophys.Biophys.Chem.* 17, 541-570.
7. Rothschild, K.J. (1992) FTIR difference spectroscopy of bacteriorhodopsin: toward a molecular model. *J.Bioenerg.Biomembr.* 24, 147-167.

8. Krimm, S. and J. Bandekar. (1986) Vibrational spectroscopy and conformations of peptides, polypeptides, and proteins. *Adv.Protein Chem.* 38, 181-364.

9. Chirgadze, Y.N., O.V. Fedorov and N.P. Trushina. (1975) Estimation of amino acid residue side-chain absorption in the infrared spectra of protein solutions in heavy water. *Biopolymers* 14, 679-694.

10. Venyaminov, S.Y. and N.N. Kalnin. (1990) Quantitative IR spectrophotometry of peptide compounds in water (H_2O) solutions. I. Spectral pararmeters of amino acid residue absorption bands. *Biopolymers* 30, 1243-1257.

11. Lin-Vien, D., N.B. Colthup, W.G. Fateley and J.G. Grasselli. (1991) *The handbook of infrared and Raman characteristic frequencies of organic molecules.* Academic Press, Boston.

12. Pinchas, S. and I. Laulicht. (1971) *Infrared Spectra of Labelled Compounds.* Academic Press Inc. London.

13. Onishi, H., M.E. McCance and N.E. Gibbons. (1965) A synthetic medium for extremely hydrophilic bacteria. *Canadian Journal of Microbiology* 11, 365-373.

14. Anonymous(1994) *Stable Isotope Applications in Biomolecular Structure and Mechanism.* Los Alamos National Laboratories, Los Alamos, New Mexico.

15. Sonar, S., C.-P. Lee, M. Coleman, N. Patel, X. Liu, T. Marti, H.G. Khorana, U.L. RajBhandari and K.J. Rothschild. (1994) Site-directed isotope labelling and FTIR spectroscopy of bacteriorhodopsin. *Struct.Biol.* 1, 512-517.

16. Ludlam, C.F.C., S. Sonar, C.-P. Lee, M. Coleman, J. Herzfeld, U.L. RajBhandari and K.J. Rothschild. (1995) Site-directed isotope labeling and ATR-FTIR difference spectroscopy of bacteriorhodopsin: the peptide carbonyl group of Tyr 185 is structurally active during the bR->N transition. *Biochemistry* 34, 2-6.

17. Lanyi, J.K. (1993) Proton translocation mechanism and energetics in the light-driven pump bacteriorhodopsin. *Biochim.Biophys.Acta* 1183, 241-261.

18. Mathies, R.A., S.W. Lin, J.B. Ames and W.T. Pollard. (1991) From femtoseconds to biology: mechanism of bacteriorhodopsin's light-driven proton pump. *Annu.Rev.Biophys.Biophys.Chem.* 20, 491-518.

19. Oesterhelt, D., J. Tittor and E. Bamberg. (1992) A unifying concept for ion translocation by retinal proteins. *J.Bioenerg.Biomembr.* 24, 181-191.

20. Ebrey, T.G. (1993) *Thermodynamics of Membranes, Receptors and Channels,* (Edited by M. Jackson), pp. 353-387. CRC Press, Boca Raton, FL.

21. Krebs, M.P. and H.G. Khorana. (1993) Mechanism of light-dependent proton translocation by bacteriorhodopsin. *J.Bacteriol.* 175, 1555-1560.

22. Siebert, F. and W. Mäntele. (1980) Investigation of the Rhodopsin/Meta I and Rhodopsin/Meta II Transitions of Bovine Rod Outer Segments by Means of Kinetic Infrared Spectroscopy. *Biophys.Struct.Mech.* 6, 147-164.

23. Siebert, F., W. Mäntele and W. Kreutz. (1981) Biochemical Application of Kinetic Infrared Spectroscopy. *Can.J.Spectrosc.* 26, 119-125.

24. Engelhard, M., K. Gerwert, B. Hess, W. Kreutz and F. Siebert. (1985) Light-driven protonation changes of internal aspartic acids of bacteriorhodopsin: An investigation by static and time-resolved infrared difference spectroscopy using [4-^{13}C] aspartic acid labeled purple membrane. *Biochemistry* 24, 400-407.

25. Siebert, F., W. Mäntele and W. Kreutz. (1982) Evidence for the protonation of two internal carboxyl groups during the photocycle of bacteriorhodopsin. *FEBS Lett.* 141, 82-87.

26. Sasaki, J., A. Maeda, C. Kato and H. Hamaguchi. (1993) Time-resolved infrared spectral analysis of the KL-to-L conversion in the photocycle of bacteriorhodopsin. *Biochemistry* 32, 867-871.

27. Sasaki, J., T. Yuzawa, H. Kandori, A. Maeda and H. Hamaguchi. (1995) Nanosecond time-resolved infrared spectroscopy distinguishes two K species in the bacteriorhodopsin photocycle. *Biophys.J.* 68, 2073-2080.

28. Dixon, A.J., P. Glyn, M.A. Healy, P.M. Hodges, T. Jenkins, M. Poliakoff and J.J. Turner. (1988) Fast time-resolved i.r. spectroscopy of biological molecules in aqueous solution: the reaction kinetics of myoglobin with carbon monoxide. *Spectrochim.Acta* 44A, 1309-1314.

29. Yuzawa, T., C. Kato, M.W. George and H. Hamaguchi. (1994) Nanosecond time-resolved infrared spectroscopy with a dispersive scanning spectrometer. *Appl.Spectrosc.* 48, 684-690.

30. Dyer, R.B., O. Einarsdóttir, P.M. Killough, J.J. López-Garriga and W.H. Woodruff. (1989) Transient binding of photodissociated CO to Cu_B^+ of eukaryotic cytochrome oxidase at ambient temperature. Direct evidence from time-resolved infrared spectroscopy. *J.Am.Chem.Soc.* 111, 7657-7659.

31. Hienerwadel, R., D. Thibodeau, F. Lenz, E. Nabedryk, J. Breton, W. Kreutz and W. Mäntele. (1992) Time-resolved infrared spectroscopy of electron transfer in bacterial photosynthetic reaction centers: dynamics of binding and interaction upon Q_A and Q_B reduction. *Biochemistry* 31, 5799-5808.

32. Hienerwadel, R., S. Grzybek, C. Fogel, W. Kreutz, M.Y. Okamura, M.L. Paddock, J. Breton, E. Nabedryk and W. Mäntele. (1995) Protonation of Glu L212 following Q_B^- formation in the photosynthetic reaction center of *Rhodobacter sphaeroides*: evidence from time-resolved infrared spectroscopy. *Biochemistry* 34, 2832-2843.

33. Griffiths, P.R. and J.A. de Haseth. (1986) *Fourier Transform Infrared Spectrometry*. Wiley-Interscience, New York.

34. Braiman, M.S., P.L. Ahl and K.J. Rothschild. (1987) Millisecond Fourier-Transform Infrared Difference Spectroscopy of BR's M412 Photoproduct. *Proc.Natl.Acad.Sci.USA* 84, 5221-5225.

35. Gerwert, K., G. Souvignier and B. Hess. (1990) Simultaneous monitoring of light-induced changes in protein side-group protonation, chromophore isomerization and backbone motion of bacteriorhodopsin by time-resolved FTIR spectroscopy. *Proc.Natl.Acad.Sci.USA* 87, 9774-9778.

36. Siebert, F. (1995) Application of FTIR spectroscopy to the investigation of dark structures and photoreactions of visual pigments. *Israel J.Chem.* 35, 309-323.

37. Fahmy, K., O. Weidlich, M. Engelhard, J. Tittor, D. Oesterhelt and F. Siebert. (1992) Identification of the proton acceptor of Schiff base deprotonation in bacteriorhodopsin: an FTIR study of the mutant Asp85->Glu in its natural lipid environment. *Photochem.Photobiol.* 56, 1073-1083.

260

38. Weidlich, O., B. Schalt, N. Friedman, M. Sheves, J.K. Lanyi, L.S. Brown and F. Siebert. (1996) Steric interactions between the 9-methyl group of the retinal and tryptophan 182 controls 13-*cis* to *all-trans* reisomerization and proton uptake in the bacteriorhodopsin photocycle. *Biochemistry* (In Press)

39. Weidlich, O., N. Friedman, M. Sheves and F. Siebert. (1995) The influenec of the 9-methyl group of the retinal on the photocycle of bacteriorhodopsin studied by time-resolved rapid-scan and static low-temperature FT-IR difference spectroscopy. *Biochemistry* 34, 13502-13510.

40. Braiman, M.S., O. Bousché and K.J. Rothschild. (1991) Protein dynamics in the bacteriorhodopsin photocycle: Submillisecond Fourier transform infrared spectra of the L,M, and N intermediates. *Proc.Natl.Acad.Sci.USA* 88, 2388-2392.

41. Souvignier, G. and K. Gerwert. (1992) Proton uptake mechanism of bacteriorhodopsin as determined by time-resolved stroboscopic FTIR-spectroscopy. *Biophys.J.* 63, 1393-1405.

42. Sonar, S., M.P. Krebs, H.G. Khorana and K.J. Rothschild. (1993) Static and time-resolved absorption spectroscopy of the bacteriorhodopsin mutant Tyr-185->Phe: evidence for an equilibrium between bR_{570} and an O-like species. *Biochemistry* 32, 2263-2271.

43. Uhmann, W., A. Becker, C. Taran and F. Siebert. (1991) Time-resolved FT-IR Absorption Spectroscopy Using a Step-Scan Interferometer. *Appl.Spectrosc.* 45, 390-397.

44. Weidlich, O. and F. Siebert. (1993) Time-resolved step-scan FT-IR investigations of the transition from KL to L in the bacteriorhodopsin photocycle: identification of chromophore twists by assigning hydrogen-out-of-plane (HOOP) bending vibrations. *Appl.Spectrosc.* 47, 1394-1400.

45. Ames, J.B. and R.A. Mathies. (1990) The role of back-reactions and proton uptake during the N -> O transition in bacteriorhodopsin's photocycle: a kinetic resonance raman study. *Biochemistry* 29, 7181-7190.

46. Schulten, K. and P. Tavan. (1978) A mechanism for the light driven proton pump of Halobacterium halobium. *Nature* 272, 85-86.

47. Váró, G. and J.K. Lanyi. (1991) Thermodynamics and Energy Coupling in the Bacteriorhodopsin Photocycle. *Biochemistry* 30, 5016-5022.

48. Brown, L.S., Y. Gat, M. Sheves, Y. Yamzaki, A. Maeda, R. Needleman and J.K. Lanyi. (1994) The retinal Schiff base-counterion complex of bacteriorhodopsin: changed geometry during the photocycle is a cause of proton transfer to aspartate 85. *Biochemistry* 33, 12001-12011.

49. Frauenfelder, H. and P.G. Wolynes. (1994) *Physics Today* 47, 58,

50. Dioumaev, A.K. and M.S. Braiman. (1996) Two bathointermediates of the bacteriorhodopsin photocycle, distinguished by nanosecond time-resolved FTIR spectroscopy at room temperature. *J.Phys.Chem.* (In Press).

51. Diller, R., M. Iannone, B.R. Cowen, S. Maiti, R.A. Bogomolni and R.M. Hochstrasser. (1992) Picosecond dynamics of bacteriorhodopsin, probed by time-resolved infrared spectroscopy. *Biochemistry* 31, 5567-5572.

52. Diller, R., M. Iannone, R.A. Bogomolni and R.M. Hochstrasser. (1991) Ultrafast infrared spectroscopy of bacteriorhodopsin. *Biophys.J.* 60, 286-289.

53. Hamm, P., M. Zurek, M. Meyer, H. Scheer and W. Zinth. (1995) Femtosecond infrared spectroscopy of reaction centers from *Rhodobacter sphaeroides* between 1000 and 1800 cm^{-1}. *Proc.Natl.Acad.Sci.USA* 92, 1826-1830.

54. Dyer, R.B., K.A. Peterson, P.O. Stoutland and W.H. Woodruff. (1991) Ultrafast photoinduced ligand transfer in carbonmonosy cytochrome c oxidase. Observation by picosecond infrared spectroscopy. *J.Am.Chem.Soc.* 113, 6276-6277.

UV RESONANCE RAMAN DETERMINATION OF α-HELIX MELTING DURING THE ACID DENATURATION OF MYOGLOBIN

Sanford A. Asher and Zhenhuan Chi
Department of Chemistry
University of Pittsburgh
Pittsburgh, PA 15260, USA

Abstract

We have studied the absorption spectrum and the 206.5 and 229 nm excited resonance Raman spectrum of aquometmyoglobin (Mb) between pH 7.5 and pH 1.5. The acid denaturation of Mb below pH 3.5 results in small absorption increases below 220 nm due to α-helix melting, and an absorption decrease at ~230 nm due to environmental changes of the trp and/or tyr residues. Dramatic heme Soret band changes indicate major alterations of the heme pocket. We examined the 206.5 nm resonance Raman spectra, which selectively enhances amide backbone vibrations and compared these results to the 229 nm excited resonance Raman spectra, which selectively enhances tyr and trp vibrations. We calculate that the Mb α-helical composition decreases from ~80% at neutral pH to ~19% below pH 3.5. The trp Raman cross sections dramatically decrease at low pH which indicates that the A helix melts. The tyr Raman bands are pH independent which indicates that the G and H helices do not melt. This result indicates that helices A, B, C, D and E melt in a concerted fashion, while the antiparallel G and H helices only partially melt. This melting of the helices framing the heme pocket is responsible for the change in heme binding; the diffuseness of the unfolded Mb heme Soret band suggests that the heme no longer has a single defined binding site.

Introduction

One central unsolved problems in the understanding of protein structure and dynamics is the mechanism by which a protein self assembles from a primary sequence of amino acids to form a three dimensional functional enzyme with its well defined, and crucial secondary tertiary and quaternary protein structure [1]. Numerous studies of protein folding, as well as the protein unfolding have utilized a host of techniques such as NMR, CD, absorption and fluorescence spectroscopy etc. [2-4]. These studies, which attempt to elucidate the rules for protein folding (and unfolding), utilize highly sensitive and selective techniques to examine protein structural changes.

G. Vergoten and T. Theophanides (eds.), Biomolecular Structure and Dynamics, 263–270.

Myoglobin has often been used in these folding and unfolding studies as a model system because of the detailed structural information available from x-ray measurements [5]. In this report we apply UV resonance Raman spectroscopy to determine myoglobin secondary structural changes during acid denaturation. We developed a new methodology to assign secondary structural changes to particular α-helical segment. We accomplish this by monitoring the UV Raman spectra of both the amide vibrations and the tyr and trp aromatic amino acid vibrations, as well as the heme Soret band. We utilized our recent PSCS methodology to determine secondary structure by fitting pure secondary component spectra obtained from 206.5 nm excited protein amide band spectra [6]. These data are combined with 229 nm excited resonance Raman spectra that selectively enhance the tyr and trp aromatic amino acids. Secondary structure changes are correlated with changes in particular aromatic amino acid environments and hydrogen bonding, to localize the secondary structural changes to particular regions of the protein. In this study we directly monitored the melting of α-helical segments of aquometmyoglobin during its reversible acid denaturation. We determine which helices melted and which were stable at low pH.

Experimental

Horse heart myoglobin (Mb) was purchased from Sigma Co. (St. Louis, MO) and was used as received; the protein was in the oxidized met form. For Raman spectra excited at 206.5 nm we utilized 18 μM Mb solutions, and for 229 nm excitation we used 91 μM Mb solutions. The pH value of the sample solutions were adjusted by adding small amounts of concentrated hydrochloric acid and potassium hydroxide solutions with rapid stirring.

The UV Raman measurements were obtained by using instrumentation described in detail previously [7,8]. 206.5 nm (2.5 mW) and 229 nm (9 mW) CW laser excitation was obtained from intracavity frequency-doubled krypton and argon ion lasers, respectively [8].

Results

Figure 1 shows the absorption spectra of aquomet Mb at various pH values between pH=7.5 and pH=1.5. The absorption spectra show the strong Mb heme Soret band between 330-450 nm, and modest tyr and trp absorption bands in the region between 260-300 nm [9]. Strong tyr and trp absorption bands occur between 210 and 230 nm [9,10]. The region below 220 nm is dominated by the π→π* absorption of the amide backbone linkages [6].

The absorption spectra are independent of pH for pH values between pH 7.5 and 3.5. However, a pH decrease below 3.5 results in a Mb absorption increase between 200-220 nm and a small absorption decrease around 230 nm. These spectral changes are correlated with the known unfolding of the dominantly Mb α-helical structure (81%); the absorption increase below 220 nm results from hyperchromism of the amide $\pi-\pi^*$ electronic transition moments due to the decreased excitonic interactions in the unfolded random coil conformations compared to that in the α-helix conformation [11].

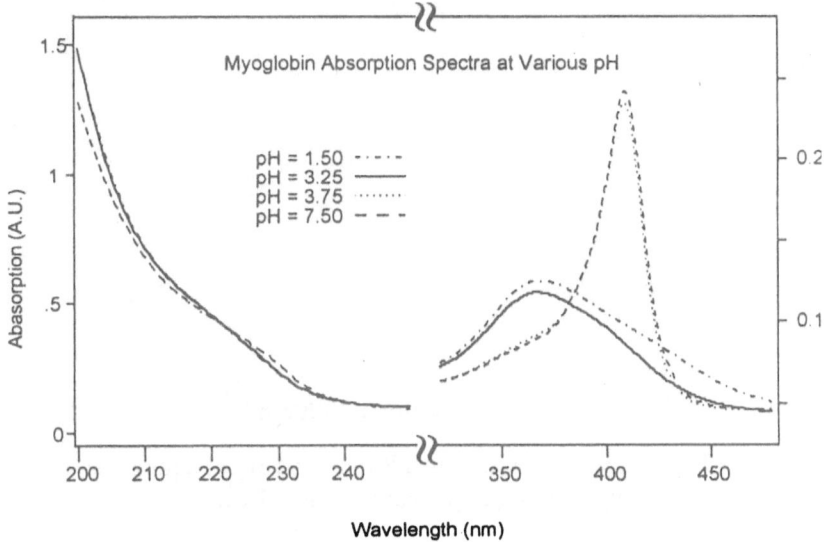

Figure 1. Aquometmyoglobin (30 μM, 0.05 cm pathlength) absorption spectra at various pH values.

The ca. 230 nm Mb absorption decrease indicates that the environment of tyr and trp residues are changing during the acid denaturation. The large 413 nm Mb Soret absorption band decrease and the replacement with a much more diffuse 363 nm Soret band indicates a major changes in binding of the heme at low pH.

Figure 2 shows the 206.5 nm excited resonance Raman spectra of Mb at pH values between 7.50 and 1.50. The pH 7.27 sample was obtained by adjusting the sample pH from 7.50 to 1.50, then returning the pH to 7.27; the fact that the pH 7.27 Raman spectrum is identical to that at pH 7.50 indicates that the protein structural changes that give rise to the low pH spectral changes are reversible.

The 206.5 nm excited Raman are dominated by the amide vibrations [6] since this excitation wavelength is resonant with the amide $\pi \rightarrow \pi^*$ transitions. The spectra show strong enhancement of the amide I vibration (~ 1660 cm^{-1}), which is predominantly carbonyl stretching; some enhancement of the ring stretching vibrations of the tyr and trp aromatic amino acid side chains (~ 1610 cm^{-1}); strong enhancement of the amide II vibration (1555 cm^{-1}), which involves all the atoms of the amide group, but contains a large contribution of C-N stretching and C-H bending; some intensity for the numerous CH$_2$ side-chain bending vibrations (1455cm^{-1}); a variable intensity of the amide C$_\alpha$-H bending vibration (1386 cm^{-1}); and strong enhancement of the amide III vibration (1240-1300 cm^{-1}), whose composition is similar to that of the amide II vibration, except that the opposite phasing occurs between the C-N stretching motion and the N-H bending motion [9, 10].

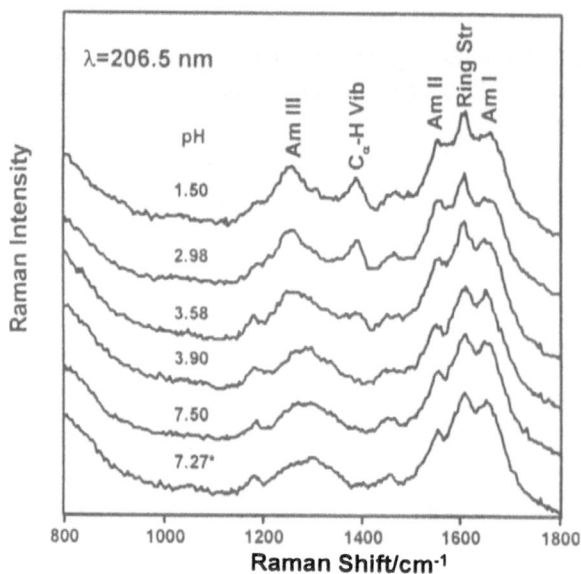

Figure 2. 206.5 nm excited Raman spectra of aquometmyoglobin (18 μM) at various pH values.

Between pH 7.50 and 3.90 the Mb amide Raman bands are essentially identical; they show a strong amide I (1655 cm^{-1}) band, and medium intensity amide II (1555 cm^{-1}) and III (1290 cm^{-1}) bands. No band near 1386 cm^{-1} is evident, which indicates that the Mb secondary structure is mainly α-helical [6, 11]. The invariance of the Raman frequencies, intensities and band shapes indicates little change of the protein secondary structure between pH 7.50 and 3.90.

Below pH 3.9, the 1386 cm^{-1} amide C_α-H bending band appears, which indicates a significant unraveling of the Mb α-helical structure. The amide III band frequency shifts from 1290 to 1268 cm^{-1} between pH 3.9 and 3.58, and further shifts to 1261 cm^{-1} at pH 2.98. The amide III band Raman cross section increases from 26 mbarn/molc str at pH 3.98 to 39 mbarn/molc str at pH 2.98. The amide I band shifts 7 cm^{-1} to lower frequency, and its Raman cross sections slightly decreases. The amide II band frequency remains constant, while its Raman cross section increases from 27 to 46 mbarn/molc·str. A further pH decrease to 1.50 causes little additional Raman spectral changes, which indicates little further change of the Mb secondary structure.

We have previous determined the amide band resonance Raman spectra of pure α-helix, β-sheet and random coil secondary structures for excitation at 206.5 nm. We also demonstrated that we can use these pure secondary structure spectra (PSCS) in a linear fit of the spectra of proteins to reliably determine protein secondary structure composition. This is a robust and sensitive determination since we simultaneously utilize the amide I, II, III bands and the C_α-H amide bending vibrations of these PSCS to determine secondary structure [6].

We used this methodology to determine the Mb secondary structure for the different pH value spectra shown in Figure 2. Table I lists our calculated Mb secondary structural compositions at the different pH values. The calculated α-helical composition remains relatively constant at ca. 75% from pH 7.50 to 3.90, but it decreases rapidly to 19% α-helix at pH 3. An abrupt transition between the folded and unfolded states occurs at pH ~3.50.

TABLE I. Myoglobin Secondary Structural
Composition at Various pH Value

pH	α-Helix %	β-Sheet %	Random Coil
1.50	19	0	81
2.98	19	0	81
3.58	44	9	46
3.90	78	0	22
4.52	77	0	23
5.05	73	0	27
7.50	73	0	27
7.27r	78	0	22

The Figure 2 bottom Mb spectrum, which is for a sample where the pH was first adjusted down from neutrality to pH = 1.50 and then adjusted back to pH = 7.27, is identical to that at pH 7.50, this clearly indicates that this acid induced unfolding of Mb is reversible.

Figure 3 shows 229 nm excited Raman spectra of Mb at various pH. This excitation occurs in resonance with the tyr and trp ring $\pi\rightarrow\pi^*$ transitions. Thus, the 229 nm Mb Raman spectra are dominated by Raman bands of tyr and trp. The 759, 857, 879, 1012, 1339, 1359, 1453, 1516 and 1557 cm^{-1} bands are assigned to trp, while the 1179, 1207 and 1258 cm^{-1} bands are assigned to tyr. The 1613 cm^{-1} bands derive mainly from the tyr Y8a band but also have a small contribution from the trp W1 band [10].

The Raman cross sections of the strongly enhanced trp 759, 879, 1012, 1360 and 1551 cm^{-1} bands decrease as the pH decreases below 4.0; the cross section decrease levels off by pH 2.50. In contrast, the strongly enhanced tyr 1129, 1207, 1258 and 1613 cm^{-1} bands show no significant frequency or Raman cross section changes.

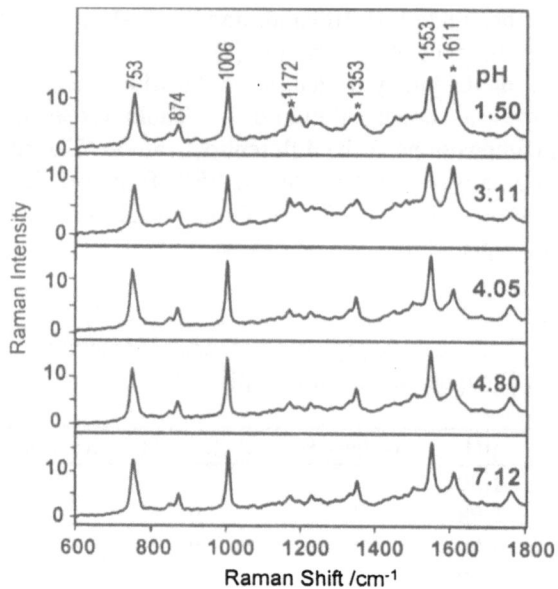

Figure 3. 229 nm excited Raman spectra of aquometmyoglobin (91 μM) at various pH values.

Discussion

Horse heart Mb is a globular, single-domain protein of 153 amino acids (Fig. 4) which possesses eight rod-like α-helix segments of variable length (labeled as A-H) [5]. Horse heart Mb contains two trp (trp7 and trp14 localized in helix A) and two tyr residues (tyr103 and tyr146 localized in helices G and H, respectively). The B, C, E, and F helices frame the heme pocket.

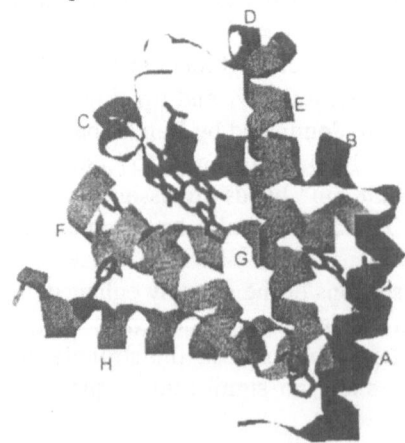

Figure 4 Structure of horse heart myoglobin showing
the helical segments, which was adapted from ref 5.

We can partially determine which helices melt by correlating changes in the trp and tyr Raman spectra with changes in the amide band Raman spectra, and with the collapse of the heme pocket which results in a change in heme binding evident from the heme Soret absorption spectral changes upon acid denaturation. The full acid denaturation indicates that the calculated ~75% α-helical structure melts to give 19% α-helical structure containing an 81% random coil structure (Table I).

The decreased trp Raman cross sections indicates a change in the trp environment from hydrophobic to hydrophilic. This indicates that the A helix melts. The invariance of tyr Raman frequencies and cross sections indicates little change in the α-helical structure of helices G and H around the tyr residues. The G and H helices alone give rise to a 29% α-helical structure which is significantly greater than the residual α-helix abundance we calculate at low pH.. Our previous determinations of α-helix abundance was accurate to within 10% for highly helical peptides. In fact, we generally overestimated the α-helical content [6]. Thus, we can conclude that parts of the G and H helices are melted and that these are localized away from the tyr residues. Finally, we can conclude the A, B, C, D, E and F helices melt. This is consistent with the destruction of the heme crevice that alters the heme binding. The breadth of the Soret band may indicate the existence of numerous modes of heme binding.

Our results indicate that there is an almost simultaneous melting of most of the α-helical segments of Mb at pH ~3.5. The origin of this concerted melting may result from destabilization of the α-helix internal hydrogen bonding at low pH values. In contrast, Helices G and H are strongly coupled, lie antiparallel to one another and the resulting helix contacts may stabilize each other to the extent that they do not completely melt at these low pH values.

Conclusions

We have demonstrated a methodology to follow the unfolding of specific helical segments by examining the alterations in the environment of specific aromatic amino acids localized within different α-helical segment. We assign the spectral changes to unfolding of helices A-F at pH ~3.5 in aquo metMb, while the helices G and H still remain some α-helical structure, especially around the tyr residues. We presume that this melting of the helices framing the heme pocket is responsible for the observed change in heme-protein binding. The diffuseness of the unfolded Mb heme Soret band indicates that the heme no longer has a single defined binding site.

Acknowledgment

We gratefully acknowledge financial support from the National Institute of Health through Grant No. R01GM30741-14 to S.A.A.

References

1. Kim, P.S. and Baldwin, R.L. (1990) *Annu. Rev. Biochem.* **59**, 631-660.
2. Johnson, W.C., Jr. (1996) In *Circular Dichroism and the Conformational Analysis of Biomolecules*; Fasman, G.D., Ed., Plenum Press, New York and London, p. 635-654.
3. Ballew, R., Sabelko, J., and Gruebele, M. (1996) Proc. *Natl. Acad. Sci. USA,* **93**, 5759-5764.
4. Hughson, F.M., Wright, P.E., and Baldwin, R.L. (1990) *Science* **249**, 1544-1548.
5. Evans, S.V. and Brayer, G.D. (1988) *J. Biol. Chem.*, **263**, 4263-4268.
6. Chi, Z., Chen, X.G., Holtz, J.S.W., and Asher, S.A., *Biochemistry*, submitted.
7. Asher, S.A., Johnson, C.R., and Murtaugh, J. (1983) *Rev. Sci. Instrum.* **54**, 1657-1662.
8. (a) Asher, S.A., Bormett, R.W., Chen, X.G., Lemmon, D.H., Cho, N., Peterson, P., Arrigoni, M., Spinelli, and Cannon, J. (1993) *Appl. Spectrosc.* **47**, 628-633;
   (b) Holtz, J.S.W., Bormett, R.W., Chi, Z., Cho, N., Chen, X.G., Pajcini, V., Asher, S.A., Arrogoni, M., Owen, P., and Spinelli, L. (1996) *Appl. Spectrosc.* **50**, 1459-1468.
9. Asher S.A., Larkin, P.J., and Teraoka, J. (1991) *Biochemistry*, **30**, 5944-5954.
10. Cho, N., Song, S., and Asher, S.A. (1994) *Biochemistry*, **33**, 5932-5941.
11. Song, S. and Asher, S.A. (1989) *J. Am. Chem. Soc.* **111**, 4295-4305.

NUCLEIC ACIDS

THE USE OF FOURIER TRANSFORM INFRARED (FT-IR) SPECTROSCOPY IN THE STRUCTURAL ANALYSIS OF NUCLEIC ACIDS

T. THEOPHANIDES AND J. ANASTASSOPOULOU

National Technical University of Athens, Chemical Engineering Department, Radiation Chemistry and Biospectroscopy, Zografou Campus, Zografou 15780, Athens, Greece.

Abstract.   Fourier Transform Infrared (FT-IR) spectroscopy is a versatile technique for studying structure and dynamics of nucleic acids. This method is particularly valuable when spectra of isotopic substitution are used and compared in order to assign the vibrational modes. Isotopic substitution provides a basis for the interpretation of the spectra and the determination of the structure. In order to do this it is important to obtain very good spectra and to interrelate and understand the changes taking place upon isotopic substitution. The current technique of FT-IR for obtaining high quality spectra of nucleotides is summarized and some qualities are discussed. This is followed by the vibrations which originate from the primary structure, i.e., the in-plane vibrations of the bases and the backbone vibrations which involve the functional groups of DNA, such as, -NH, $-NH_2$, C=O, C-H, -OH. Finally the interpretation of the spectra leads to diagnostic bands or characteristic bands for the bases, the sugar and the phosphate.

1. Introduction

Big advances and progress in science come by way of individual efforts which are seldom predictable. In the last twenty years we have seen an extraordinary and unexpected progress of techniques and experimentation of molecular spectroscopy not only in apparatus and development, but also in electronics and theoretical understanding. Most significant in many of these spectroscopic developments have been the refinement of the electronics, the laser spectroscopy and the arrival of Fourier transform spectrometers. FT-IR

G. Vergoten and T. Theophanides (eds.), Biomolecular Structure and Dynamics, 273–284.

274

microanalysis assures accuracy in particular when an analytical microscope has been attached, which makes more flexible the IR instruments.

Fourier Transform Infrared (FT-IR) spectroscopy has been widely used by nucleic acid scientists in the determination and identification of nucleobases, nucleosides and nucleotides [1,3]. Some diagnostic peaks for each component have been made [4-7]. The structure of nucleic acids has been extensively studied since Watson and Crick [8] elucidated and proposed the helical structure of right-handed B-form DNA in the early 1950's (Figure 1).

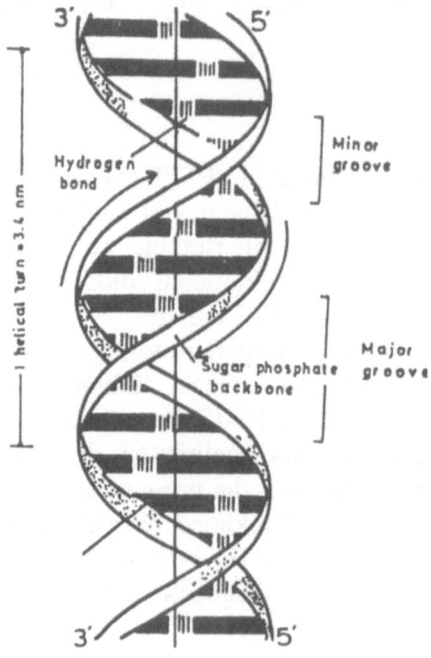

Figure 1. The helical structure of B-DNA given by Watson and Crick [8]

Much of the structural information and particularly on the conformation of nucleic acids come from X-ray diffraction of fibers and crystals, nuclear magnetic resonance, and other spectroscopic methods. These studies have revealed that the structure of DNA is polymorphic. The various conformations A-, B- and Z- are classified according to the notions of handiness helical parameters, and backbone geometry. These nucleic acid conformations are due principally to sequence of bases, cation type and concentration, temperature and solvent or the ionic environment.

2. Fourier Transform Spectroscopies

In all these spectroscopies the common thing is the use of Fourier transform,

$$f(t) = \int_{-\infty}^{+\infty} F(v) e^{2pv\,ti} dv \qquad F(v) = \int_{-\infty}^{+\infty} f(t) e^{-2pvti} dt \qquad\qquad [1]$$

The basic apparatus here is the Michelson interforometer [9] (Figure 2).

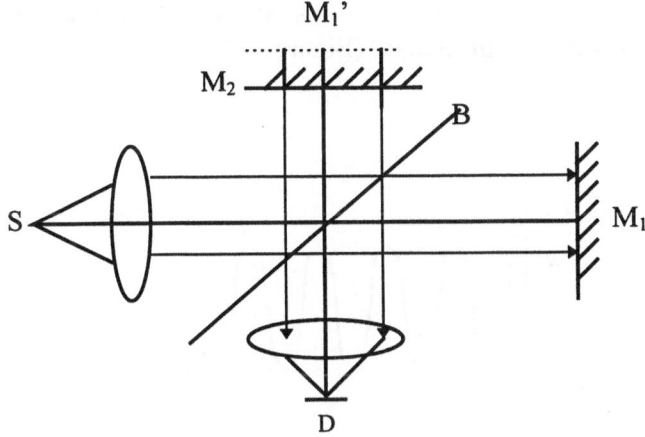

Figure 2. Schematic diagram of the double beam interferometer of Michelson. S = source, M_1= fixed mirror, M_2= movable mirror, B = beam divider, D = detector, M_1' = image of M_1 in M_2 as "seen" from D.

The interferometer consists essentially of two mirrors and the beam divider. The beam divider allows the detector to see what is coming from both mirrors. The beam reflected by mirror M_1 and that reflected by M_2 do not travel the same distance. There is thus a phase difference between the two beams. This phase difference is given by multiplying by 2 the distance between the two mirrors (aller-retour) divided by the wave length of the beam [10]. We can show that,

$$n\lambda = 2d\sin\theta \quad \sin\theta = n\lambda/2d \qquad\qquad [2]$$

$$n = 1,2,3,... \qquad \text{or} \qquad 1/\sin\theta = 2d/n\lambda$$

$$D = 1000 \ l/cm$$

θ= phase difference

if $\dfrac{(2n+1)/2}{\lambda}$: destructive interference

if $2n/\lambda$:   constructive interference

Amplitude = $0.5(1\pm\cos2\pi\theta/\lambda)$

The observed signal for a given wavelength is sinusoidal for a given point of observation with a given speed of displacement (Figure 3).

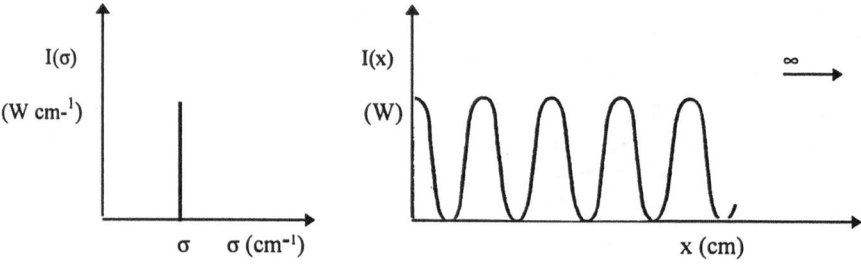

Figure 3. The sinusoidal shape of the signal and its intensity

The mathematical analysis of this phenomenon shows that the received signal from a detector placed at the exit named the interferogramme is equal to the Fourier transform. So what is left then to do is the inversed numerical transform of the interferogramme. The received signal is then the Fourier transform and the spectral distribution of the transmission of the sample.

3. Nomenclature and Diagnostic Bands of Nucleic Acids

Nucleic acids are made of Adenine (A), Guanine (G), Cytosine (C), Thymine (T) bases and Uracil (U) for RNA, as well as pentoses and the phosphate backbone [11].
Adenine and guanine are called purines, while cytosine and thymine are called pyrimidines. Ribose is for RNA or deoxy-ribose is for DNA. Finally, the phosphate groups are PO_2^-, OPO and COP (sugar-phosphate).

Adenosine -5'- monophosphate   (AMP)
Guanosine - 5'- monophosphate (GMP)
Cytidine -5'- monophosphate   (CMP)
Thymidine -5'- monophosphate (TMP)
Uridine- 5'- monophosphate   (UMP) for RNA.

The assignment of vibrational bands in polynucleotides as diagnostics from FT-IR spectra is made according to the position of the bands, intensity, shape and relation to known base, sugar and phosphate vibrations. The ring vibrations occur in the molecular planes of the bases, whereas the sugars involve C-C and C-O stretchings and the phosphate groups involve symmetric stretching of PO bonds [12,13].

4. Presentation and Interpretation of FT-IR Spectra

The characteristic bands for the nucleotides 5'-AMPNa$_2$, 5'-GMPNa$_2$, 5'-CMPNa$_2$ and 5'-TMPNa$_2$ and their assignments are given in Table 1.

Table 1. FT-IR absorption bands of the nucleotides in cm^{-1}.

5'-AMPNa$_2$	5'-GMPNa$_2$	5'-CMPNa$_2$		Assignments
	1693		1696	$vC6=O$
1653	1627	1653		δNH_2, $vC=N$, $vC=C$
1607		1612		
1576	1599		1590	δNH_2, $vC=N$, $vC=C$,
	1582			δNH
1506	1537	1528	1519	
1479	1492	1492	1479	$vC=N$, C-N, C=C
1110	1128	1115	1112	vC-O sugar ring
	1105			
1091	1084	1086	1092	vPO_3^{2-} deg. Str.
976	982	982	977	vPO_3^{2-} sym
		820	816	vPO,
797	781	785	786	ribose -phosphate
721	696		617	ring breathing

In this review we discuss the FT-IR spectra of the four nucleotides 5'-AMPNa$_2$, 5'-GMPNa$_2$, 5'-CMPNa$_2$ and 5'-TMPNa$_2$ [5,6]. The FT-IR spectra of 5'-AMP (disodium salt) in solid state, water and deuterated water are given in Table 2. The solution spectra were taken at pH(D)=6.5.The spectra of 5'-AMPNa$_2$ and 5'-GMPNa$_2$ are shown in Figure 4.

278

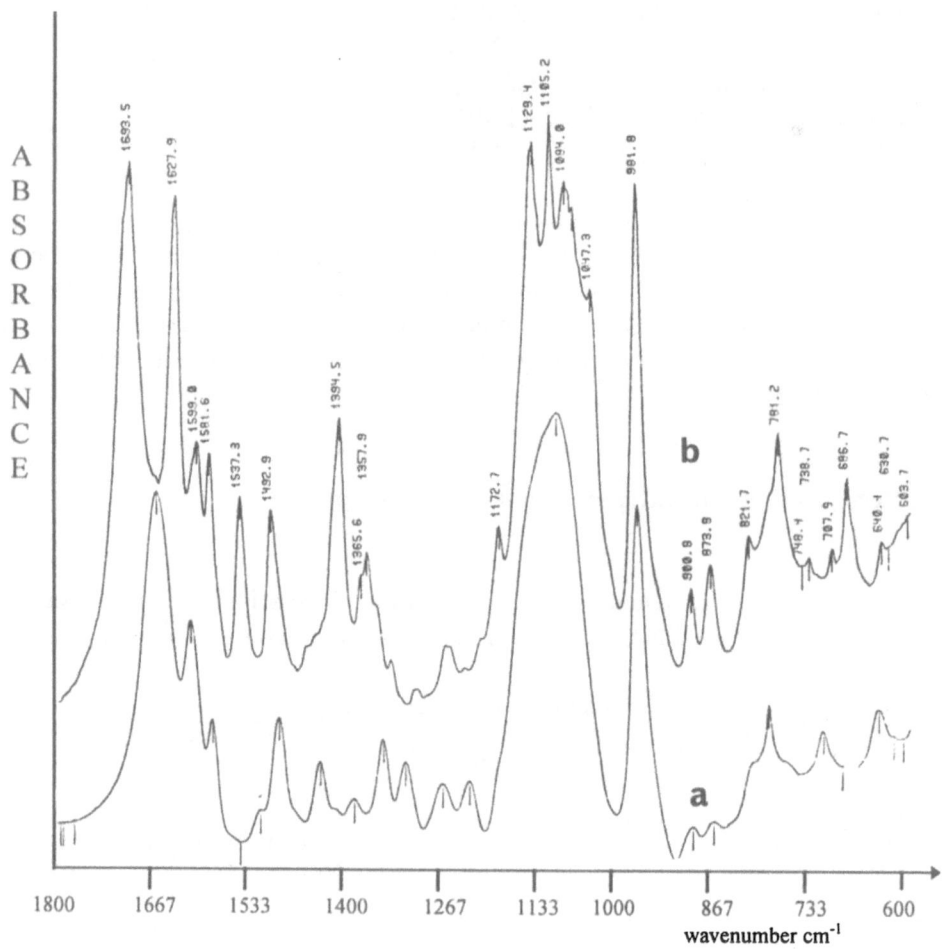

Figure 4. FT-IR spectra of purines a) 5'-AMPNa₂ and b) 5'-GMPNa₂ in KBr pellets. Resolution 4 cm⁻¹.

Table 2. FT-IR absorption bands of 5'-AMPNa₂ in solid state, water and D₂O solutions in cm⁻¹.

5'-AMPNa₂	5'-AMPNa₂-in H₂O	5'-AMPNa₂-in D₂O	Assignments
1653	1655	1626	δNH_2
1606	1605		$\nu C=C$, δNH_2
1575	1578	1576	$\nu C=C$,
			$\nu N=C4=C5$
1100	1092	1105	C-O sugar, PO_3^{2-} deg.
976	978	974	PO_3^{2-} sym.
817	818		ribose hosphate
796	799	795	$\nu P-O$
721	723	733	purine ring breathing

Table 3. FT-IR absorption bands of 5'-GMPNa$_2$ in solid state, water and D$_2$O solutions in cm^{-1}.

5'-GMPNa$_2$	5'-GMP Na$_2$- in H2O	5'-GMPNa$_2$- in D$_2$O	Assignments
1693	1693		vC6=O,
1628	1655	1665	δNH$_2$, vC2-N2 , vC4=C5
1600	1602		vC4-N3, vC4=C5, vC4-N3
1582	1575	1580	vC4=C5, δN1-H, vC6-N1
1183	1176		vC8=N7, N9-sugar
1128			C-O sugar ring
1105		1117	C-O sugar ring
1084	1092	1082	C-O sugar ring vPO$_3$$^{2-}$ deg.
1048			vPO$_3$$^{2-}$ sym.
982	978	974	vPO$_3$$^{2-}$ sym.
822	818		sugar phosphate
781	802	789	vP-O, O-P-O
696	690	690	purine ring breathing

In Figure 5 are shown the FT-IR spectra of pyrimidines 5'-CMPNa$_2$ and 5'-TMPNa$_2$ in KBr pellets. The assignments and the bands of these pyrimidines in water and deuterated solutions are given in Tables 4 and 5 for 5'-CMPNa$_2$ and 5'-TMPNa$_2$, respectively.

Table 4. FT-IR absorption bands of 5'-CMPNa$_2$ in solid state, water and D$_2$O solutions in cm^{-1}.

5'-CMPNa$_2$	5'-CMP Na$_2$- in H$_2$O	5'-CMP Na$_2$- in D$_2$O	Assignments
1653	1657	1651	vC6=O, vC6-C5
1612	1603	1620	δNH2, vC2-N2
1086		1524	vC4-N3, vC4-C5
		1504	
1072	1107	1105	vC4-C5, δN1-H
	1090	1092	
982	978	974	vC8-N7
820	802		C-O sugar ring
785	794	791	vP-O

Table 5. FT-IR absorption bands of 5'-TMPNa$_2$ in solid state, water and D$_2$O solutions in cm^{-1}.

5'-TMP	5'-TMP- in H2O	5'-TMP- in D$_2$O	Assignments
1696	1690	1663	vC6=O, vC6-C5
	1666	1632	
1519			δNH$_2$, vC2-N2
1095	1090	1088	vC4-N3, vC4-C5
977	978	976	vPO$_3$$^{2-}$ sym
	812	804	sugar phosphate
786		777	vP-O
768			

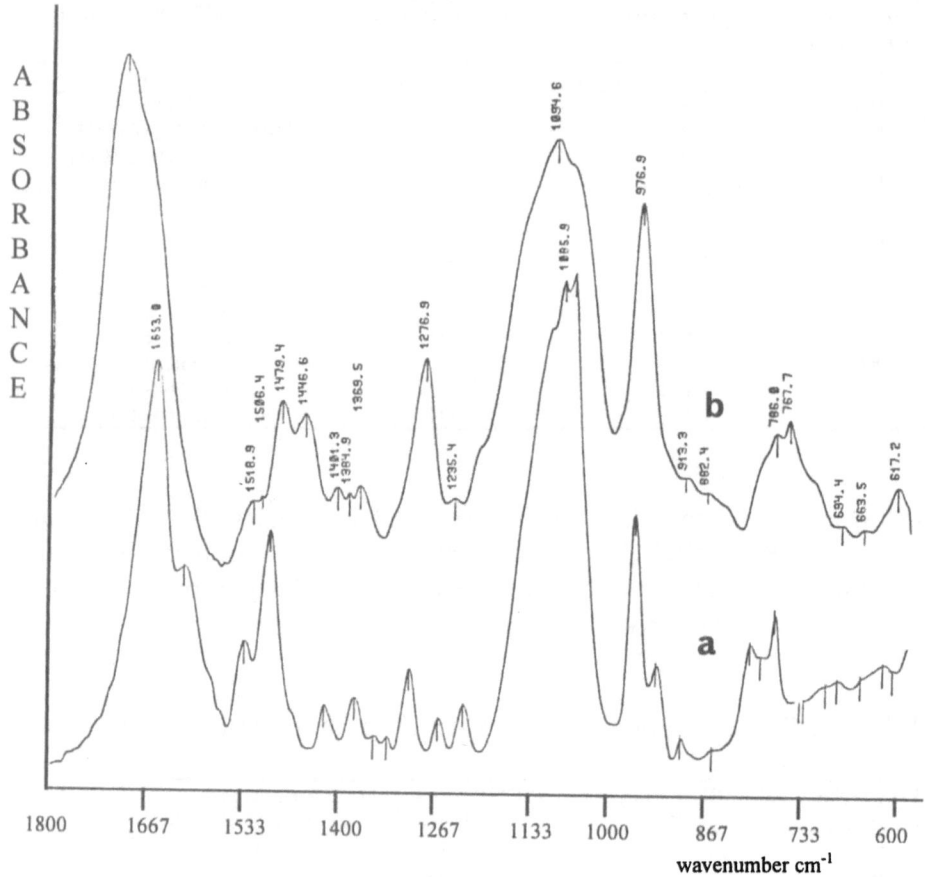

Figure 5. FT-IR spectra of pyrimidines a) 5'-CMPNa₂ and b) 5'-TMPNa₂ in the region 600 - 1800 cm⁻¹. The spectra were taken in KBr pellets with 4cm⁻¹ resolution.

As we see the vibrations of nucleic acids can be classified in two categories [14].

1) The vibrations which originate from the primary structure (covalent bonds). These can be divided into two parts (i) the in-plane vibrations of the bases, (ii) furanose-phosphate, backbone vibrations.
2) The vibrations of the secondary structure of the double helix of DNA involving the functional groups, -NH, -NH₂, C=O, C-H, OH, and the hydrogen bonds between each pair of bases.

In the solid spectra two regions of strong absorptions are visible. These are approximately the 1700-1500 cm⁻¹ and the 1200-900 cm⁻¹ regions. The first region is the region of double bonds of groups C=O, C=C, C=N in-plane vibrations of the base purine together with the functional groups -NH, and -NH$_2$ [5]. The second region is the phosphate backbone region which involves principally the phosphate vibrations (PO$_3$ $^-$, and O-P-O groups) together with the ribose and the ribose-phosphate vibrations (C-O). The solution spectra in H$_2$O and D$_2$O again display strong absorption of these vibrations in the above regions and a displacement of these bands to lower frequencies due to deuteration of the -NH$_2$ and the sugar -OH groups. The solvent (H$_2$O, D$_2$O) spectra have been subtracted. The assignments of the bands are given in Table 2.

The spectra of 5'-GMP (disodium salt) in solid state, H$_2$O and D$_2$O are displayed in Figure 4. The solution spectra are taken at pH(D)=2. The solid state spectra show strong absorptions in the above two regions and medium or weak absorptions elsewhere. The shape of the bands differ from that of the solution spectra in the case of the solid spectra (1693.5, 1627.9, 1599.0, 1581.6 cm^{-1}) in the region 1700-1550 cm^{-1} and (1172.7, 1128.4, 1105.2, 1084.0, 1047.3, 981.8 cm^{-1}) in the case of ribose and ribose-phosphate. The solution spectra do not show these splittings which are expected, whereas the spectra taken in D$_2$O show the vibrations of the groups -NH, NH$_2$ and the ribose -OH displaced upon deuteration. These changes allow us to assign some of the bands of the characteristic group vibrations (see Table 2).

The pyrimidine nucleotide FT-IR spectra (5'-CMP and 5'-TMP) are displayed in Figure 5 in the region 1800-600 cm^{-1}. The solid, H$_2$O and D$_2$O solution spectra are taken at a pH(D)=6.4. A similar pattern of absorptions is followed for all three pyrimidine derivatives. If we compare the spectra, in particular in the regions 1700-1500 cm^{-1} and 1200-900 cm^{-1} of these two pyrimidine derivatives in the solid state and in both H$_2$O and D$_2$O solutions we arrive at the assignments given in Table 2.

The ribose phosphate (disodium salt) FT-IR spectra in the solid state in H$_2$O and D$_2$O solutions at pH(D)=6 have also been studied.

One predominant region of strong absorption is shown in the FT-IR spectra of D-ribose-phosphate in the solid state and H$_2$O and D$_2$O solutions taken in pH(D)=6.0 in the 1200-900 cm^{-1} region, which is due to the phosphate group. In addition, a medium strength absorption is also displayed in the 850-750 cm^{-1} region. The latter absorptions upon deuteration do change drastically indicating the involvement of hydrogen in deuteration. The assignment of these bands was proposed previously [6,14] as C2'-endo, anti and C3'-endo, anti of the sugar pucker, involving also the hydroxyl groups at C2'-OH and C3'-OH.

5. FT-IR Spectral Analysis of Conformational Transitions as a Function of Temperature

It is well known that the conformational transitions of DNA B→A→Z are salt concentration and temperature dependent [15,16]. The B→ Z conformation occurs at low salt concentration and at high temperatures, whereas the B conformation dominates at low salt concentrations and at low temperatures. The FT-IR spectra in the region 1000-500 cm^{-1} show characteristic changes, which correspond to these conformations (Figure 6).

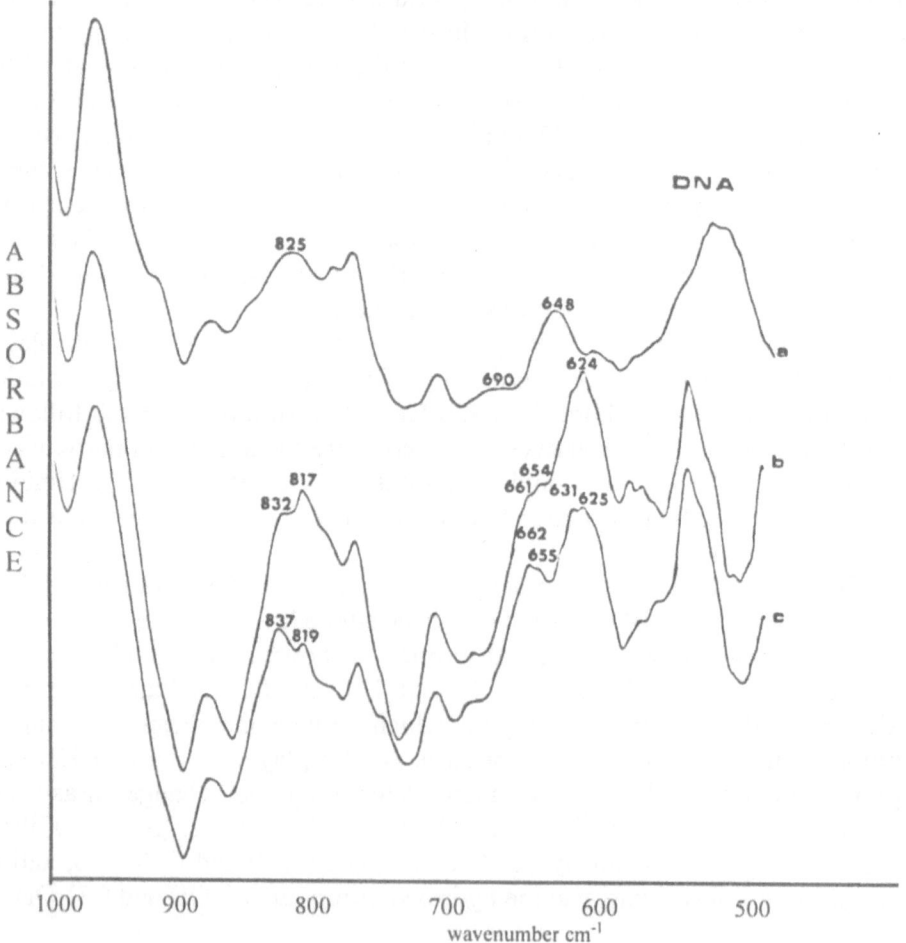

Figure 6. FT-IR spectra of DNA samples obtained from freeze-drying at different temperatures in the region 1000-500 cm-1. a) at room temperature, b) solid CO_2 temperature and c) liquid nitrogen temperature.

The natural B-DNA crystallized at room temperature shows characteristic infrared absorptions at 825 cm^{-1}, which are assigned to sugar-phosphate (C2'-endo) vibrational modes, as well as characteristic absorptions near 690 cm^{-1}, which is assigned to guanine ring breathing mode [14,15].

The infrared spectra of freeze-dried samples obtained at solid CO_2 temperatures (-70 °C) exhibit two strong absorptions at 832 and 817 cm^{-1}, as well as two bands of medium intensity at 662 and 655 cm^{-1} (Figure 6). The four bands in these regions are indicative of both B and Z conformations. In the spectra of Figure 6 the bands for Z conformation are dominant which means that from the two conformations present the Z dominates over the B. Similarly, with the liquid nitrogen temperatures again the two conformations are present, but the B conformation predominates now over the Z conformation. In DNA-metal ion adducts (16) with Co(II), Pt(II) and Cu(II) the strongest features in their infrared spectra are shown in 812-809 cm^{-1} and 625-600 cm^{-1} regions, indicating the stabilization of the Z-form. In the case of RNA the analogous characteristic sugar-phosphate absorption are at 811 cm^{-1} and the guanine breathing mode at 675 cm^{-1} and do not change either with temperature changes or RNA-metal adduct formation, where the A conformation is always, present, which is in accordance with structural data reported on different RNA crystals.

References

1. Shimanuchi, T., Tsuboi, M. and Kyogoku, Y. (1964), in J. Duschene(ed.), The structure and Properties of Biomolecules and Biological Systems, Interscience, London, Chap. 12, p. 435.
2. Tsuboi, M. (1974), Infrared and Raman spectroscopy in P.O.P. Tso (ed.), Basic Principles in Nucleic Acids, Academic Press, New York, Vo 1, pp. 399-452.
3. Theophanides, T. (1981) Fourier transform infrared spectra of calf thymus DNA, and its reactions with the anticancer drug cisplatin, J. Appl. Spectrosc., 35, 461-465.
4. Theophanides, T. and Anastassopoulou, J. (1988) Metal binding and conformational changes, in E.D. Schmid, F.W. Schneider and F. Siebert (eds.), Spectroscopy of Biological Molecules- New Advances, John Wiley & Sons, Chichester, pp. 433-438.
5. Theophanides, T. (1989) Metal ion-nucleic acid interactions as studied by Fourier transform infrared spectroscopy, in T. Theophanides (ed.), Spectroscopy of Inorganic Bioactivators Theory and Applications- Chemistry, Physics, Biology, and Medicine, Kluwer Academic Publishers, Dordrecht, pp. 265-272.
6. Theophanides, T. and Tajmir-Riahi, H.A (1984) Spectroscopic properties of metal-nucleotide and metal-nucleic acid interactions, in C. Sandorfy and T. Theophanides (eds.), Spectroscopy of Biological Molecules, Kluwer Academic Publishers, Dordrecht pp. 137-152.
7. Taillandier, E., Liquier, J., Taboury, J. and Chomi, M. (1984) Structural transitions in DNA (A,B,Z) studied by IR spectroscopy, in C. Sandorfy and T. Theophanides (eds.), Spectroscopy of Biological Molecules, Kluwer Academic Publishers, Dordrecht pp. 171-189.
8. Watson, J.D. and Crick, H.C. (1953) A structure for deoxyribose nucleic acid, Nature, 171, 737-738.
9. Michelson, A.A. (1891) On the application of interference methods to spectroscopic measurements, I. Phil. Mag., (5), 31, 338

284

10. Michelson, A.A. and Stratton, S.W. (1898) A new harmonic analyser, *I. Phil. Mag.* **(5), 45,** 85.
11. Theophanides, T. (1979) Vibrational spectroscopy of metal nucleic acid systems, in T. Theophanides (ed.) *Infrared and Raman Spectroscopy of Biological Molecules*, Reidel Publishing Co, Dordrecht pp. 205-223.
12. Bailey, L.E., Hernanz, A., Navarro, and Theophanides, T. (1996) Normal coordinate analysis and vibrational spectra of 9β-D- arabinofuranosyladenine hydrochloride (ara-A.HCl), *European Biophysics J.*, **24**, 149-158.
13. Bailey, L.E., Hernanz, A., Navarro, R., Huvenne, J.P., and Legrand, P. (1995), J.-C. Merlin, S. Turrell and J.P. Huvenne (eds.), Kluwer Academic Publishers, Dordrecht pp.291.
14. Theophanides, T. and Tajmir-Riahi, H.A. (1985), FT-IR spectroscopic evidence of C2'-*endo, anti,* C3'-*endo, anti* sugar ring pucker in 5'-GMP and 5'-IMP nucleotides and their metal-adducts, in E. Clementi and G. Gorongiu (eds.), *Structure & Motion: Membranes, Nucleic Acids & Proteins*, M.H. Sarma & R.H. Sarma, ISBN Adenine Press, pp. 521-530.
15. Theophanides, T. and Tajmir-Riahi, H.A. (1985), Flexibility of DNA and RNA upon binding to different metal cations. An investigation of the B to A to Z conformational transition by Fourier transform infrared spectroscopy, *J. Biomol. Struct. & Dynam.* **2**, 995-1004.
16. Theophanides, T. (1984) FT-IR spectra of nucleic acids and the effect of metal ions, in T. Theophanides (ed.), Fourier Transform infrared Spectroscopy, D. Reidel Publishing Co, Dordrecht, pp. 105-124

GEOMETRIES AND STABILITIES OF G·GC, T·AT, A·AT AN C·GC NUCLEIC ACID BASE TRIPLETS

TIZIANA MARINO, NINO RUSSO, ANNA SARUBBO
and MARIROSA TOSCANO

*Dipartimento di Chimica, Università della Calabria,
I-87030 Arcavacata di Rende (CS), Italy*

ABSTRACT. Geometrical structures and stabilities of G·GC, T·AT, A·AT and C·GC nucleic acid triplets have been computed by using the AM1 and PM3 advanced semimpirical methods. Results indicate that the third nucleic acid base binds with the corresponding Watson-Crick base-pair via Hoogsteen hydrogen bonds. Where possible comparisons have been made with previous ab initio investigations.

1. Introduction

During the last 20 years, the interest for triplex DNA has grown considerably because of its possible biological function (e. g. in genetic regulation) and its potential therapeutics applications (e. g. as antisense drug) [1-12].
Triple-helix is usually formed between a double-stranded homopurine-homopyrimidine and a single-stranded homopyrimidine or a homopurine tract [13-15]. The third homopyrimidine single-stranded DNA binds to the major groove of the double-stranded DNA, in parallel with the purine strand, via Hoogsteen hydrogen bonding [16,17], whereas the third homopurine strand binds to the major groove antiparallel to the other purine strand [15].
Typical nucleic acid base triplets are formed between thymine (T) and adenine-thymine (AT) base pairs (T·AT triad) or between

285

G. Vergoten and T. Theophanides (eds.), Biomolecular Structure and Dynamics, 285–297.
© *1997 Kluwer Academic Publishers.*

protonated cytosine (C) and guanine-cytosine (GC) base pairs (C$^+$·G C triad) [1,13,14,18].   Other base combinations (G·TA, G·GC, T·CG, A·AT) are known to form stable triplets [19,20].

Despite the renewed interest for triple helices (especially in the area involving synthetic modifications), structural and energetic information on these systems is scarce and a better understanding of these aspects is required to interpret and rationalize the experimental findings.

In order to contribute to the knowledge of the mechanism of triplex formation we have undertaken a theoretical study on a series of triads (G·GC, T·AT, A·AT and C·GC) by using the advanced AM1 and PM3 semiempirical methods [21,22]. The choice of these methods is essentially due to their high computational speed with modern workstations and to the possibility to take into account the correlation effects via empirical parametrization. Previous works have demonstrated that both the methods can be used with confidence also to study hydrogen bonding containing systems [23-25]. The present work is a further test on their reliability in this field.

2.   Computational   details

All the computations have been performed by using the MOPAC 6 code [26] implemented to run on the Convex C220 of the Computer Centre of the University of Calabria.

The structures have been optimized at both AM1 and PM3 levels without any constrain by using the keyword PRECISE.

The interaction energies have been computed as

$$\Delta E = E_{tot} \text{ (Triad)} - \sum E_{tot} \text{ (single base)}]$$

E_{tot} (Triad) and E_{tot} (single base) have been obtained from full energy minimization of the involved systems.

For G·GC, T·AT, A·AT triplets we have considered both the Hoogsteen (H) and reverse Hoogsteen (rH) positions of the third base that would lead to parallel and antiparallel orientations respectively of the third chain with respect to the Watson-Crick paired purine chain.

3. Result and discussion

First of all, we have optimized the AT and GC Watson-Crick and non-Watson-Crick base pairs. Results show that the PM3 hydrogen bond distances are shorter than those coming from the experiments.

The AM1 data are more consistent with previous ab-initio 4-31 G theoretical studies [10] and with the available experimental ones [27-29].

3.1. GEOMETRIES AND HYDROGEN BONDS IN BASE TRIPLETS

The terms Hoogsteen and reverse Hoogsteen are used here in a more general sense with respect to the convention, and denote pairing between any purine and any pyrimidine in which N7 and N6 of A, or N7 and O8 of G are forming hydrogen bonds.

The AM1 and PM3 optimized geometries for T·AT(H) ,T·AT(rH), A·AT(H), A·AT(rH), G·GC(H), G·GC(rH) and C·GC triads are reported in Figures 1-4. Hydrogen bond distances are reported in Tables 1 (Z·AT) and 2 (Z·CG) together with previous HF [10] and experimental data [30].

As a general trend we note that for the studied systems AM1 and PM3 structures have the same topologies, but the two methods give different values for the hydrogen bond distances.

All the triads show nearly planar structures with the exception of the C·GC triplet in which one cytosine lies in a plane that is nearly perpendicular to that of the CG base pair.

In both the AM1 and PM3 computations the Watson-Crick base pairs in the Z·AT and Z·GC base triads retain essentially the same geometry of the isolated base pairs. Thus, the third base is bound without disturbing the original Watson-Crick base pairs. The same trend has been suggested by an HF study employing minimal basis sets [10].

The Hoogsteen and reverse Hoogsteen base hydrogen bond lengths and angles have nearly the same values in Z·AT studied systems. In the case of Z·GC triplets structural differences, especially in the inter-bases torsional angles, can be noted between C·GC and G·GC triads. An analysis of the Table 1 shows that the hydrogen bond distances obtained by the AM1 parametrization

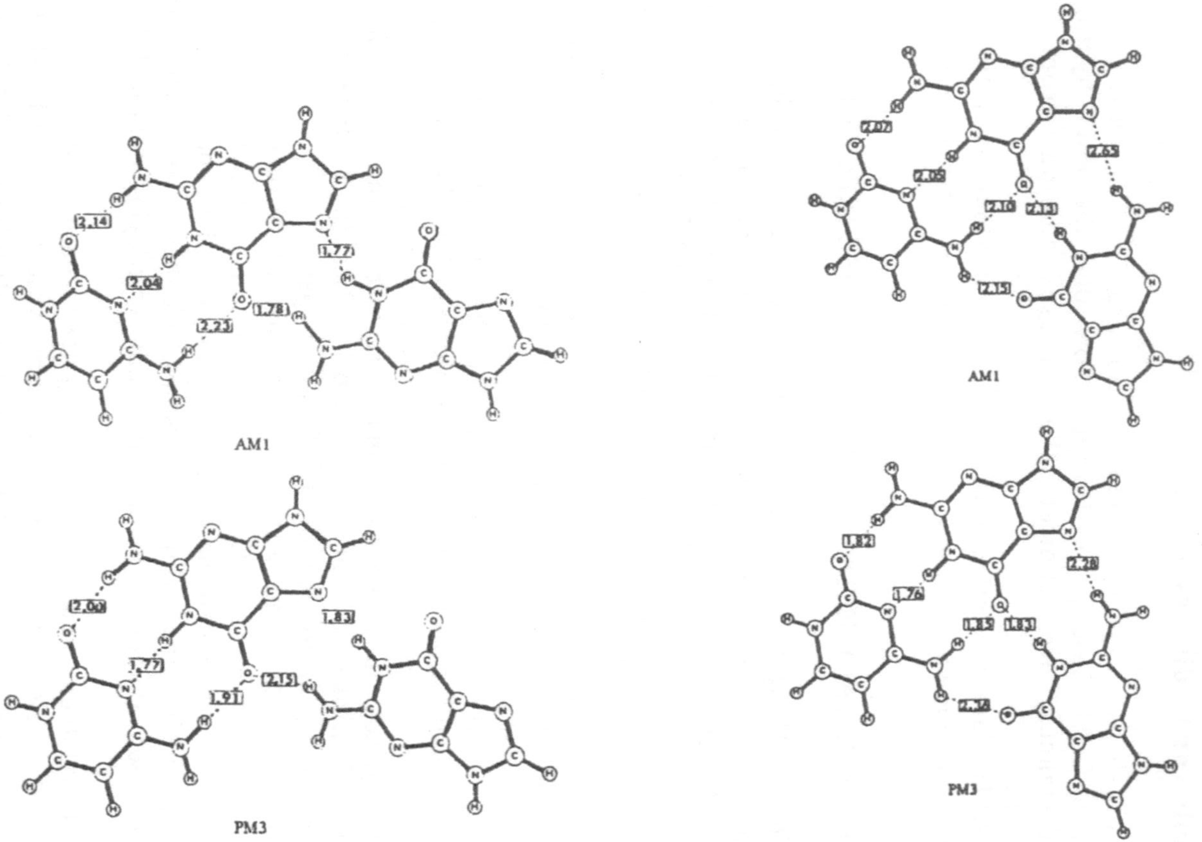

Figure 1. AM1 and PM3 optimized structures of G·GC(rH) (left side) and G·GC(H) (right side) triads.

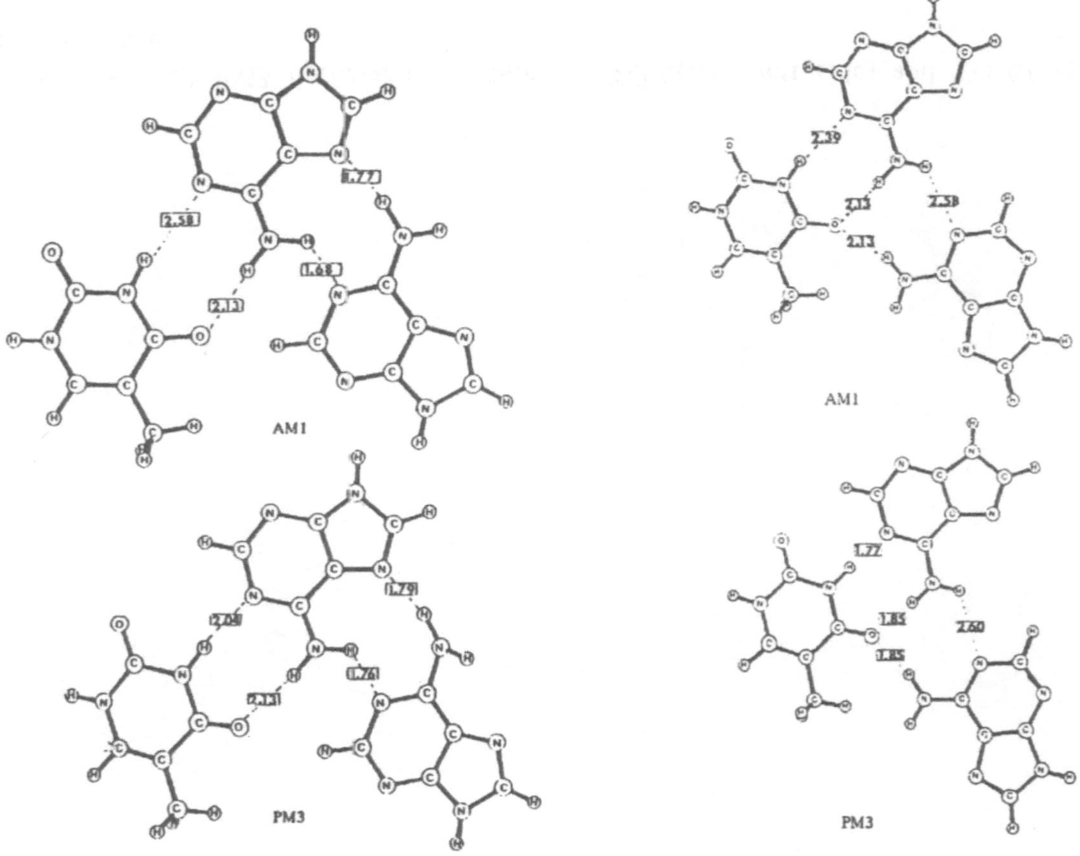

Figure 2. AM1 and PM3 optimized structures of A·AT(rH) (left side) and A·AT(H) (right side) triads.

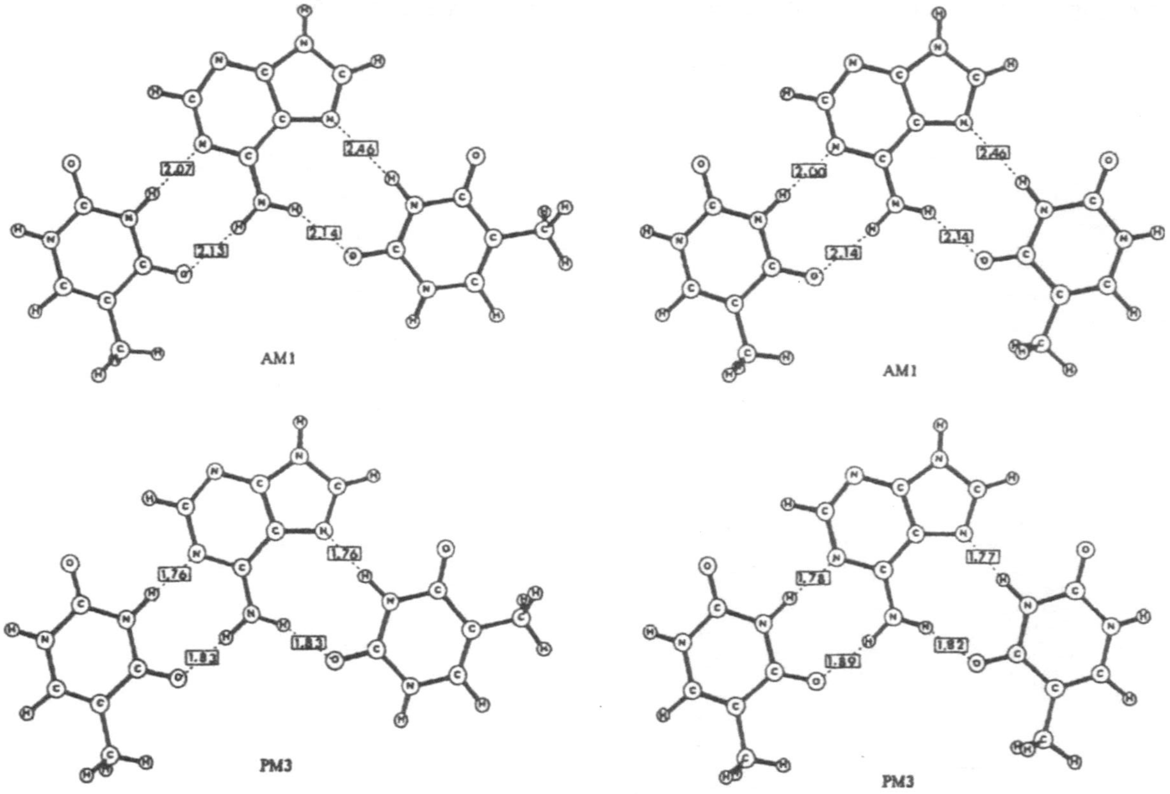

Figure 3. AM1 and PM3 optimized structures of T·TA(rH) (left side) and T·TA(H) (right side) triads.

Table 1. AM1, (PM3), [HF-STO-3G] (ref. 10) and {experimental} (ref. 30) hydrogen bond lengths (Å) for Z·AT (Z=A,T) base triplets

H-bond	T·AT(H)	T·AT(rH)	A·AT(rH)
	Watson-Crick		
N-H---N	2.06	2.07	2.58
	(1.78)	(1.78)	(2.04)
	[2.78]	[2.77]	[2.77]
	{1.99}		
NH$_2$---O=C	2.13	2.13	2.13
	(1.89)	(1.83)	(2.13)
	[2.73]	[2.74]	[2.74]
	{1.95}		
	Hydrogen bonds involving the third base		
NH$_2$---O=C	2.14	2.14	/
	(1.82)	(1.83)	
	[2.79]	[2.79]	
	{1.82}	{1.82}	
N-H---N	2.46	2.46	/
	(1.77)	(1.76)	
	[2.80]	[2.80]	
	{1.88}	{1.88}	
NH$_2$---N	/	/	1.68
			(1.76)
			[2.88]
			{1.82}
N---H$_2$N	/	/	1.77
			(1.79
			[2.86]
			{1.88}

Table 2. AM1, (PM3), [HF-STO-3G] (ref. 10) and {experimental} (ref. 30) hydrogen bond lengths (Å) for Z. GC (Z=C,G) base triplets.

H-bond	G·GC(H)	G·GC(rH)	C·GC
Watson-Crick			
N-H---N	2.05	2.04	2.04
	(1.76)	(1.77)	(1.78)
	[2.70]	[2.71]	
	{1.99}	{1.99}	
NH$_2$---O=C	2.10	2.23	2.09
	(1.85)	(1.91)	(1.99)
	[2.68]	[2.68]	
	{1.95}	{1.95}	
C=O---NH$_2$	2.07	2.14	2.13
	(1.82)	(2.00)	(1.85)
	[2.71]	[2.73]	
	{1.95}	{1.95}	
Hydrogen bonds involving the third base			
C=O---NH$_2$	2.15	1.78	2.25
	(2.36)	(2.15)	289
		[2.81]	(2.71)
		{1.82}	(2.80)
N-H---O	2.13	/	/
	(1.83)		
	[2.80]		
	{1.88}		
NH$_2$---N	2.65	/	/
	(2.28)		
	[2.88]		
NH---N	/	1.77	/
		(1.83)	
		(2.81)	

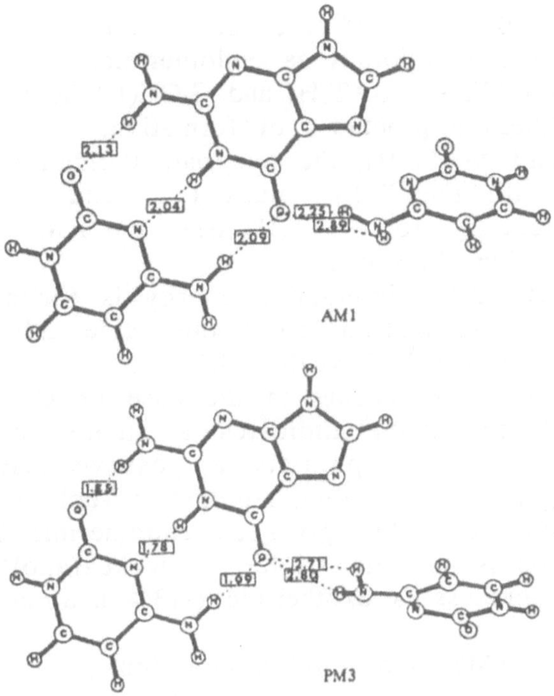

Figure 4. AM1 and PM3 optimized
structure of base triplet.

result close to the average crystallographic [30]. Ab-initio HF
hydrogen bond lengths are strongly overestimated [10]. This is
essentially due to the use of minimal STO-3G basis set and to the
lack of correlation effects in the calculations.

3.2. INTERACTION ENERGY, STABILITY AND COOPERATIVITY OF BASE TRIPLETS

The AM1 and PM3 interaction energies (DE) of the considered
base triplets are shown in the Table 3.
From this table it is evident that the two methods give different
energetic behaviour. In addition, the sign of ΔE indicates that the
energy of the base triplet is not perfectly pairwise additive and
exhibits some cooperativity. Firstly we can observe that both AM1
and PM3 methods predict a positive interaction energy for the

A ·AT(rH) triad. Because of the definition of the ΔE, the thermodynamics of this topology is endothermic.

For A·AT(H), A·AT(rH), A·AT(H) and G·GC(H) both methods agree in predicting exothermic processes of formation.

For A·AT(H) and A·AT(rH), the two parametrizations give almost the same trend with the PM3 values that tend to stabilize the triplets. On the contrary, for A·AT(H) triad the AM1 ΔE is higher than the corresponding PM3 one.

For G·GC(H) the AM1 interaction energy is found to be -27.8 kcal/mol. PM3 parametrization gives the same ΔE sign but the absolute value is lower (-21.5 kcal/mol).

Very different situation occurs in the case of G·GC(rH) base triplet for which the AM1 indicates a stable complex (-21.8 kcal/mol), while the PM3 predicts a positive ΔE value (0.8 kcal/mol). Looking at the ΔE values reported in Table 3 for the C·G C complex , we note that AM1 gives an endothermic behaviour for the formation process of this triplet (22.8 kcal/mol) while PM3 indicates that the process is exothermic (-13.4 kcal/mol).

Table 3. AM1 and PM3 interaction energies (kcal/mol) of Z·XY base triplets.

System	Method	ΔE
C·GC	AM1	22.8
	PM3	-13.4
T·AT(H)	AM1	-9.1
	PM3	-11.7
T·AT(rH)	AM1	-9.1
	PM3	-12.5
A·AT(H)	AM1	-9.0
	PM3	-8.4
A·AT(rH)	AM1	59.8
	PM3	49.4
G·GC(H)	AM1	-27.8
	PM3	-21.5
G·GC(rH)	AM1	-21.8
	PM3	0.8

The stability order derived from the calculated ΔE through the two methods is:

AM1

$$G \cdot GC(H) > G \cdot GC(rH) > T \cdot AT(H) = T \cdot AT(rH) \cong A \cdot AT(H) > C \cdot GC(rH) >$$
$$A \cdot AT(rH)$$

PM3

$$G \cdot GC(H) > G \cdot GC(rH) > T \cdot AT(rH) > T \cdot AT(H) > A \cdot AT(H) > C \cdot GC(rH) > A \cdot AT(rH)$$

At HF STO-3G level, the stability order found for the base triplet containing AT Watson-Crick base pair is $T \cdot AT(H) > T \cdot AT(rH) > A \cdot AT(H)$ [10].

The agreement with our AM1 results is satisfactory. Concerning the order of stability of base triplet containing the Watson-Crick base pair GC, the ab-initio result show that the $G \cdot GC(H)$ is more stable than the corresponding reverse Hoogsteen base triad. In this case the agreement is good with both our AM1 and PM3 data.

4. Concluding remarks

Full geometry optimization have been performed for $G \cdot GC$, $T \cdot AT$, $A \cdot AT$ and $C \cdot GC$ base triplets employing the advanced semiempirical AM1 and PM3 methods. Results show that:
i. both the Watson-Crick and Hoogsteen hydrogen bond distances are reproduced better by the AM1 parametrization. In fact, the PM3 hydrogen bond lengths appear to be underestimated;
ii. with the exception of CGC triad all other structures show little deviations from the planarity between the three bases;
iii. base pairing in triplets shows both positive and negative cooperativity;
iv. the PM3 and AM1 stability orders are different. AM1 trend is more similar to that obtained at ab initio level.

Acknowledgements

We acknowledge the CNR (Comitato Nazionale Scienze Chimiche) and the MURST for the financial support.

References

1. R. D. Wells, D. A. Collier, J. C. Hanvey, M. Shimizu and F. Wohlrab, FABES J. 2 (1988) 2939.
2. S. F. Singleton and P. B. Dervan, J. Am. Chem. Soc. 114 (1992) 6957.
3. G. Duval-Valentin, N. T. Thuong and C. Helene, Proc. Natl. Acad. Sci. USA 89 (1992) 504.
4. T. Ito, C. L. Smith and C. R. Cantor, Proc. Natl. Acad. Sci. USA 89 (1992) 495.
5. D. S. Pilch, C. Levenson and R. H. Dchafer, Biochemistry 30 (1991) 6081.
6. J. N. Davidson and W. E. Cohn (eds) Preogress in Nucleic Acid Research and Molecular Biology, Academic Press, New York, vol. 10, 1970.
7. A. Rich and V. L. Raybahnday, Ann. Rev. Biochem. 45 (1976) 805.
8. P. A. Beal and P. B. Dervan, J. Am. Chem. Soc. 114 (1992) 4976.
9. J. S. Koh and P. B. Dervan, J. Am. Chem. Soc. 114 (1992) 1470.
10. S.-P. Jiang, R. L. Jernigan, K.-L. Ting, J.-L. Syi and G. Raghunathan, J. Biomol. Struct. Dyn. 12 (1994) 383.
11. S. Weerasinghe, P. E. Smith, V. Mohan, Y.-K. Cheng and B. M. Pettitt, J. Am. Chem. Soc. 117 (1995) 2147.
12. M. Kamiya, H. Torigoe, H. Shind and A. Sarai, J. Am. Chem. Soc. 118 (1996) 4532.
13. H. E. Moser and P. B. Dervan, Scienze 238 (1987) 645.
14. T. Le Doan, L. Perronault, D. Prasenth, N. Habhoub, J. L. Decant, N. T. Thuong, J. Lhomme and C. Helene, Nucleic Acid Res. 15 (1987) 7749.
15. P. A. Bell and P. B. Dervan, Science 251 (1991) 1360.
16. P. Rajagopal and J. Feigon, Nature 339 (1989) 637.
17. C. de los Santos, M. Rosen and D. Patel, Biochemistry 28 (1989) 7282.
18. D. H. Live, I. Radhakrishnan, V. Misra and D. Patel, J. Am. Chem. Soc. 113 (1991) 4687.
19. K. Yoon, C. A. Hobbs, K. Koch, M. Sardaro, R. Kutny and A. L. Weis, Proc. Natl. Acad. Sci. USA 89 (1992) 3840 and references cited therein.
20. J.-S. Sun and C. Helene, Curr. Opin. Struct. Biol. 3 (1993) 345 and references cited therein.

21. M. J. S. Dewar and W. Thiel, J. Am. Chem. Soc. 98 (1977) 4899.
22. J. J. P. Stewart, J. Comp. Chem. 10 (1989) 209, ibidem 221.
23. M. J. S. Dewar and C. Cone, J. Am. Chem. Soc. 99 (1977) 372
24. M. J. S. Dewar, E. G. Zoebish, E. F. Healy and J. J. P. Stewart, J. Am. Chem. Soc. 107 (1985) 3902.
25. P. Davis, L. W. Burggraf and D. M. Storch, J. Comp. Chem. 12 (1990) 350.
26. J. J. P. Stewart, QCPE 455, Department of Chemistry, Indiana University, Bloomington, Indiana.
27. L. Katz, K. Tomita and A. Rich, Acta Cryst. 21 (1966) 754.
28. F. S. Matthews and A. Rich, J. Mol. Biol. 8 (1964) 89.
29. E. V. Haschemeyer and H. M. Sabel, Acta Cryst. 18 (1965) 525.
30. W. Saenger, Principles of Nucleic Acid Structure, Springer-Verlag, New York, 1984.

VIBRATIONAL CIRCULAR DICHROISM OF NUCLEIC ACIDS.
Survey of Techniques, Theoretical Background, and Example Applications

TIMOTHY A. KEIDERLING
Department of Chemistry, University of Illinois at Chicago, 845 W. Taylor St. (m/c 111), Chicago, IL 60607-7061 USA

1. Introduction

Electronic circular dichroism (ECD) of transitions in the ultraviolet for biomolecular structural studies has been one of the dominant applications of that technique due to its exquisite sensitivity to molecular conformation which is often manifested in complete sign reversals for various structural changes. For nucleic acids, ECD measurements depend solely on the π-π^* transitions of the bases which are spread over a modest range of the near-uv spectrum resulting in severe overlap of these broad electronic excitations [1]. Interactions among these transitions yield information about the nucleotide base stacking convoluted with the local chirality effects of base-sugar interaction. Other structural aspects of nucleic acids have less impact in ECD due to the difficulty of accessing spectral transitions centered on other parts of the molecule. Furthermore since the accessible electronic excitations are relatively delocalized and involve changes in the π-bonding electron configurations, the resulting transitions are susceptible to significant frequency shifts and intensity variations due to environmental or local perturbations.

In recent years a growth of Fourier transform infrared (FTIR) and Raman spectroscopic applications for determining nucleic acid structural variations has occurred [2,3]. In the FTIR, base deformation modes again overlap (in the region from 1800-1500 cm^{-1}). However, due to the ability of vibrational spectroscopic techniques to probe many different types of motion, FTIR and Raman can sense aspects of ribose and phosphate backbone conformations independently by utilizing data from several other distinct spectral regions. However both of these techniques are dependent on "diagonal" terms of the perturbation interaction in that they evidence an intensity that is a function of the squared amplitude of the local motion in a vibrational mode convoluted with its dipolar or polarizability derivatives, respectively. Thus all

G. Vergoten and T. Theophanides (eds.), Biomolecular Structure and Dynamics, 299–317.
© 1997 *Kluwer Academic Publishers.*

transitions contribute "positive" intensity making overlapping contributions difficult to separate.

This contrast of FTIR and ECD sensitivities has stimulated the technical development of vibrational (or infrared) CD (VCD) and of Raman optical activity (ROA) in our and other laboratories. Only the former measurement will be addressed here; but reviews of ROA abound [4]. The key impetus for moving to the vibrational region of the spectrum is that it is rich with resolved transitions which are characteristic of localized parts of the molecule and that by making a VCD measurement one can impart distinct stereochemical sensitivity to each of them. The chromophores needed in the molecule for VCD measurement are simply the bonds themselves as sampled by their stretching and bond deformation excitations. Chiral interaction of these bonds will then be manifest in the spectral bandshape that results from a VCD measurement. Furthermore these vibrational excitations are part of the ground state of the molecule for which one normally wishes to gain structural insight. VCD has the three dimensional structural sensitivity of a chiroptically sensitive technique but distributed over a large number of localized probes of the structure [5]. In other words, VCD is to IR what ECD is to uv absorption spectra, nothing is lost but much is gained.

Of course this benefit comes at a cost, which arises from significantly reduced signal to noise ratio (S/N) and some theoretical interpretive difficulty as compared to IR. Developments on the latter front are fast bringing the theoretical capability for prediction of VCD spectra for small molecules to a level that is demonstrably superior to that for ECD spectra [6]. These *ab initio* quantum mechanical methods are severely limited in their applicability to large molecules, such as the nucleic acids of interest in this talk. Experimentally, instrumentation has reached a stage where VCD spectra for most systems of interest can be measured under at least some sampling conditions [7-9]. However, most studies of biomolecules in aqueous solutions, naturally those of prime interest, are restricted to high concentration samples. Experimental aspects of VCD will be addressed next in this talk, and a brief survey of theoretical methods will follow. Applications of VCD for nucleic acid conformational studies, primarily based on empirical correlation with structure, will comprise the balance.

The theory and experimental aspects of VCD have been extensively reviewed previously with a focus on instrumentation and small molecule applications [6,7,9] and more recently for biomolecular studies [8]. This talk will focus on VCD applications for nucleic acids and my next one will review the VCD aspects of our peptide and protein structure studies. These are meant to introduce the biomolecular structurally oriented student to new developments related to the VCD field by discussion of the general principles and methods of VCD and by example applications.

In summary, unlike ECD, VCD can be used to correlate data for several different spectrally resolved features; and, unlike IR and Raman spectroscopies, each of these features will have a physical dependence on stereochemistry. But from another point of view, the combination of these techniques can compensate and balance for each other. The prime questions remaining in the VCD field now relate to application and interpretation of the method. It is clear that, despite claims of fundamental advantages of any one technique, progress in understanding of biomolecular structures will come from synthesizing all the data gathered from various techniques. In our biomolecular work, different types of spectral data are used to place bounds on the reliability of structural inferences that might be drawn from any one technique.

2. Experimental Techniques

This new dimension in optical activity comes at some cost in that the rotational strengths of vibrational transitions as detected in VCD are much weaker than are those of electronic transitions detected in ECD. Similarly since VCD is a differential IR technique, its S/N can never approach that of FTIR which represents a summed response. Several research groups have developed instrumentation that makes the measurement of VCD reasonably routine over much of the IR region [7,9] and commercial FTIR vendors are just now entering the market with VCD accessories. In this section those designs are briefly summarized and compared.

2.1 FT-VCD vs. DISPERSIVE VCD.

Instrumentation that has been developed over the past two decades makes the measurement of VCD reasonably routine over much of the IR region down to ~700 cm^{-1} on most samples. Development of a VCD instrument is normally accomplished by extending a dispersive IR or an FTIR spectrometer to accommodate, in terms of optics, time-varying modulation of the polarization state of the light and, in terms of electronics, detection of the modulated intensity that results from a sample with non-zero VCD. Our instruments and those of others are described in the literature in detail as referenced in recent reviews [7,8]. A detailed review contrasting these designs and detailing components needed to construct either type of instrument has been published by this author [7]. Here only a brief survey of the important components is given.

VCD instruments share several generic elements with "normal" CD instruments, as schematically outlined in Figure 1. All current instruments use a broad band source of light, typically utilizing black-body radiation from something like a ceramic or graphite-based glower, to allow sampling of a spectrum over the IR region. The method chosen for encoding the optical frequencies divides VCD instruments into two styles. Dispersive VCD instruments use a monochromator, based on grating technology, and are optimized for efficient light collection. On the other hand, Fourier transform (FT) VCD instruments use a Michelson interferometer that encodes the optical frequencies as an interferogram and gains efficiency through the multiplex advantage. Both styles of instrument then provide for linear polarization of the light beam, normally with a wire grid polarizer, and modulation of it between (elliptically) right- and left-hand states, with a photo-elastic modulator (PEM), before passing the beam through the sample and onto the detector, typically a cooled $Hg_{1-x}(Cd)_x Te$ (MCT) photoconducting diode. After preamplification, the electrical signal developed is divided to measure the overall transmission spectrum (I_{trans}) of the instrument and sample in one channel and the modulation intensity (I_{mod}, which is related to the VCD intensity) in the other.

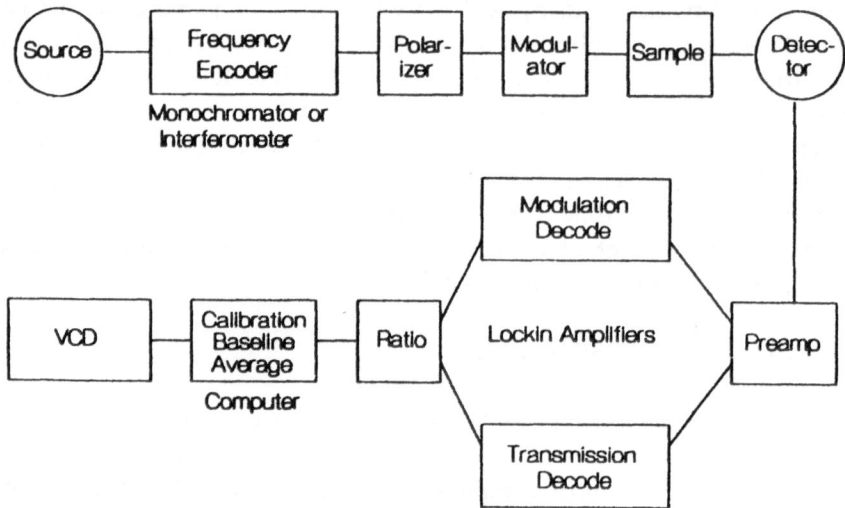

Figure 1. Schematic diagram of a generic CD spectrometer

After A-to-D conversion, these signals are ratioed which results in the raw VCD signal. Since VCD is a differential absorbance measurement, $\Delta A = A_L - A_R$, $A_{L/R}$ being the absorbance of left/right circularly polarized light, it is necessary to ratio these two intensities to normalize out any dependence on the source intensity and instrument transmission characteristics. In the limit of small ΔA values,

$$I_{mod}/I_{trans} = 1.15\ \Delta A\ J_1(\alpha_0)\ \text{[instrument gain factor]} \qquad (1)$$

where $J_1(\alpha_0)$ is the first order Bessel function at the maximum retardation of the modulator, α_o. Evaluation of this term and elimination of the gain factor are obtained by calibration of the VCD using a pseudo sample composed of a birefringent plate and a polarizer pair or by measuring the VCD of a known sample [7,9]. Further processing of the computer stored VCD spectrum involving calibration, baseline correction and spectral averaging or smoothing, as desired, and conversion to molar quantities, *e.g.* $\Delta\varepsilon = \Delta A/bc$ where b is the path length in cm and c is the concentration in moles/L, completes the process. Some details of the UIC dispersive and FTIR based instruments are given below to flesh out these somewhat abstract ideas.

Our original dispersive instrument is configured around a 1.0 m focal length, ~f/7 monochromator (Jobin-Yvon, ISA) that is illuminated with a home built carbon rod source [7,10]. A mechanical chopper provides the modulation necessary for detecting the instrument transmission with an MCT detector. The monochromator output is filtered with a long wave pass interference filter (OCLI) to eliminate light due to higher order diffraction from the grating and is manipulated using mirrors to focus the beam on the sample. A more compact design has been shown to have advantages in terms of S/N and baseline stability [11].

The light is linearly polarized with a wire grid polarizer on a BaF_2 substrate and modulated between left and right circularly polarized states with an AR coated ZnSe PEM. Following the sample a ZnSe lens focuses the beam onto an MCT detector chosen to match the slit image for optimal operation down to ~800 cm^{-1}. Very high sensitivity in the near-IR (~5000-1900 cm^{-1}) is possible with InSb photovoltaic detectors, CaF_2 modulators and lenses, and the proper grating for that region.

To process the signal, a lock-in amplifier is used to detect the transmission intensity of the instrument and sample as evidenced by the signal developed in phase with the chopping frequency. The polarization modulation intensity is detected as that component of the detector signal in phase with the modulator frequency. Since the VCD is only detectable when the light is on, it is also modulated by the chopper whose effect on the signal can be demodulated by using a second lock-in following the first. This scheme effects an added stage of amplification, added protection from amplifier overload, and discrimination against ground loops or other sources of unchopped signals at the modulator frequency. Dynamic normalization uses the signal from the transmission detecting lock-in to vary the amplification gain such that the transmission signal is constant. Applying the same gain to the polarization modulated signal assures a normalization much as is accomplished in "normal" CD instruments. In an alternate design, both signals can be A-to-D converted and normalization effected by division in the data computer [11].

Our FTIR-VCD spectrometer uses a Digilab (BIORAD) FTS-60A FTIR as its core [12] but the choice of FTIR is fully open to the user, since these optics seem to impart little limitation for practical VCD operation. In a separate compartment, an external beam goes through the same type of linear polarizer, stress optic modulator, lens and relatively large area MCT detector as described above but very weak focussing is used at the sample which leads to the flattest baseline characteristics of any of our VCD instruments. MCT detectors can easily saturate to a non-linear regime due to the high light levels which can be aided by using optical filters (*e.g.* 1900 cm^{-1} cut-off low-pass) to isolate the spectral region of interest and by controlling the preamp gain. For aqueous, biological samples, the spectral band pass of the solvent is limited, so high light level is not a big problem. The optical frequency at which the modulator retardation α_0 has its maximum is selected to lie in the spectral range of interest.

The raw VCD is obtained by ratioing the spectrum of the polarization modulated signal with the normally developed transmission single beam spectrum using the FTIR computer processing software. In a rapid scan instrument, the modulated signal is obtained using a lock-in amplifier to detect the modulation frequency and yield the sidebands as its output forming an ordinary interferogram which the FTIR electronics can process. Since the near-IR corresponds to higher frequency sidebands, the lock-in output attenuates them at moderate scan speeds, meaning that most rapid scan FTIR VCD spectra have concentrated on mid-IR bands. By going to slow- or step-scan operation, better response for higher frequency, near IR, components of the spectrum can be obtained [13]. The optical frequencies are also encoded through correlation to the mirror position, as measured by use of laser fringe counting, but the time element is removed.

While the instrument throughput (equivalent to I_{trans}) is an ordinary IR intensity measurement for which all instruments are adequately programed to process, the polarization modulation signal (I_{mod}) is not so simply processed. Often the integral of the modulated spectrum is very small having a rough balance between positive and negative going VCD. This results in their being only a very weak center burst in the interferogram. For purposes of interferometer alignment and phase correction, this can pose difficulties [7,12,13]. Normally, software is required that permits use of a transferred phase file, resetting of the centerburst position before transformation, simultaneous or sequential measurement from two independent detectors (I_{trans} and I_{mod}) inputs, and a variety of arithmetic manipulations of the "single beam" spectra (I_{trans} and I_{mod}). Software is perhaps the biggest limitation in choosing an FTIR instrument for modification to VCD operation.

While FT-VCD has many advantages, the restriction to measurement only in the spectral windows of water and the relatively broad bands seen in biopolymer IR spectra can nullify the multiplex and throughput advantages of FTIR and, all other things being equal, favor use of dispersive VCD. Until now, the expected FTIR advantages have <u>not</u> been experimentally realized in terms of the S/N for low resolution biomolecular (aqueous) FTIR-VCD spectra as compared to what can be measured with the dispersive instrument over a similar time span [13,14]. FT-VCD measurement, which always encompasses the full spectrum, can be quite inefficient as compared to concentrating one's effort and maximizing S/N in dispersively scanning just one or two spectral bands. This is particularly true if one uses D_2O based solvents to measure the base deformation modes in nucleic acids near 1650 cm^{-1} and shifts to H_2O solution to measure the lower frequency ribose and phosphate modes [2,8].

To get adequate signal to noise ratio (S/N) and determine scan-to-scan reproducibility, the dispersive spectra are averaged for several scans, often using time constants of the order of 10 sec and resolutions of ~10 cm^{-1}. This means a typical IR band can take about 1/2 hour for a single scan. FTIR measured VCD spectra can sample a much wider spectral region and take ~1/2 hour to collect an adequate number of scans for detecting the features of interest at 8 cm^{-1} resolution for a rigid, chiral organic molecule in a highly transparent solvent, but would require much more extensive averaging to match the S/N available using the dispersive instrument for single bands in aqueous phase biopolymers. In both cases, these VCD scans must be coupled with equally long collections of baseline spectra to correct for instrument and sample induced spectral response.

While VCD, just as ECD, is inherently a single beam measurement, due to its much weaker characteristic, it is subject to artifacts which must be corrected by more careful baseline subtraction. The best baseline is determined using racemic material, which is impractical for most biological materials. However, satisfactory baselines for spectral corrections can often be acquired with carefully aligned instruments by measuring VCD spectra of the same sample cell filled with just solvent. However there exists no satisfactory theory of these artifacts that can be used to control baselines. Rather there is an empirical body of evidence that parallel or slowly converging beams, few reflections and uniform detector surfaces give the best results. Finally, most artifacts can be minimized by careful optical adjustments which, at least in our instruments, are very stable, not requiring corrections for months.

2.2 SAMPLING TECHNIQUES.

Most biomolecular systems are best studied in an aqueous environment. This poses difficulties for IR techniques due to water being a very strong absorber whose fundamental transitions strongly overlap regions of interest in biomolecules such as the N-H and C=O stretches. Consequently, nucleic acid base deformation VCD in the range of 1800-1400 cm^{-1}, which is dominated by the C=O and aromatic ring C=N stretch contributions, has normally been measured in D_2O based solution. On the other hand, phosphate centered modes in the 1250-1000 cm^{-1} region are best studied in H_2O based solution.

Nucleic acid samples in D_2O can be prepared at concentrations in the range of 10-40 mg/ml. An aliquot of the solution is placed in a standard cell consisting of two BaF_2 windows separated by a 25-50 μm Teflon spacer. At a path of 50μ, these yield an absorbance of ≤0.1 in the 1700-1600 cm^{-1} region yet still produce acceptable VCD, at least for duplex samples [15]. Nucleic acids are best studied in H_2O in the PO_2^- region at concentrations of >50 mg/ml and pathlengths of 15μ giving absorbances of ~0.5 [16]. To date little progress has been made in the measurement of VCD for DNA ribose-based transitions.

Final VCD curves are obtained by subtraction of a baseline VCD scan (which for nucleic acids, due to their typically low absorbance, can be obtained with just of the same cell filled with the buffer solution) from the sample spectrum and by calibration as noted above.

Typically, after obtaining baseline and sample VCD scans, single beam IR transmission spectra of the sample and of the solvent are recorded in the same cell to obtain an absorbance spectrum taken on the same instrument and under identical conditions as were the VCD spectra. It is often useful to obtain FTIR spectra at higher resolution and optimal S/N on the same samples for purposes of comparison and for Fourier self-deconvolution resolution enhancement [17]. The FTIR spectra can also be used to frequency correct the dispersive VCD spectra by shifting the observed dispersive absorbance to align it with the FTIR absorbance. Ideally, VCD should be plotted in molar units such as ε and Δε as is done commonly with ECD measurements. Since concentration and path lengths are rarely known to an accuracy comparable to that used in ECD studies, VCD spectra of biomolecules are often normalized to the absorbance to effect a comparison between the spectra of different molecules. Because the absorbance coefficients for different molecules studied will vary, this is only a first order correction for concentration. Hence, VCD must be viewed as having some magnitude error intrinsic to the conditions required for IR study of biomolecules.

In our laboratory, ECD spectra are additionally measured for the samples studied using a commercial instrument (Jasco J-600). These spectra are usually obtained under more dilute conditions using strain-free quartz cells (NSG Precision Cells) obtained with various sample path lengths from 0.1 to 10 mm, the shorter path length cells being somewhat difficult to clean. Since relatively small amounts of biopolymer can give rise to significant ECD signals, it is very important to thoroughly clean sample cells between uses. Concentrations used in our laboratory for ECD are often of the order of magnitude of 0.1 mg/ml. For comparison of data obtained under comparable conditions, it is possible to measure ECD on sample of similar concentration as that used for VCD by employing a 15μ path cell constructed with quartz windows and a teflon spacer [18].

3. Theoretical basis for VCD

The theory of VCD has been a challenge since before the first VCD was measured for any real samples. This side of VCD research has continued to develop and has yielded valuable tools for the study of small molecules in particular. Since most such theoretical models do not easily apply to large biomolecules, they will only be touched on here as background for the examples given in this and the following talk on protein and peptide VCD. Extensive reviews of theoretical methods of VCD and applications have appeared [6].

Biopolymer oriented calculations of VCD have mostly been based on exciton coupling concepts, whereby local vibrations are dipole coupled yielding a pattern of oppositely signed VCD for coupled modes whose frequency dispersion results from electric transition dipolar coupling [19]. Schellman [20] modified the exciton method to simulate VCD for polypeptides in α-helices and β-sheets. Holzwarth and Chabay [21] put forth a dipole-based exciton model for the VCD of dimers, much as has been used successfully in ECD studies for biopolymers, which has been revived by Diem and coworkers [22], and have termed their result the extended coupled oscillator (ECO) model. By use of comparison to more exact theoretical methods [23], coupled oscillator approaches have been shown to be valid for weakly interacting (non-bonded) dipolar vibrations.

More accurate means of computing VCD spectra have been developed in the last decade. These involve use of quantum mechanical force fields and *ab initio* calculation of the magnetic and electric transition dipole moments, usually in the form of parameters termed the atomic polar and axial tensors (APT and AAT, respectively). These computations normally involve use of relatively large basis sets and some approximation to represent the magnetic dipole term. The magnetic field perturbation method (MFP) of Stephens and coworkers [6,24] effectively evaluates transition matrix elements due to the perturbation of the ground state wave function by a magnetic operator. While successful for a number of small molecules, the MFP model is difficult to extend to large molecules. This is particularly true if one carries out the MFP calculations at higher levels of *ab initio* theory to incorporate correlation effects [6,25] such as MP2 or using DFT methods (the latter being much more efficient). None-the-less some model calculations for dipeptides [26], and others coupling *ab initio* methods for computing the VCD arising from local interaction with dipolar methods for longer range interactions [27] and coupling *ab initio* results from fragments of nucleic acids [28] have been carried out. In particular, density functional theory [6,25,29] used for force field development could provide a means of applying *ab initio* methods to larger molecules and of sorting out the overlapped bands in polymeric molecules through improved precision in the frequencies so obtained [30].

At this juncture, theory for small molecule VCD is routinely carried out at a higher level of precision than is even possible for ECD. Thus the younger VCD field can be viewed to be progressing well in terms of both experiment and theory in its efforts to match the status of the older, established spectroscopic tools, such as ECD and FTIR.

4. Nucleic Acid Studies

VCD measurements on RNAs and DNAs in buffered aqueous solution yield quite large signals in terms of $\Delta A/A$ for a variety of modes. VCD can access the in-plane base deformation modes to study interbase relative disposition, and stacking interactions, the phosphate P-O stretches to sense backbone stereochemistry, and coupled C-H or C-O motions to monitor the ribose conformation. The VCD of the sugar-based modes has not proven very useful to date, due to their having little spectral definition in the C-H region, where such characteristic sugar modes are isolated, and to overlap of the C-O stretches with other nucleic acid modes in the mid-IR. Other diagnostic modes have not been adequately studied to date.

4.1 EMPIRICAL STUDIES

The first report of nucleic acid VCD spectra was for synthetic RNA-based systems [15]. Single stranded RNA samples give rise to a positive VCD couplet, which is typically centered over the most intense in-plane base deformation band which lies in the range of 1600-1700 cm^{-1}. In most cases, this band arises from a C=O stretching mode of the planar base itself. Dinucleotides and random copolymers have similar but weaker VCD, while duplex RNAs have similar but more intense patterns. Monophosphates have little or no detectable VCD in the base deformation or phosphate centered modes, so that the spectra observed for the polymers must be a direct consequence of the regular interactions between near-degenerate oscillators that is a property of the polymer or oligomer structure. Since base stacking is a dominant structural characteristic of nucleic acids, VCD provides a measure of these interactions. The generality in VCD bandshape found for simple polynucleotides has been attributed to their regularity in the helix twist [15]. From the dipole coupling point of view, the relative stereochemistry is the same for all of these RNAs, and is basically the same for DNAs, since the degree of turn and the interbase separation between planes are fairly independent of the type of base. In Figure 2, the spectrum for an A-form DNA is compared to that of a t-RNA. These are very similar except in terms of magnitude and temperature dependence, since the stacking interaction dominates [31,32].

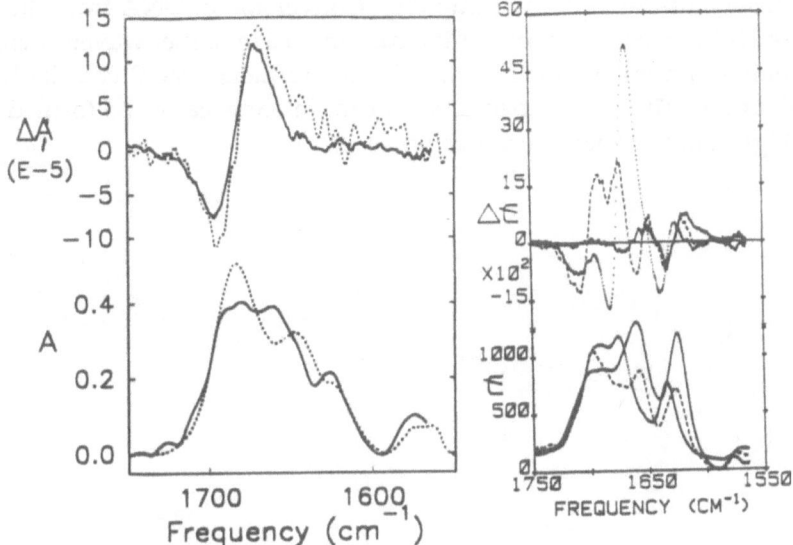

Figure 2. VCD and absorption of t-RNA (——) and A-form DNA (- - -) in 80% TFE.

Figure 3. Poly(A)·poly(U) temperature variation. 25 (· · ·), 56 (- - -), 80°C (——).

Somewhat of an exception to the consistent RNA VCD pattern described above is that shown in Figure 3 of the duplex, poly(rA)•poly(rU), which has a more complex pattern. At 55°C, the spectrum abruptly changes to another complex pattern that can be assigned to formation of a triple helical form, polyU*polyA•polyU [33]. Finally at higher temperatures the triple helix melts to single strands. This intermediate spectral form has general characteristics that can be found in all pyrimidine-purine-pyrimidine A(T)U DNA, RNA or mixed triplexes (Figure 4) and provides a much more definitive diagnostic that was previously available for this conformation with just FTIR or ECD data [34]. The G(I)C system also can form triplexes but these must be stabilized by ionizing one C strand to form C^{+}*G•C type triplexes. The pattern seen in the VCD is different due to the differences in frequency for the underlying base modes, but the changes from duplex to triplex are consistent with those of the A(T)U system and offer promise of generalizing the method for analytical purposes. VCD can discriminate among various stacking and complexing conformations of nucleic acids to a degree which is difficult with ECD or FTIR techniques.

In general, DNA VCD in the base stretching modes is very similar to that of the RNAs allowing for the helical conformational differences. Natural DNAs give somewhat broader and weaker VCD spectra patterns, since they are heteropolymers, than do their homopolymer, synthetic analogues. While

RNAs are mostly in an A-like form, DNAs are in the B form in aqueous solution and can be transformed to A-form with dehydration accomplished by use of alcohol solvents for solution sampling. Conversion of DNA from the B- to A-form results in a frequency shift of the base modes to higher wavenumbers and a sharpening and intensifying of the highest frequency VCD couplet [35] as seen in Figure 5. The left-handed Z-form can, in some cases, be formed in solution at high ionic strengths (*vide infra*).

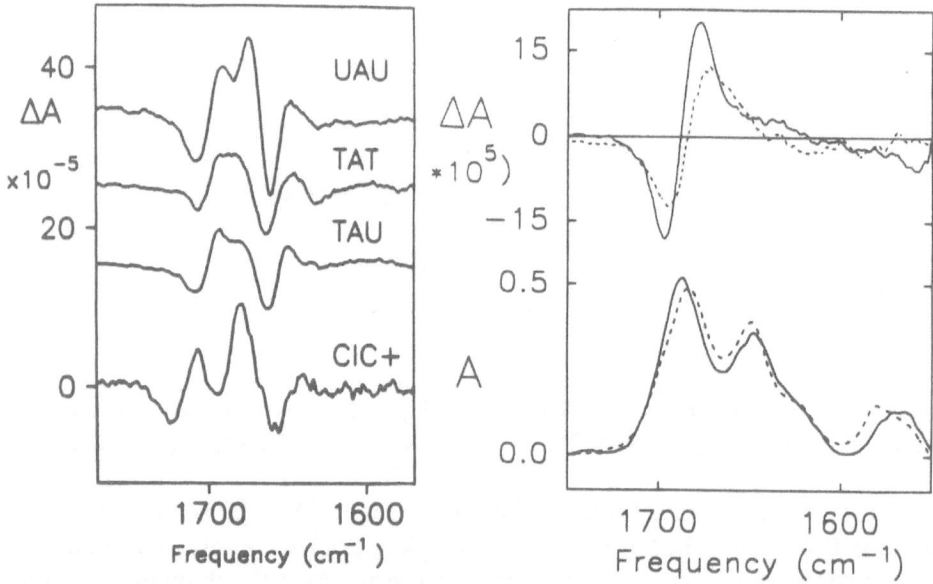

Figure 4. Comparison of VCD for Triplexes   Figure 5. A (——) and B (- - -) form DNA

As for RNAs, the poly(dA-dT) type species give distinctly different patterns from the high G-C content DNAs [35]. Consequently, base deformation VCD, though C=O stretching dominated, is sequence dependent and offers a means of typing DNA base content, at least in a qualitative sense [16]. A comparison of the VCD bandshapes for base modes of DNAs of varying base content, all B-form, is given in Figure 6. By contrast, to the considerable shape variation for the base centered modes, the PO_2^- modes (Figure 7) are virtually constant in form for these same molecules.

The VCD spectra of poly(dG-dC) (Figure 8) and related DNA oligomers in the B- and Z-forms have distinctly different bandshapes for the base stretching modes [22,32,33,36]. Both are dominated by a VCD couplet, but are significantly shifted in frequency and have opposite sign patterns which seems to reflect the handedness of their duplex helices. While the detailed shape of the Z-form base deformation VCD may be dependent on the specific conditions used for the conformational transformation and on the nature of the

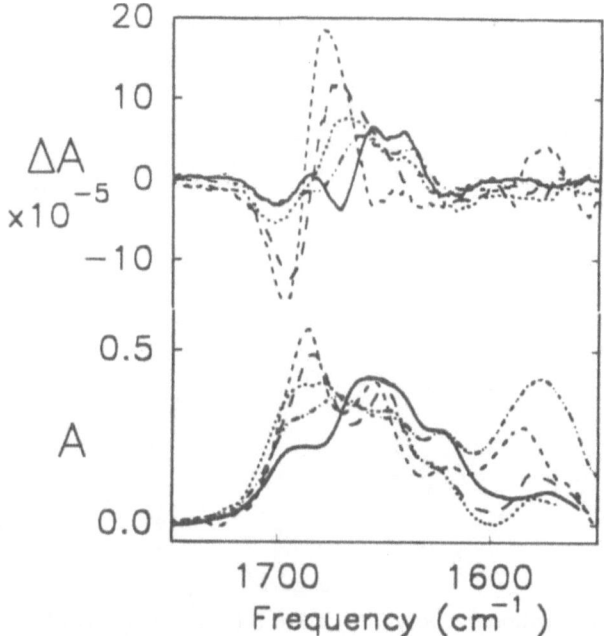

Figure 6 Comparison of VCD and absorption of duplex DNAs with 100 (- - -), 72 (– –), 44 (. . .), 26 (- . . -), and 0% (___) G-C content.

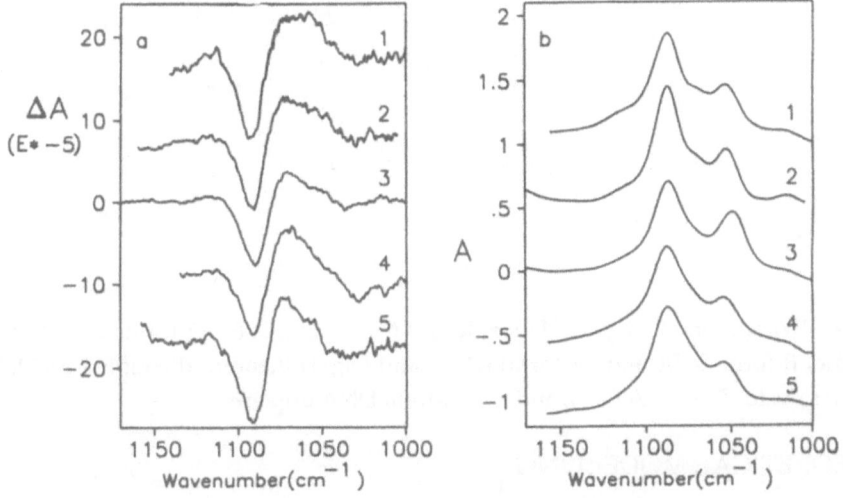

Figure 7 VCD and absorption in the PO_2^- region of the same DNAs as in Figure 6.

counter ions present, the Ż, Z* and Z' forms that have been postulated based on ECD studies have not been identified in VCD. Similarly, the generic Z-forms resulting from Na^+ and alcohol induced transitions lead to very similar VCD spectra [35]. VCD of Z-form DNA is also opposite in sign that of B-form in the symmetric PO_2^- stretch [16], as shown in Figure 9.

312

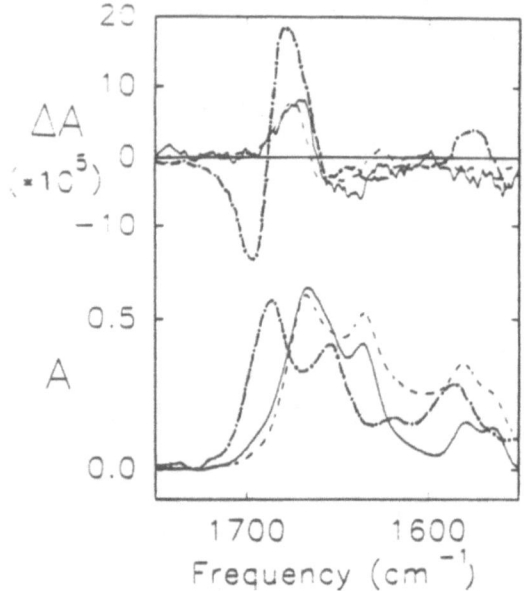

Figure 8. VCD and absorption for poly (dG-dC) in B-form (– · –), 0.1M NaCl, and Z form, 4.7M NaCl (——) and 80% TFE/D$_2$O (- - -), solution conditions

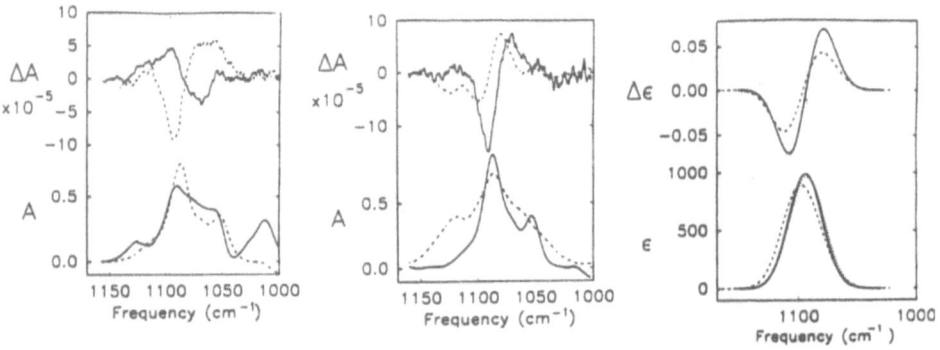

Figure 9. Experimental PO$_2^-$ VCD for (left) B (- - -) and Z (——) forms of (dG-dC), (middle) B-form DNA and (A-form) tRNA and (right) theoretical coupled oscillator VCD results for B (——), A (- - -), and Z (· · ·)-form DNA duplexes.

4.2 THEORETICAL MODELLING

VCD spectra of A-form DNA change little, other than sharpening and increasing in intensity, as compared to B-form DNA VCD and are very much like that of typical RNA VCD. These observations are consistent with a model for VCD of base deformations (primarily C=O stretch) arising through stacking interactions. The similarity of RNA and DNA in these transitions is due to the local nature of the interacting oscillators and the independence of their chirality from perturbations by the ribose moiety. The bases are spectroscopically the same in both types of molecules, the helicity is similar, but the angle with respect to the helix axis and the spatial relationship of the bases to the axis are different. The pairwise relationship is quite conserved. While the two right handed forms, B and A, give very similar spectra, there is no possibility of confusing them with the spectra of the Z form as has happened using ECD [37]. For example, we have measured the VCD of poly(dI-dC) in both the base stretching and phosphate regions and in both cases a clear spectral pattern results consistent with the right-handed B-form results.[31]. There is no possibility that these spectra could result from a Z-form. However, in ECD, poly(dI-dC) has the opposite pattern to typical B-form DNA in the near uv that has led to a series of confused conformational assignments. This results from the bases having slightly different electronic structures which affects the π-π^* transitions but has very little effect on the vibrations which are studied with VCD. Such a clear dependence on helicity is indicative of the role of dipolar coupling in the interaction between bases that leads to the observed VCD. This is also why model calculations using the simple coupled oscillator concept as a basis, have had such success in interpreting DNA spectra [16,22,31,35,36,33,39].

The phosphate, PO_2^-, modes dominate the IR spectrum in the 1250-1050 cm^{-1} region and yield VCD signals with strong couplets for the symmetric PO_2- stretch at 1070 cm^{-1} that reflect the helical sense (positive for B and A form, negative for Z form) [16]. Overlap with ribose modes potentially can complicate interpretation, but the patterns seen to date are systematic. The ribose modes overlapping the PO_2^- modes seem to affect the VCD in a primarily additive manner that only involves broadening and frequency/intensity shifts in higher order. However the asymmetric stretch at 1250 cm^{-1} has vanishingly weak VCD (Figure 10). The patterns, intensities and relative frequencies seen in the PO_2^- VCD, both asymmetric and symmetric stretches can be quite satisfactorily calculated using the dipolar coupling model [16]. As shown in Figure 9, while A and B forms are not practically distinguished in this manner, Z forms are opposite in sign and reduced in magnitude much as seen experimentally, and as shown in Figure 10, the asymmetric mode has a much weaker anticipated VCD intensity. The high dipole moments for these transitions (involving motions of charged groups) and the weak mechanical coupling of the PO_2^- groups makes them ideal candidates for coupled oscillator theory. Actually, the VCD of A-form DNA cannot be studied well in this region due to interference from the alcohol solvent used to dehydrate the DNA to transform to A form, but RNAs which have a conformation very close to that of the DNA A form can be used for rough comparison. The important aspect of the PO_2^- modes is their relative

314

independence of sequence. These modes sense the helical twist directly and are all identical for each nucleotide, thus providing a probe of helicity independent of sequence.

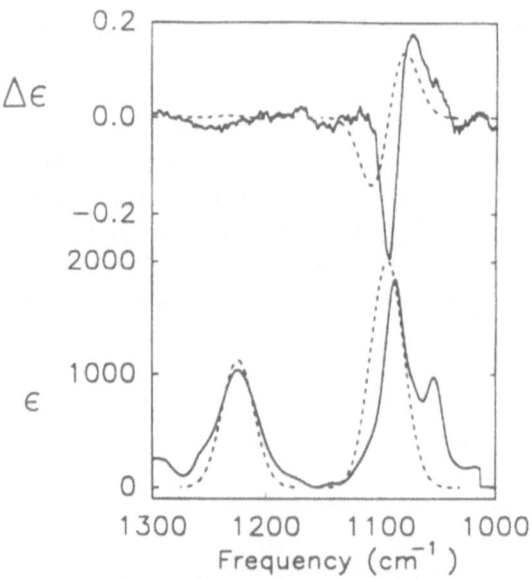

Figure 10. PO$_2^-$ asymmetric and symmetric stretching VCD and absorption for B-form DNA, experimental (——) and theoretical (- - -) results.

4.3 SUMMARY

While most nucleic acid studies have been oriented at characterizing the spectra of known structures, some applications have arisen. The previously noted triplex studies have identified structural phase transitions [33,34] and more recent oligonucleotide studies have put constraints on possible conformations for di- and tetra-nucleotides [39]. The ideal dipole coupling environment for the components of these molecules has spawned conformational studies utilizing approximate theoretical approaches. Finally a whole unexplored area of drug and protein or peptide-DNA interaction awaits study. The local sensitivity and relatively rapid analysis of VCD should offer new insight into the stereochemistry of such interactions not now available with other techniques.

Acknowledgement. This work was primarily supported by a continuing grant from the National Institutes of Health (GM 30147) for which we are most grateful. Instrumentation and theoretical development have been supported in the past by the National Science Foundation.

5. References

1. J. Brahms, and S. Brahms, (1974) In P. O. P T'so (ed.)*Basic Principles of Nucleic Acid Chemistry*, Academic Press, New York, p. 191-270. C. A. Bush,(1974) In P. O. P T'so (ed.) *Basic Principles of Nucleic Acid Chemistry*, Academic Press, New York, **Vol. II**, pp. 92-172. I. Tinoco, C. Bustamante, and M. F. Maestre, (1980)*Ann Rev. Biophys. Bioeng.* **9**, 107-141 . W. C. Johnson, (1995) in G. D. Fasman (ed.),*Circular Dichroism and the Conformation of Analysis of Biomolecules*. Plenum, New York, p. 433-468.

2. F. S. Parker, (1983) *Applications of Infrared, Raman, and Resonance Raman Spectroscopy*, Plenum, New York. Taillandier, E., Liquier, J. (1992) *Meth. Enzymol.* **211**, 307-335

3. Tsuboi, M., Nishimura, T., Hirakawa, A. Y. and Peticolas W. L. (1987) in T. G. Spiro (ed.), *Biological Applications of Raman Spectroscopy*, J. Wiley, New York, p. 109-79.

4. L. D. Barron, (1989) In J. R. Durig (ed) *Vibrational Spectra and Structure*. **17B**, 343-368. Barron, L. D. and Hecht, L. 1993 in Clark, R. J. H. and Hester, R. E. (ed.),*Biomolecular Spectroscopy Part B*, Wiley, Chichester, pp 235-266. Nafie, L. A. and Che, D. (1994), *Adv. Chem. Phys.* **85**, (Modern Nonlinear Optics, Part 3, Ed. Evans, M. and Kielich, S.) Wiley, New York pp 105-206. Polavarapu, P. L., (1989) in "Raman Spectroscopy Sixty Years On" *Vibrational Spectra and Structure*. **17B** 319-342.

5. Nafie, L. A. (1984) *Adv. Infrared Raman Spectr.* **11**, 49- .Freedman, T. B. and Nafie, L. A., (1987) *Top. Stereochm.* **17**, 113-206. L. A.Nafie, G. S. Yu, X. Qu, T. B. Freedman (1994) *Faraday Discuss.* **99**, 13-34 (1994).

6. Freedman, T. B. and Nafie, L. A. (1994), *Adv. Chem. Phys.* **85**, (Modern Nonlinear Optics, Part 3, Ed. Evans, M. and Kielich, S.) Wiley, New York pp 207-263. P. J. Stephens, F. J. Devlin, C. S. Ashvar, C. F. Chabalowski and M. J. Frisch (1994)*Faraday Discuss.* **99**, 103-120.

7. Keiderling, T. A. (1981) *Appl. Spectr. Rev.* **17**, 189-226. Keiderling, T. A. (1990) in *Practical Fourier Transform Infrared Spectroscopy* (Ferraro, J. R., and Krishnan, K., Eds.) Academic: San Diego, pp. 203-284.

8. Keiderling, T. A. (1995) in G. D. Fasman (ed.),*Circular Dichroism and the Conformation of Analysis of Biomolecules*. Plenum, New York, p. 555-598. Keiderling, T. A. (1994) in Nakanishi, K., Berova, N., Woody, R. W. (ed.), *Circular Dichroism Principles and Applications*, VCH Publishers, New York, pp 497-521. Keiderling, T. A. and Pancoska, P. (1993) in Clark, R. J. H. and Hester, R. E. (ed)*Biomolecular Spectroscopy Part B*, Wiley, Chichester, pp 267-315. Freedman, T. B., Nafie, L. A.,

316

and Keiderling, T. A. (1995), *Biopolymers (Peptide Sci.)* **37**:265-279.

9. Nafie, L. A., Keiderling, T. A., Stephens, P. J. (1976)*J. Am. Chem. Soc.* **98**, 2715- . Diem, M. (1991) in J. Durig (ed.)*Vibrational Spectra and Structure* **19**, 1-54 , Elsevier, Amsterdam. Polavarapu, P. L., (1985) in Ferraro, J. R. and Basile, L. Ed. *Fourier Transform Infrared Spectroscopy Vol.4* Academic New York. Diem, M. 1994 inPurdie, N., Brittian, H. G. (ed.) *Techniques and Instrumentation in Analytical Chemistry.* Elsevier, Amsterdam, pp 91-130.

10. Su, C. N., Heintz, V., Keiderling, T. A. (1981) *Chem. Phys. Lett.* **73**, 157-159.

11. G. Yoder, (1996), PhD Thesis, University of Illinois at Chicago. Diem, M., Roberts, G. M., Lee, O., Barlow, A. (1988) *Appl. Spectrosc.* **42** 20- .

12. Malon, P., Keiderling, T. A. (1988) *Appl. Spectr.* **42**, 32-38. Malon, P., Kobrinskaya, R. and Keiderling, T. A. (1988) *Biopolymers* **27**, 733-746. Yoo, R. K., Wang, B., Croatto, P. V., Keiderling, T. A., (1991) *Appl. Spectr.* **45**, 231-236.

13. Wang, B. and Keiderling, T. A. (1995) *Appl. Spectr.* **49**, 1347-1355 . Marcott, C., Dowrey, A. E., Noda, I., (1993) *Appl. Spectr.* **47**, 1324-1328.

14. Pancoska, P., Yasui, S. C. and Keiderling, T. A. (1989) *Biochemistry* **28**, 5917-5923.

15. Annamalai, A. and Keiderling, T. A. (1987) *J. Am. Chem Soc.* **109** 3125

16. Wang, L., Yang, L. and Keiderling, T. A. (1994) *Biophys. J.* **67**, 2460-2467

17. Kauppinen, J. K., Moffat, D. J. Mantsch. H. H. and Cameron, D. G. (1981) *Appl. Spectr.* **35** 271-276.

18. Baumruk, V., Huo, D., Dukor, R. K., Keiderling, T. A., Lelievre, D. and Brack, A. (1994) *Biopolymers* **34**, 1115-1121.

19. Tinoco, I. (1963) *Radiation Res.* **20**, 133.

20. Snir, J., Frankel, R. A., Schellman, J. A. (1974) *Biopolymers* **14**, 173.

21. Holzwarth, G., Chabay, I. (1972) *J. Chem. Phys.* **57** 1632.

22. Gulotta, M., Goss, D. J. and Diem, M., (1989) *Biopolymers,* **28**, 2047-58. Zhong,W., M. Gulotta, D. J. Goss, M. Diem, (1990) *Biochemistry* **29** 7485-91.

23. Bour, P. and Keiderling, T. A. (1992) *J. Amer. Chem. Soc.* **114**, 9100-9105.

24. Stephens, P. J. (1985), *J. Phys. Chem.* **89**, 784. Stephens, P. J. (1987), *J. Phys. Chem.* **91**, 1712. Stephens, P. J. and Lowe, M. A. (1985) *Ann. Rev. Phys. Chem.* **36**, 213-241.

25. F. J. Devlin, P. J. Stephens (1994) *J. Am. Chem. Soc.* **116**, 5003- . P. J. Stephens, F. J. Devlin, K. J. Jalkanen, (1994) *Chem. Phys. Lett.* **225**, 247- P. J. Stephens, K. J. Jalkanen, F. J. Devlin, C. F. Chabalowski (1993) *J. Phys. Chem.,* **97**, 6107.

26. Bour, P. and Keiderling, T. A. (1993) *J. Amer. Chem. Soc.* **115**, 9602-9607.

27. Bour, P. (1993) PhD Thesis, Academy of Science, Prague, Czech Republic.

317

P. Bour, J. Sopkova, L. Bednarova, P. Malon, T. A. Keiderling, (1996) *J. Comp. Chem.* (submitted).

28. P. Bour, H. Wieser (unpublished results).
29. Stephens, P. J., Devlin, F. J., Chabalowski, C. F., Frisch, M. J., (1994) *J. Phys. Chem.* **98**, 11623-11627. P. Bour, C. N. Tam, T. A. Keiderling (1996) *J. Phys. Chem.* (in press).
30. C. N. Tam, P. Bour, T. A. Keiderling (1996) *J. Amer. Chem. Soc.* (in press).
31. Wang, L. & Keiderling, T. A. (1993) *Nucl. Acids Res.* **21**, 4127-4132.
32. Keiderling, T. A., Pancoska, P., Dukor, R. K. and Yang, L. 1989 *Biomolecular Spectroscopy*, (Birge, R. R. and Mantsch, H. H., Eds.) *Proc. SPIE* **1057**, 7-14.
33. Yang, L. and T. A. Keiderling (1993)*Biopolymers*, **33**, 315-327.
34. Wang, L., Pancoska, P., and Keiderling, T. A. (1994) *Biochemistry* **33**, 8428-8435.
35. Wang, L. & Keiderling, T. A. (1992) *Biochemistry* **31**, 10265-10271.
36. Wang, L., Yang, L. and Keiderling, T. A. (1991) in Hester, R. E. and Girling R. B. Eds. *Spectroscopy of Biological Molecules* Royal Society of Chemistry, Cambridge, pp. 137-138. Birke, S. S., Zhong, W., Goss, D. J., Diem. M. 1991 in Hester, R. E. and Girling R. B. Eds. *Spectroscopy of Biological Molecules* Royal Society of Chemistry, Cambridge, pp. 135-136.
37. J. C. Sutherland and K. P. Griffen (1983)*Biopolymers* **22** 1445-1448.
38. Xiang, T., Goss, D. J., Diem, M., (1993) *Biophys. J.* **65**, 1255-1261.
39. Birke, S. S., Agbadje, I., and Diem, M. (1992) *Biochemistry* **31**, 450-455, Birke, S. S., Moses, M., Gulotta, M., Kagarlovsky, B., Jao, D., Diem, M. (1993) *Biophys. J.* **65**, 1262-1271, Birke, S. S., Diem, M. (1995) *Biophys. J.* (in press).

SUBJECT INDEX

- A -

A.AT	286,287,291,294
A-form DNA	312
ab initio	180,300,307
ab initio FEP	68
affinity	4,5,8,9,13,14
alanine	151,153,159,160,161,162,163,173
algorithms	99,111
alpha-helical	264
alpha helix bundles	124,219
alpha helix melting	263
Altona equation	183
AM1	285,286,287,291,292,293,294
amide-I	245,246
amide-II	245
antigenic determinants	122
aquometmyoglobin(Mb)	263,264,265,266,268
asparagine	102
association	10

- B -

barnase	5
B-form DNA	312
backbone	83,102
bacteriorhodopsin	246,247,249,253,257
band assignment	230
base	273
base triplets	285,293
binding energy	16
biological processes	17
biomolecules	243,248

- C -

catalytic power of enzymes	47
(+)-catechin	181
characteristic infrared absorption	283
Chou-Fasman method	126,129
C.GC	285,286,287,291

chromophore   252
circular dichroism   124,299,303
computer simulation   217
conformational behavior   166,173
conformational energy map   166
conformational transitions   281
conformations   274,283
conformer ensemble   183
CONTIN   132
continuum dielectric   49
cooperativity   293,294
cysteine bridges   110
cytochrome C   249

- D -

density functions, DF   151,152,155,160,162,163,173,174
D-glucitol   182
diagnostic bands   273,276
diamagnetic susceptibility   153,162
diffuse scattering   36
1,4-dihydronicotinamide   151,153,164,167
dipole moment   153,162,168,173
DNA   79,273,274,306,307,308,310,312,313
DNA-binding protein   190,202
docking   17,18
drug design   21
DSSP   126,130

- E -

egg-white lysozyme   40
elafin   141
elastin   124,140
electron affinity   153,162
electrostatic potential   223
energy   16
energy minimization   110,201
entropy   96,98
enzymatic reaction   61
enzymes   244
enzyme catalysis   47
(-)-epiciatechin-4β phloroglucinol   182
epitopes   121,123,137
equilibrium constant   173
extended Karplus method   184

- F -

far infrared	35
far infrared spectroscopy	41,45
flavan 3-ol	181
flexibility	122,138
fluidity	240
5-fluorouracil	153,169,170,171,172,173
folding	250
Fourier transform infrared spectroscopy, FT-IR	251,273,274
free energy	22,23,28,32,33
free energy curve	66
free energy surface	58
free energy transfer	128
FT-IR instruments	248,252,303,304

- G -

gas phase acidity	162,173
GEA	156
G-GC	285,286,287,291
generalized gradient approximations, GGA	152
globular proteins	121
glutamic acid	102
glycine	103,162,164,173
GMMX	179
GOR methods	126,131

- H -

Hartree-Fock	151,153
heat of hydration	163
helix bundles	212,213,214,221,223,224
helix-helix association	226
hemoglobin	90,254
heteropolymer	309
HF	154,155,159,160,162,171
homopolymer	309
Hoogsteen hydrogen bond	285,287,295
horse heart myoglobin	268
human myoglobin	254
hybrid orbital	64
hybrid QM	49,68
hydration free energy	48,153
hydrogen bonding	10,29
hydrophobic effect	213
hydrophobic interaction	212,223
hydrophobic mapping	214,217,219
hydrophobic organization	213,214,221,225

hydrophobic properties   218,221,223,224,227
hydrophobic templates   213
hydrophobicity   122,137
1,4-hydroxynicotinamide   165,166,173

- I -

information   128
infrared   229,233,300,301,302
infrared spectroscopy   124
interaction energy   293,294
interface area   16
interferogram   251
interferometer   275
ion channels   217,221
ionization potentials   153,162,173
iron-superoxide dismutase   143
isotopic substitution   273

- J -

J-coupling data   189

- K -

Karplus equation   179
Kohn-Sham, KS   152,153,155,156,158

- L -

Langevin dipole models   48
LD approach   68,152,156
ligand binding   21,27
LINK   123,132,133
lipids   229,231,235,236,238,239
local functional   157
loops   100
lysozymes   5,7,10,11,12,13,16,102

- M -

Mb secondary structure   265
membrane   229,230,231,236
membrane bundle   217,226
membrane domain   212,213
membrane helices   215
metal adduct   283
MHP method   224
MHP-templates   223

mid-IR	304
modeling	82,83,107
modeling enzymatic reactions	48
molecular dynamics	35,105
molecular dynamics simulations	41
molecular hydrophobicity potential, MHP	214,215,217,219,220 ,221,225,226,227
molecular modeling	121,212,213,227
Monte Carlo	21,33,92,96
multi helix bundles	225
mutagenesis	251
mutant	252,253,254,257
mutation	48,103,104,253
myoglobin	263,264,266

- N -

NADH	166
NADH/NAD$^+$	164
near-IR	304
NLSD	158,160 ,163,169,171,172
NMR-spectroscopy	189
NOE	189,180,194,196,202
NOE forces	198
NOE intensities	197
nucleic acids	273,274,276,285,306,307
nucleic acid triplets	285
nucleobases	274
nucleophilic attack	48,66,67
nucleotides	276,281

- O -

Onsager approach	158
optical spectroscopies	122

- P -

PDB	126,129,132
P.E.O.P.L.E.	123,137
peptide backbone	255
peptide group	247
phase transition	243,235
phenylalanine	98
phosphate	273,281,283,305,313
phosphate vibration	281
phosphoryl transfer	48
photolysis	255
PM3	285,286,287,291,292,293,294

polar solution   171
polynucleotides   276
potential surfaces   66
preference   128,131
probability   127
proline   103
propensity   127
protein   6,21,22,82,84,95,99,100
protein data set   126
protein lipid interactions   225
protein-protein   10,12,17
protein-protein complexes   13
protein-protein interface   10
protein-protein recognition   17
protein-solvent interaction   18,224
protein structure   107,111
protein structure determination   121
proteolysis reaction   48
proteolysis cleavage   110
proton affinity   153,162,173
proton transfer   163
proton transfer path   153,164

- Q -

quadrupole moment   153,162
quantum mechanical   49

- R -

Raman spectroscopy   124,265
reaction field, SCRF   157
recognition   17,18
RNA   276,283,307,308,312
rotational barrier   167,168
relative energies   60
resonance Raman   263

- S -

Schiff base   246,248
second moment   162
secondary structures   111,121,122
strategy 1D=>3D   121
self consistent   157,173
semiempirical methods   47
simulated annealing, SA   200
solution phenomena   173
solute-protein-solvent model   47

solvation energies	52,53
specific binding	15
specific recognition	17
specificity	14
spin polarized, LSD	152
stability	293,294
stability order	295
structural biology	126
subtilisin	48
sugar	273,283
sugar pucker	281
sulfonamide	31,32
surface accessibility	138
surface probability	122

- T -

T.AT	286,287
tautomeric equilibrium	171,173
tautomerization processes	153,168,169
temperature factor	36
template	89
template recognition	85
theoretical VCD	307
4-thiouracil	151,153,168,170,171,172,173
threonine	102
time resolved IR	243,248
time resolved step scan	253
time resolved UV	244,251
thrombin	31,32
transmembrane domains	219
transition	281
tropoelastin	124,136,140
trypsin	5,28
trypsin-benzamidine	27
tyrosin	98,103

- U -

unfolding	249,250
uracil	153,169,170,171,172,173
UV resonance Raman	263,264
UV-Vis spectroscopy	256

- V -

Van der Waals	92
VCD	180,181,300,301,304
vibrational spectroscopy	299,300

vicinal proton coupling   179

- W -

- X -

X-ray   35
X-ray diffuse scattering   40,41

- Y -

- Z -

Z conformation   283
Z form   310,311

ABBREVIATIONS

ASA	Accessible Surface Area
Brh	Bacteriorhodopsin
CD	Circular Dichroism
DF	Density Functional
EP	Environment Profile
FT-IR	Fourier Transform Infrared
2D,3D	two, three dimensional
GGA	Generalized Gradient Approximations
KS	Kohn-Sham
LSD	Spin Polarized
Mb	Aquometmyoglobin
MC	Monte Carlo method
MHP	Molecular Hydrophobicity Potential
NOE	Nuclear Overhauser Effect
P.E.O.P.L.E.	Predictive Estimation Of Protein Linear Epitopes
QM/CD	Quantum Mechanical/Continuum Dielectric
RC	Photoreaction Center Rhodopseudomonas Viridis
SA	Simulated Annealing
TM	transmembrane
TMS	transmembrane segment
VCD	Visible Circular Dichroism

AUTHOR INDEX

Alix A.J.P.   121
Anastassopoulou J.   273

Arrondo L.R.J.	229
Asher S.A.	263
Blake J.F.	21
Boelens R.	189
Bonvin A.M.J.J.	189
Chi Z.	263
Duffy E.M.	21
Efremov R.G.	211
Essex J.W.	21
Florian J.	47
Georg H.	243
Goni M.F.	229
Hauser K.	243
Hery S.	35
Janin J.	3
Jones-Hertzog D.K.	21
Jorgensen W.L.	21
Kaptein R.	189
Keiderling T.A.	299
Lamb M.L.	21
Marino T.	151,285
Mineva T.	151
Muller R.P.	47
Rödig C.	243
Rodriguez R.	79
Russo N.	151,285
Sarubbo A.	285
Severance D.L.	21
Siebert F.	243
Smith J.C.	35
Souaille M.	35
Theophanides T.	273
Tirado-Rives J.	21
Tobiason F.	179
Toscano M.	151,285
Vergoten G.	179,211
Vriend G.	79
Warshel A.	47
Weidlich O.	243